Conversion Factors (continued)

Temperature - see below			
Temperature difference	degree F (or R)	5/9	K (degree Kelvin)
	degree C	1	K
Energy (work, Heat)	ft lb_f	1.36	J (Joule = W·s = N·m)
	Btu	1.055×10^3	
	hph	2.68×10^6	
	erg	10^{-7}	
	$kg_f m$	9.81	
	kcal	4.19×10^3	
	kWh	3.60×10^6	
Power	Btu/hr	0.293	W (Watt = J/sec)
	ft lb_f/s	1.36	
	hp	746	
	erg/s	10^{-7}	
	kcal/hr	1.163	
	cal/s	4.19	
Heat flux	Btu/ft²hr	3.15	W/m² (= J/sec·m²)
	cal/cm²s	4.19×10^4	
	kcal/m²hr	1.163	
Specific heat	Btu/lb°F	4.19×10^3	J/kg·K
	kcal/kg°C	4.19×10^3	
Latent heat	Btu/lb	2.33×10^3	J/kg
	kcal/kg	4.19×10^3	
Thermal conductivity	Btu/ft hr°F	1.73	W/m·K
	cal/cm s°C	4.19×10^2	
	kcal/m hr°C	1.163	
Heat transfer coefficient	Btu/ft²hr°F	5.68	W/m²·K
	cal/cm²s°C	4.19×10^4	
	kcal/m²hr°C	1.163	

Temperature Conversion Formulas

T,°C = 5/9 (T,°F −32.0)
T,K = (T,°C) + 273.15
T,K = 5/9 (T,°R)
T,°R = (T,°F) + 459.67

Economics of Solar Energy and Conservation Systems

III

Volume III

Energy Management and Conservation

Editors

Frank Kreith, D.Sc., P.E.

Chief
Thermal Conversion Branch
Solar Energy Research Institute
Golden, Colorado

Ronald E. West, Ph.D., P.E.

Professor
Department of Chemical Engineering
University of Colorado
Boulder, Colorado

CRC PRESS, INC.
Boca Raton, Florida

Library of Congress Cataloging in Publication Data

Main entry under title:

Economics of solar energy and conservation systems.

Includes bibliographical references and index.
1. Solar energy industries. 2. Solar energy.
3. Energy conservation—United States. I. Kreith,
Frank. II. West, R. E.
HD9681.A2E27 338.4'3 78-20910
ISBN 0-8493-5231-2 (v. 3)

Direct all inquiries to CRC Press, Inc., 2000 N.W. 24th Street, Boca Raton, Florida 33431.

International Standard Book Number 0-8493-5231-2

Library of Congress Card Number 78-20910
Printed in the United States

PREFACE

Energy is universally acknowledged to be the mainstay of an industrial society. Without an adequate supply of energy the stability of the social and economic order, as well as the political structure, are in jeopardy. As the world supply of inexpensive, but nonrenewable, fossil energy sources decreases, the need for conservation as well as for developing other sources, such as solar energy and fusion, becomes ever more critical.

America has made two major energy transitions in its relatively short history (see Figure 1). Before the Civil War American industry depended mainly on wood, wind, and water for energy, heat, and power. After the Civil War these sources gave way to coal, which was more economical and technically more convenient for railroad transportation, industrial process heat, and home heating. Coal supplied the major part of U.S. energy need between 1885 and the start of World War II. Changes in American technology after World War II and increased availability of oil and gas paved the way for the second energy transition. During the 1950s a transition from coal to oil and natural gas occurred.

The second transition, as the first, was a transition from one abundant type of fuel to another; both resulted from technological advantages, lower cost, greater versatility, and ease of handling of the new source. Neither of these two transitions was the result of shortages. In fact, coal still remains today a vast energy resource for the future.

The energy crisis facing America today results from the divergence between the historically increasing energy demand of America and its decreasing supplies of oil and gas. To meet this crisis, America must make a new kind of energy transition which is quite different from the previous two. It is a transition from an infrastructure with abundant and cheap oil and gas to one in which these resources will be in ever decreasing supply. At one time it was thought that nuclear energy would be the major energy source of the future. However, the cost of nuclear energy has proved to be higher than was expected 20 years ago and there are also still unresolved problems regarding the safe long-term deposition of nuclear waste products. At this point in time it is not believed that any single energy source will take the place of oil and gas. Therefore, we must prepare for a condition in which many different sources of energy will play a role.

It is generally accepted that beginning in the late 1980s a diversity of energy sources will be required to supply the energy needs of the U.S. This makes it necessary to develop renewable energy sources in addition to liquid and gaseous fuels, and that safe nuclear power and fusion be aggressively pursued. However, it is inevitable that the quality of the feedstock for liquid fuels will deteriorate as coal, oilshale, tar-sands, and biomass take the place of oil and gas in the energy infrastructure later in this century. The process of phasing these new feedstocks into the energy infrastructure will be slow and the cost of energy will continue to increase for some time to come at a faster rate than the general level of inflation.

All of the new energy technologies, including solar energy, are more complex and expensive than the old oil and gas-based technologies. Therefore, energy conservation will become increasingly important. Also, a more sophisticated approach to investment decision in energy supply and conservation systems will be required. The three volumes of the CRC Uniscience Series *The Economics of Solar Energy and Conservation Systems* (SE&C) are designed to provide the economic, as well as the technical background necessary to make such investment decisions wisely.

The structure and information flow of the three volumes is displayed schematically in Figure 2. The first volume presents an overview of the U.S. energy system today, the principles of economics relevant for investment decisions in SE&C systems, basic

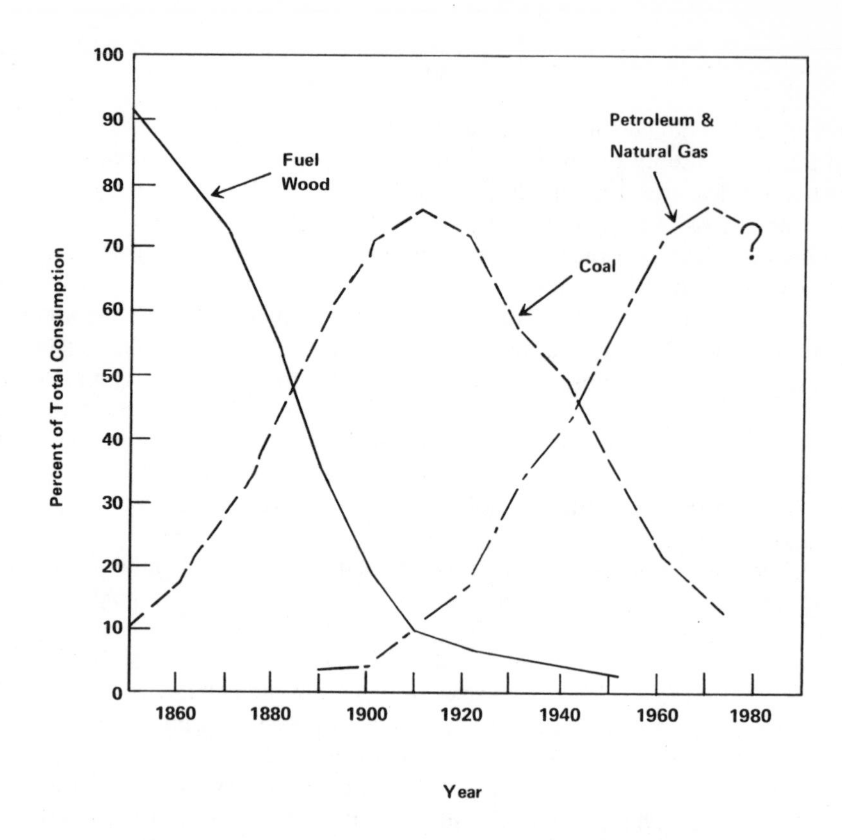

FIGURE 1. U.S. energy consumption patterns by major energy source.

background information in thermal sciences, as well as a likely scenario for the price of energy during the rest of this century. Since the economic viability of solar energy and conservation systems will depend on the tax structure and incentives, such as special write-off provisions or low interest loans, Volume I, Chapter 4 presents a summary of the economic incentives and special tax measures which had been enacted in each state before September of 1978, the time the three volumes went to press. Since that time, the long awaited National Energy Bill was signed into law in October 1978. Because of its importance in defining future energy policy, an addendum summarizing the provisions in this national law which are relevant to investment decisions in solar and conservation systems is presented in a special appendix to Volume I.

In many situations a conservation device and a solar system will compete with each other. Since for both measures large initial investments are required in order to reduce consumption of other fuels in the future, the economic trade offs in the investment decisions are similar. By applying the same criteria to all available options, a decision can be reached as to which option is the most favorable in a specific case. Chapter I-6, which is devoted to the "a priori" decision process for investments in conservation and/or solar energy system, can provide a comparative viewpoint for the decision maker.

The second and third volumes apply the principles presented in the first volume to specific solar energy and conservation systems. In both volumes only those applications for which hardware is commercially available at a reasonable price have been included. For each case the reader is provided with a step-by-step procedure on how to calculate the life cycle cost of energy delivered to the load, given basic data such as the term of the loan, the interest rate, the maintenance cost, the price of the auxilary

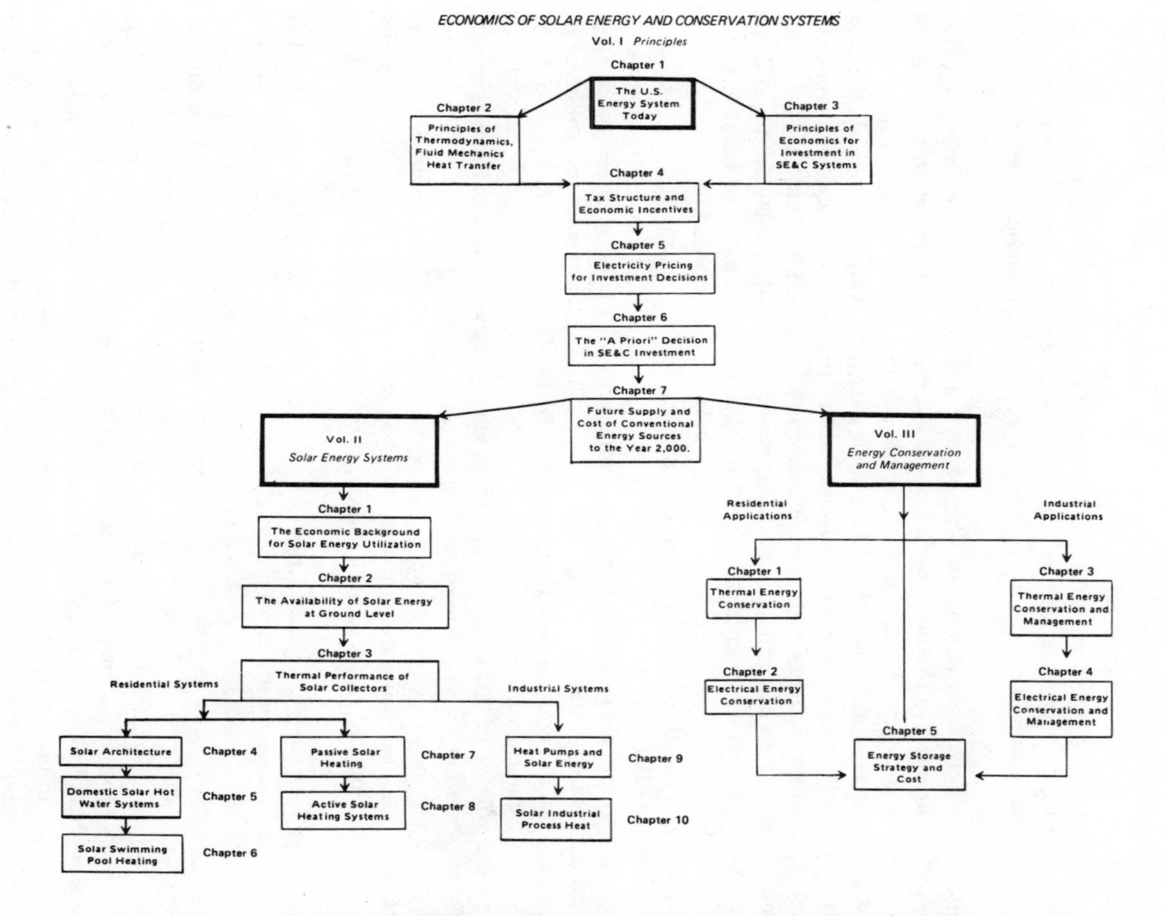

FIGURE 2. Outline of *Economics of Solar Energy and Conservation Systems.*

and the competing fuel source, and the inflation rate. Wind systems and photovoltaic (or direct) solar energy conversion systems have not been included in this edition because for the former little recent U.S. experience with commercial units is available while the cost of the latter is still prohibitive. However, we plan to add these in a later edition.

The second volume deals with solar energy systems. Solar energy, although the most promising renewable source, still has serious problems. It is environmentally more benign than fossil and nuclear fuels, but faces a number of technical and economic barriers, some of which are inherent in the nature of the source itself. Solar energy is a diffuse and intermittent source of energy and, hence, not capable of delivering an uninterrupted flow of energy on demand without adequate storage or fossil back-up. Since solar energy can not meet our needs at present and there is uncertainty as to how much it can supply in the near future, a mixture of conventional and new sources, as well as conservation measures, will be required in most cases to meet the demand. Just what this mixture should be and how it can be achieved economically are the crucial issues facing us.

Volume II, "Solar Energy Systems" begins with a discussion of the economic background of solar energy, followed by a chapter on the availability of solar energy at ground level and a general discussion of solar collector performance; then it treats separately the main applications for which solar energy is likely to be used in the next 25 years. The following chapters are arranged according to their probable sequence of market penetration, starting with solar architecture, followed by domestic hot water heating, swimming pool heating, passive solar heating, and active solar heating systems in the residential sector. Solar cooling systems have not been included because so far no technically and economically viable units are commercially available.

In the industrial sector solar assisted or stand alone heat pumps and solar process heat systems are likely to have the greatest impact, with heat pumps also having near term applications in the residential sector. Commercial solar cooling systems with large absorption or Rankine units are closer to commercialization than solar residential cooling systems, but not enough operating experience exists to consider this application of solar energy technically viable.

The third volume in this series is devoted to energy conservation and management. Energy conservation is a problem of national urgency. Americans use more energy per person than any other people in the world. With only 6% of the world population, Americans use about ⅓ of the energy consumed on the earth. Our total national energy costs in 1975 amounted to about $170 billion, and each year this cost is rising by about 10%.

The amount of energy wasted in the U.S. is enormous. If waste is defined as that portion of the prime energy input which is not converted to its intended use, Table 1 shows the percentages of energy wasted in transportation, utilities, manufacturing and industrial processes, and in residential/commercial heating and cooling. A considerable portion of the 75% of energy loss in transporation and the 65% loss by utilities are due to heat rejections inherent in the thermodynamic cycles used. However, the 25% of energy lost in the industrial sector and in heating of buildings is largely due to unnecessary heat leakage and can be substantially reduced by conservation measures. Of course also in the transportation sector conversion from Otto to Diesel cycles and reduction in aerodynamic drag and friction could substantially reduce the percentage of energy lost in that sector, while use of mass transport could reduce the total amount of energy consumed in transportation. Using the heat rejected by power plants for home heating and industrial processes could reduce the energy waste in that sector. However, conservation measures for the transportation and utility sectors are not the

TABLE 1

Annual Energy Consumed and Lost in the U.S.

Sector	Energy Consumed			Energy Lost			
	10^{18}J	10^{15}Btu	Percentage of total	10^{18}J	10^{15}Btu	Percentage of sector	Percentage of total
Transportation	17.3	16.3	24	12.9	12.2	75	17.9
Utilities	16.0	15.1	22	10.4	9.8	65	14.4
Manufacturing and in- dustial processes	22.2	21.0	31	5.5	5.2	25	7.6
Commercial/residential heating & cooling	16.9	15.9	23	4.2	4.0	25	5.9
Total	72.4	68.3	100	32.8	31.2	—	45.8

Note: Extracted from *Energy and the Future*, American Association for the Advancement of Science, Washington, D.C., 1973 and Pinkus, O., Decker, O., and Wilcock, D. F., *Mech. Eng.*, 32, Sept. 1977. Data are for 1970. Although total energy consumption has increased the percentages are essentially unchanged.

topic of this volume which deals with the potential for reducing energy waste in manufacturing and industrial processes and in heating of residential or commercial buildings.

The industrial energy consumption pattern in the U.S. is displayed graphically in Figure 3 for the year 1972, the last year for which complete information is readily available. The total industrial energy consumption was 28.4 Quads (1 Quad = Q = 10^{15} Btu). Of that total, 13.5 Q were used in manufacturing processes and more than 3.9 Q were lost. Figure 3 shows how the energy input was distributed among the various industries and what part of the energy was used for process steam or hot water, what part was used for direct heat, and what part was wasted. The total industrial sector energy use in Figure 3 differs substantially from that in Table 1 because losses from electricity generation for the industrial sector are included only in the former.

Inspection of Figure 3 demonstrates several important points. First, the total amount of energy wasted by the manufacturing industry is larger that its energy requirements for direct heat. Also, the amount of energy wasted is nearly equal to the energy requirements for hot water and process steam. Therefore, it is obvious that conservation measures could reduce the total energy consumption appreciably. If, in addition, solar energy were utilized to supply process heat wherever technically feasible, the total energy consumption in the manufacturing sector could even be further reduced. Of course, direct solar energy utilization and energy conservation measures will have to be judged from economic as well as technical points of view. Deciding where to invest capital for most economic reduction in energy consumption requires careful analysis.

Volume III, "Energy Conservation and Management," is subdivided into two parts: one treats applications to residential and commercial buildings, the other deals with industrial applications. Each of these two parts is subdivided further into thermal and electrical conservation procedures. All of these are tied into the chapter on energy storage and cost. This chapter, III-5, is also relevant to solar applications when the time of availability of the resource does not fit the demand, as for example, in heating of buildings. The chapter dealing with energy storage attempts not only to present current technology, but also to look ahead because currently available energy storage methods are limited and will most likely undergo substantial change and technical development in the coming decade.

At the present time conservation measures, properly designed and executed, are the most economical way of conserving energy in heating residential and commercial buildings. However, in many places solar heated domestic hot water systems are already competitive with electrically heated domestic hot water systems and they could compete with gas heating if the price of natural gas were to be placed at its marginal replacement value sometime in the future. After conservation and solar hot water heating it appears that industrial process heat will be the next economically viable solar application for future capital investment. The reason for the economic superiority of investment in industrial process heat over other direct solar applications, such as home heating, is the fact that the collectors can be used all year round and will, therefore, deliver more energy per dollar invested. Swimming pool heating may not always be economically viable, but due to laws in many states forbidding the use of gas for that purpose, it may become a necessity.

Although the chapter on heat pumps appears in Volume II, their potential as a conservation measure with or without solar assistance should not be overlooked. Heat pumps can be used in the building heating and cooling sector or for industrial applications. Their full potential depends on environmental conditions as well as on future technical developments which could improve the year-round efficiency of these devices. The performance calculation procedures presented in Volume II are applicable

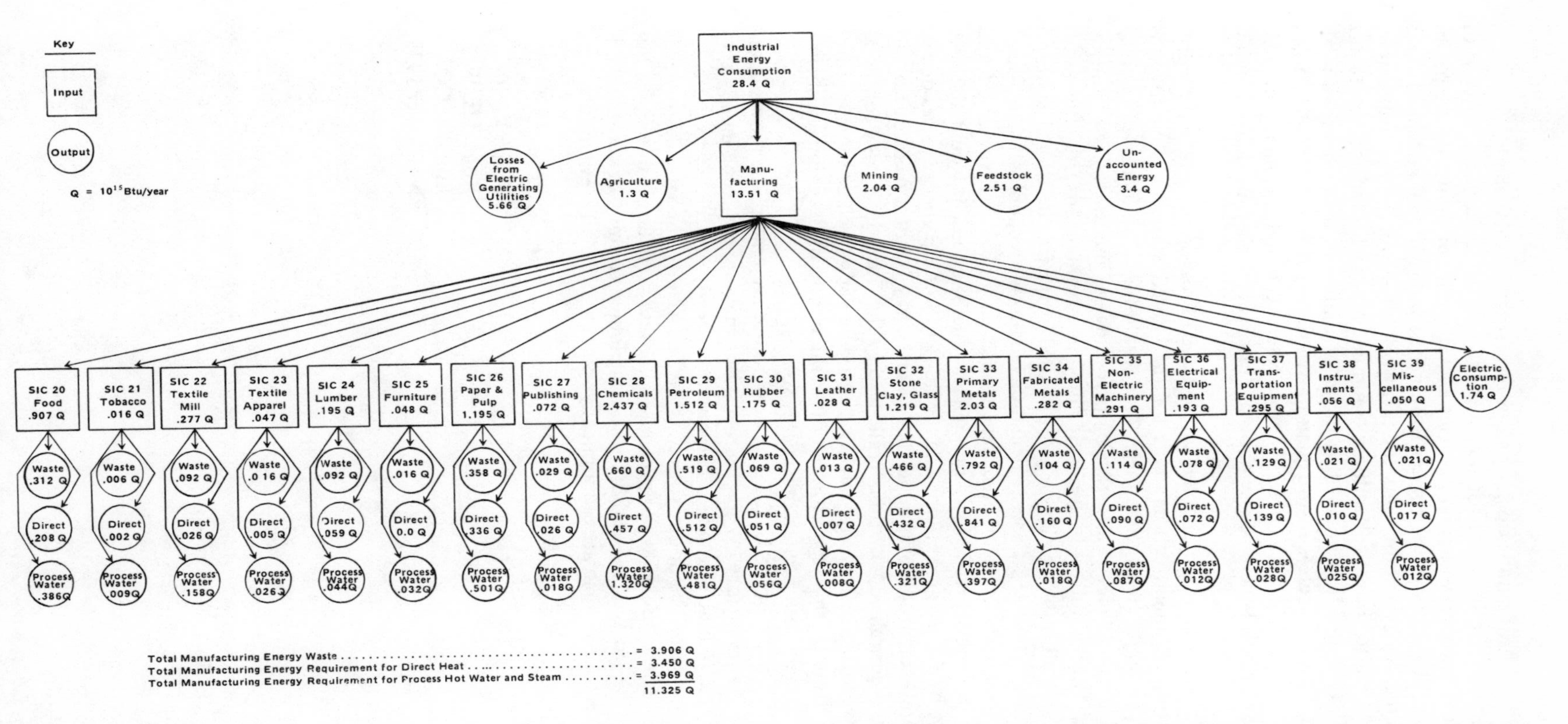

FIGURE 3. Industrial energy consumption - 1972. Details may not add to totals due to independent rounding. Data for agriculture, mining, feedstock, electric generating losses, and unaccounted energy from Planning Research Corp. / Energy Analysis Co. from meeting at D.O.E. (report to be released soon). Data for individual SIC's are from Survey of Manufacturers, U.S. Bureau of the Census, 1972. Fuels and Energy Consumed - Special Report, July, 1973. Percentages for the breakdown of individual SIC energy consumption into waste, direct heat, and process hot water and steam were taken from Industrial Waste Energy Data Base / Technology Evaluation, CONS 2862-1, Drexel University, December, 1976.

to any conditions, i.e., heat pumps working alone or assisted by solar energy.

These three volumes are intended for use over the next 10 to 20 years. In view of the fact that United States government agencies as well as major industries are beginning to require the SI system in their specification, the international system of units has been used as the key system of units for this work. However, wherever possible, engineering units are also shown to facilitate communication with those practitioners still dealing in those units. It is hoped this blend of units will make it easy for all potential users to avail themselves of the quantitative reference material and perform necessary calculations. All of the material is cross-referenced and a common nomenclature is used throughout. Where new symbols are introduced, they are defined within the text.

Both in terms of responsible national self-interest and of responsibilities to the world community, a strong argument can be made for the United States to move towards lower growth of per capita fossil energy use. This would imply an increase in energy conservation measures as well as an increasing percentage of our energy use from renewable resources. For a number of reasons, it would be advantageous from the standpoint of both the overall resource picture and of world image for the United States to supply a significant fraction of its energy demands from solar technology. The use of solar technology would result in lowered dependence on global reserves of oil and would extend our limited domestic fossil and nuclear fuel supplies. It would also free those fuels for use by future generations and would help to keep energy prices down. It is important, however, to keep in mind that solar energy is not a panacea for our energy problem; but combined with an overall strategy that includes energy conservation, cascading, hydroelectric and safe nuclear electric power as well as some sacrifice of convenience in lifestyle, this scenario does offer the option of developing a society which can operate largely with a renewable energy source.

The energy choices made in the U.S. in the near future are among the most important of any choices in its history. While the questions may appear to be mostly economic, they are actually much broader. The questions really are related to the basic assessments of major priorities in our society today, as well as our attitudes towards future generations. These books do not pretend to address these larger questions, but it is our hope that the work of the authors who contributed to this work will make it easier to arrive at decisions which are economically sound and socially responsible.

Frank Kreith
Ronald E. West
Boulder, CO.
November 1978.

THE EDITORS

Dr. Frank Kreith, P.E., is Chief of the Thermal Conversion Branch at the Solar Energy Research Institute, Golden, Colorado. He is also Professor of Chemical Engineering (on Leave) at the University of Colorado in Boulder, Colorado.

Dr. Kreith graduated from the University of California at Berkeley with a B.S. in mechanical engineering in 1945. He received an M.S. in engineering in 1949 from the University of California, Los Angeles, and a Doctorate from the University of Paris in 1963.

From 1945 to 1949, Dr. Kreith was a Research Engineer at the California Institute of Technology Jet Propulsion Laboratory. From 1949 to 1951 he was associated with the Aeronautical Engineering Center at Princeton University. In 1951 he was appointed Instructor in Mechanical Engineering at the University of California (Berkeley) and promoted to Assistant Professor the following year. In 1953 he became an Associate Professor at Lehigh University and in 1959 was appointed Professor of Engineering at the University of Colorado, where he remained until 1977, when he accepted his present position at the Solar Energy Research Institute.

Dr. Kreith's research interests are in the area of heat transfer and thermodynamics, with applications to energy conservation and solar energy systems. He has published over 100 articles in the area of his specialty and is the author of several textbooks, including *Principles of Heat Transfer, Solar Heating and Cooling,* and *Principles of Solar Engineering.* His current research interests are in developing methods for reducing consumption of nonrenewable energy resources by developing energy conservation and economically viable solar energy systems.

Dr. Kreith has been the recipient of two Guggenheim Fellowships, several Fulbright Grants, two University Fellowships, and a Senior NATO Research Fellowship. He is a member of many professional societies, including the ASME, AAAS, Pi Tau Sigma, and Sigma Xi. He is a Fellow of ASME and Technical Editor of the Solar Energy Division of that Society. Dr. Kreith has served as a consultant to industrial organizations, as well as to government agencies and to the Academy of Sciences.

Ronald E. West, Ph.D., P.E., is Professor of Chemical Engineering at the University of Colorado, Boulder, Colorado.

Dr. West graduated from the University of Michigan, Ann Arbor, with a B.S. in chemical engineering in 1954. He received his M.S. in 1955 and Ph.D. in 1958 in chemical engineering from Michigan.

Dr. West has been on the University of Colorado faculty since 1957. He was Year-In-Industry Professor at E. I. duPont de Nemours & Co., Inc. in 1965-66 and was a Visiting Research Engineer at the University of California, Santa Barbara in 1971.

His research interests are in the control of air and water pollution and the conservation of energy in industrial processes. He has authored several papers and monographs in these areas. He is a member of the American Institute of Chemical Engineers, the American Chemical Society, and the Water Pollution Control Federation. He presently serves as a consultant to the Solar Energy Research Institute on industrial process heat.

CONTRIBUTORS

Stanley W. Angrist, Ph.D.
Professor of Mechanical Engineering and Public Policy
Carnegie-Mellon University
Pittsburgh, Pennsylvania

Jerold W. Jones, Ph.D.
Associate Professor of Mechanical Engineering
Director of Conservation Programs
Center for Energy Studies
University of Texas
Austin, Texas

Wesley M. Rohrer, Jr., M.S.M.E.
Associate Professor of Mechanical Engineering
Director
Energy Analysis and Diagnostic Center
University of Pittsburgh
Pittsburgh, Pennsylvania

Craig B. Smith, Ph.D.
Principal
ANCO Engineers, Inc.
Santa Monica, California

Charles E. Wyman, Ph.D.
Group Manager
Thermal Conversion Branch
Solar Energy Research Institute
Golden, Colorado

TABLE OF CONTENTS

Volume III

Chapter 1

THERMAL CONSERVATION

J. W. Jones

TABLE OF CONTENTS

I. INTRODUCTION

Buildings are constructed to provide a means of maintaining a safe and comfortable internal environment despite variations in external environmental conditions. The extent to which the desired interior conditions can be economically maintained is one important measure of the success of the building project. Although environmental control usually involves an active heating and cooling system as well as the building envelope, energy-conscious design must start with an examination of the thermal performance of the envelope. If the envelope is designed so as to optimize heat gain from, or heat loss to, the exterior, both equipment capacity and energy consumption may be minimized. This consideration is of critical importance in solar heating and cooling applications.

This chapter examines procedures for evaluating the impact of the thermal characteristics of the building envelope on the design of solar energy systems.

II. EVALUATION OF THERMAL PERFORMANCE

Heat transfer through a building envelope is influenced by the materials used; by geometric factors such as size, shape, and orientation; by the existence of internal heat sources; and by climatic factors. Energy conserving design requires each of these factors be examined and the impact of their interactions with one another carefully evaluated during the design process. An overall framework for analysis must be established in order for this task to be performed. In the case of solar energy systems, it is necessary to go beyond the usual heat gain/heat loss calculations to the point of being able to estimate annual energy requirements. It is only on this basis that the economics of a design can be properly judged.

A. Calculation of Heat Loss/Heat Gain

The primary function of heat loss/heat gain calculations is to estimate the capacity that will be required for the various heating and air conditioning components necessary to maintain comfort within a space. These calculations are, therefore, based on peak load conditions for heating and cooling and correspond to environmental conditions which are near the extremes normally encountered. Standard outside design values are usually specified for the dry bulb temperature and coincident wind velocity for calculating heating load.

The set of conditions specified for cooling load calculations includes dry bulb temperature, humidity, and solar load. Peak load conditions during the cooling season usually correspond to the maximum solar load rather than to the peak outdoor air temperature. Thus it is often necessary to make several calculations to actually fix the appropriate maximum cooling capacity requirements. The choice of the sun time for which a cooling load calculation is made will depend on the geographic location and on the orientation of the space being considered. For example, peak solar loading on an east-facing room may occur at 8 a.m., while for a west room maximum load may occur at 4 p.m. Peak solar loads for south-facing rooms will occur during the winter rather than the summer. Of course, when a cooling system serves several spaces with different orientations, the peak system load may occur at a time other than the peak for any of the several spaces. Fortunately, after making a number of such calculations, one begins to recognize likely choices for times when the peak load may occur.

A number of load calculation procedures have been developed over the years. Although the procedures differ in some respects, they are all based on a systematic evaluation of the basic components of heat loss/heat gain. Loads are generally divided into the following four categories:

TABLE 1

Thermal Resistance of Common Building Materials
(at Mean Temperature of 24°C)
Resistance = (thickness, m) × (1/k, m°C/W) = 1/C m²°C/W

	1/k m°C/W	1/C m²°C/W
Exterior materials		
Brick		
Face	0.77	
Common	1.39	
Stone	0.55	
Concrete block		
Sand and gravel aggregate 200mm		0.20
Lightweight aggregate 200mm		0.35
Stucco	1.39	
Siding		
Asbestos-cement 6mm/lapped		0.04
Asphalt insulating 13mm		0.14
Wood drop 25mm × 200mm		0.14
Wood plywood 10mm		0.10
Aluminum or steel, insulation board backed (10mm)		0.32
Architectural glass		0.02
Building board		
Asbestos-cement	1.73	
Plywood	8.66	
Fiberboard, regular density 13mm		0.232
Hardboard, medium density	9.49	
Particle board, medium density	7.35	
Insulating materials		
Blanket and batt		
Mineral fiber 75—90mm		1.94
Mineral fiber 135—165mm		3.32
Board and slab		
Glass fiber, organic bond	27.7	
Expanded polystyrene, extruded	27.7	
Expanded polyurethane	43.8	
Loose fill		
Mineral fiber 160mm		3.35
Vermiculite	14.7	
Perlite	18.7	
Cellulosic	21.7—25.6	
Interior materials		
Gypsum or Plaster Board 13mm		0.08
16 mm		0.10
Plaster materials		
Cement plaster	1.39	
Gypsum plaster, lightweight 16mm		0.066
Gypsum plaster, sand aggregate		0.02
Wood		
Softwood (fir, pine, . . .)	8.76	
Hardwood (maple, oak, . . .)	6.3	
Roofing		
Asphalt shingles		0.08
Built-up roofing 10mm		0.06
Wood shingles 13mm		0.17
Slate 13mm		0.01

TABLE 1 (continued)

Thermal Resistance of Common Building Materials
(at Mean Temperature of 24°C)
Resistance = (thickness, m) × (1/k, m°C/W) = 1/C m²°C/W

	1/k m°C/W	1/C m²°C/W
Concrete		
Sand and gravel aggregate	0.55	
Lightweight aggregate	1.94	
Surface resistance		
	Direction of heat flow	1/C m²°C/W
Still air (ε = 0.9)		
	Horizontal, upward	0.11
	Horizontal, downward	0.16
	Vertical, horizontal	0.12
Moving air		
6.7 m/s		0.029
3.4 m/s		0.044
Air space (ε = 0.8)		
	Horizontal	0.14
	Vertical	0.17
(ε = 0.2)		
	Horizontal	0.24
	Vertical	0.36

1. Transmission, heat loss/heat gain due to a temperature difference across a building element
2. Solar, heat gain due to transmission of solar energy through a transparent building component or absorption by an opaque building component.
3. Infiltration, heat loss/heat gain due to the infiltration of outside air into a conditioned space
4. Internal, heat gain due to the release of energy within a space (lights, people, equipment, etc.)

In the following paragraphs we consider procedures for evaluating each of these load components. This presentation attempts to provide the basic information but cannot be complete. The interested reader is referred to the *ASHRAE Handbook and Product Directory, Fundamentals Volume* for a more complete discussion.[1]

In addition to being used for determining equipment capacity requirements, load calculations can be used for estimating seasonal energy requirements and evaluating the extent to which various design decisions can influence thermal energy conservation.

1. Transmission Loads

The general procedure for calculating heat loss/heat gain due to thermal transmission is to apply Fourier's conduction equation in the form of Equation 36 of Volume 1, Chapter 2.

$$q = UA(T_o - T_i) \tag{1}$$

As discussed in Volume I, Chapter 2, the overall heat transfer coefficient U is a function of the thermal resistance. The above Table 1 and Tables 5 and 8 through 15 of

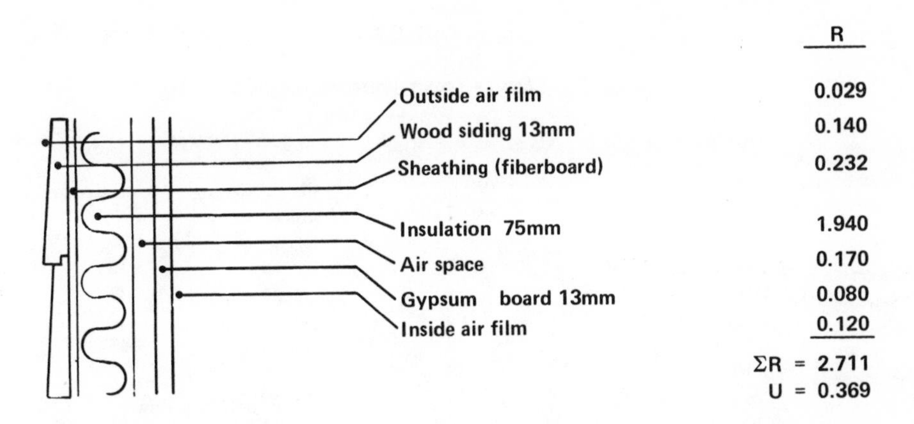

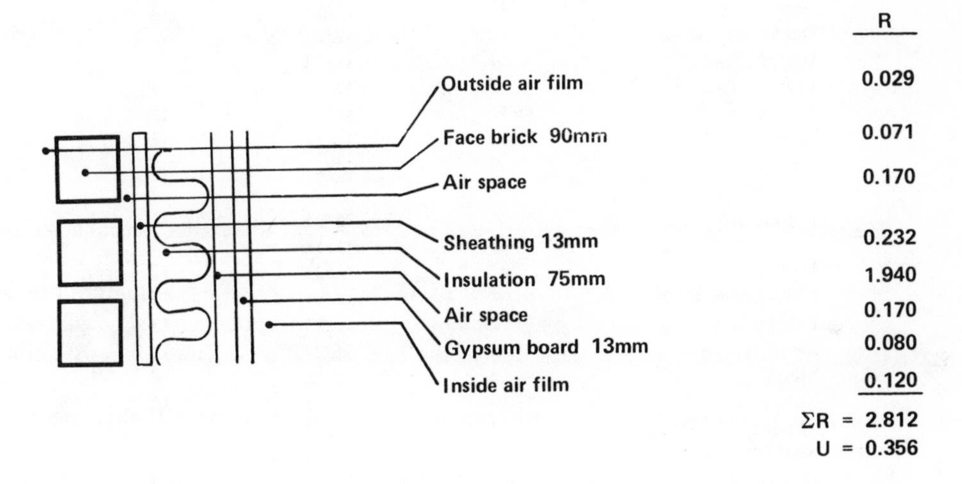

FIGURE 1. Wall U values.

Volume I, Chapter 2, provide values of thermal resistance for common building materials, enclosed air spaces, and building envelope boundaries. Figure 1 illustrates the determination of U values for typical wall and roof cross sections. The areas used in transmission calculations are nominal inside areas (linear dimensions rounded to nearest 10 cm). The temperature difference is the difference between the inside and outside design temperatures. Values for summer and winter design temperatures are given in Table 2.

Example 1 — Select outside and inside design temperatures for a home to be built in Omaha, Nebraska.

From Table 2 for summer conditions

- Summer dry bulb temperature = 33°C
- Coincident wet bulb temperature = 24°C
- Daily temperature range = 12°C

TABLE 2

Design Temperature Data

		Summer	
	Winter 97.5% db°C	2.5% db°C/ Coincident wb°C	Mean daily range°C
Albuquerque, N.M.	−9	33/16	15
Atlanta, Ga.	−6	33/23	11
Birmingham, Ala.	−6	34/24	12
Bismarck, N.D.	−28	33/20	15
Boise, Idaho	−12	34/18	17
Boston, Mass.	−13	31/22	9
Charlotte, N.C.	−6	34/23	11
Cheyenne, Wyo.	−18	30/14	17
Chicago, Ill.	−18	33/23	11
Cincinnati, Ohio	−14	32/22	12
Cleveland, Ohio	−15	31/22	12
Columbus, Ohio	−15	32/23	13
Dallas, Tex.	−6	28/24	11
Denver, Colo.	−17	33/15	15
Des Moines, Iowa	−21	33/23	13
Detroit, Mich.	−14	31/22	11
El Paso, Tex.	−5	37/18	15
Fort Wayne, Ind.	−17	32/22	13
Great Falls, Mont.	−26	31/16	15
Houston, Tex.	0	34/25	10
Jackson, Miss.	−4	35/24	12
Kansas City, Mo.	−14	36/23	11
Las Vegas, Nev.	−2	41/18	17
Little Rock, Ark.	−7	36/25	12
Los Angeles, Calif.	4	32/21	11
Louisville, Ky.	−12	34/23	13
Memphis, Tenn.	−8	35/24	12
Miami, Fla.	8	32/25	8
Milwaukee, Wis.	−20	31/23	12
Minneapolis, Minn.	−24	37/23	12
Nashville, Tenn.	−10	34/23	12
New Orleans, La.	−4	33/26	9
New York, N.Y.	−9	32/23	9
Oklahoma City, Okla.	−11	36/23	13
Omaha, Neb.	−19	33/24	12
Philadelphia, Pa.	−10	32/23	12
Phoenix, Ariz.	1	42/22	15
Pittsburgh, Pa.	−14	31/22	11
Portland, Maine	−18	29/22	12
Portland, Ore.	−4	30/20	12
Richmond, Va.	−8	33/24	12
Sacramento, Calif.	0	37/21	20
Salt Lake City, Utah	−13	35/17	18
San Antonio, Tex.	−1	36/23	11
San Francisco, Calif.	+4	22/17	8
Seattle, Wash.	−3	28/19	11
Shreveport, La.	−4	36/24	11
Spokane, Wash.	−17	32/17	15
St. Louis, Mo.	−13	34/24	10
Syracuse, N.Y.	−17	31/22	11
Tampa, Fla.	4	33/25	9
Washington, D.C.	−8	33/23	10
Wichita, Kan.	−14	37/23	13

Assuming no special requirements exist, an inside design temperature of 25°C is chosen.

Winter conditions, winter dry bulb temperature = −19°C.

Again, assuming no special requirements exist, an inside design temperature of 20°C is chosen.

It should be noted that the choice of the inside design temperature only limits the conditions that can be maintained in extreme weather. When the outside temperature is above the outside design value, an inside temperature greater than 20°C could be maintained if desired.

In the case of spaces adjacent to an attic it is necessary to estimate the attic temperature in order to calculate the heat loss/heat gain through the ceiling. The attic temperature T_a is estimated on the basis of an energy balance on the attic space which results in:

$$T_a = \frac{T_i(A_1U_1 + A_2U_2 + ...) + T_o(\dot{m}\,c_p + A_aU_a + A_bU_b + ...)}{A_1U_1 + A_2U_2 + ... + \dot{m}\,c_p + A_aU_a + A_bU_b + ...} \quad (2)$$

where T_i = inside temperature; T_o = outside temperature; A_1, A_2, etc. = areas of surfaces adjacent to conditioned space; U_1, U_2, etc. = heat transfer coefficients for surfaces 1, 2, etc.; A_a, A_b, etc = areas of surfaces exposed to outside conditions; U_a, U_b, etc. = heat transfer coefficients of surfaces a, b; $\dot{m}$ = mass flow rate of infiltration (or ventilation) air into attic; c_p = the specific heat of air.

An important exception to the use of the outdoor design temperature to calculate transmission heat gain is in the case of sunlit walls and roofs. This situation is discussed in the next section.

2. Solar Loads

Heat gain due to solar energy incident on a surface will depend upon the physical characteristics of the surface. Surface optical properties are described by Equation 53 (Volume I, Chapter 2): $\tau + \varrho + \alpha = 1$; τ = transmittance; ϱ = reflectance; α = absorptance. The value of each of these has a pronounced effect on solar heat gain.

For transparent surfaces, with the total solar irradiation reaching the surface being I_t, the energy passing through is

$$q_{sg} = A(\tau I_t + N\alpha I_t) = AI_t(\tau + N\alpha) \quad (3)$$

where N = the fraction of the absorbed radiation that is transferred by conduction and convection to the inside environment = $U\alpha/h_o$; U = air-to-air heat transfer coefficient; h_o = outside heat transfer coefficient. Values of the term $(\tau + U\alpha/h_o)$ have been calculated for different transparent materials. The product of $(\tau + U\alpha/h_o)$ for a single sheet of clear window glass and I_t is frequently referred to as the solar heat gain factor (SHFG). Values for the SHGF are presented in tabular fashion for various latitudes and orientations in Table 3. A shading coefficient (SC) is used to adjust these values for other types of glass. This coefficient is simply;

$$SC = \frac{(\tau + U\alpha/h_o)}{(\tau + U\alpha/h_o)_{\text{single sheet clear glass}}}$$

Typical values of the shading coefficient for several types of glass with and without

TABLE 3a

Solar Position and Intensity; Solar Heat Gain Factors for 32°N. Latitude

Date	Time A.M.	Solar position		Direct normal irradiation, (Btu/hr ft²)	Solar Heat Gain Factors, Btu/(hr)(sq ft)[a]									Time P.M.
		Alt.	Azimuth		N	NE	E	SE	S	SW	W	NW	Hor.	
Jan 21	7	1.4	65.2	1	0	0	1	1	0	0	0	0	0	5
	8	12.5	56.5	202	8	29	160	189	103	9	8	8	32	4
	9	22.5	46.0	269	15	16	175	246	169	16	15	15	88	3
	10	30.6	33.1	295	19	20	135	249	212	45	19	19	136	2
	11	36.1	17.5	306	22	22	67	221	238	110	22	22	166	1
	12	38.0	0.0	309	23	23	25	174	246	174	25	23	176	12
			Half day totals		75	91	529	974	834	262	75	75	509	
Feb 21	7	6.7	72.8	111	4	47	102	95	26	4	4	4	9	5
	8	18.5	63.8	244	12	64	205	217	95	12	12	12	63	4
	9	29.3	52.8	287	19	32	199	248	149	19	19	19	127	3
	10	38.5	38.9	305	23	24	151	241	189	31	23	23	176	2
	11	44.9	21.0	314	26	26	76	208	213	87	26	26	207	1
	12	47.2	0.0	316	27	27	29	155	221	155	29	27	217	12
			Half day totals		97	207	749	1091	780	227	98	97	689	
Mar 21	7	12.7	81.9	184	9	105	176	142	19	9	9	9	31	5
	8	25.1	73.0	260	17	107	227	209	62	17	17	17	99	4
	9	36.8	62.1	289	23	64	210	227	107	23	23	23	163	3
	10	47.3	47.5	304	27	30	158	215	144	29	27	27	211	2
	11	55.0	26.8	310	30	31	82	179	168	58	30	30	242	1
	12	58.0	0.0	312	31	31	33	122	176	122	33	31	252	12
			Half day totals		122	368	891	1054	588	191	123	122	872	
Apr 21	6	6.1	99.9	66	9	54	65	37	3	3	3	3	7	6
	7	18.8	92.2	206	17	147	201	136	15	14	14	14	61	5
	8	31.5	84.0	256	23	144	228	178	30	22	22	22	130	4

	9	43.9	74.2	278	28	103	206	188	58	27	27	27	189	3
	10	55.7	60.3	290	32	52	156	173	87	33	32	32	234	2
	11	65.4	37.5	296	34	36	83	135	108	40	34	34	263	1
	12	69.6	0.0	298	35	35	38	82	115	82	38	35	272	12
			Half day totals		159	559	965	898	359	174	150	149	1022	
May 21	6	10.4	107.2	118	32	108	116	55	8	8	8	8	21	6
	7	22.8	100.1	211	35	170	204	118	18	18	18	18	81	5
	8	35.4	92.9	249	29	165	220	149	27	25	25	25	146	4
	9	48.1	84.7	269	32	128	198	155	37	30	30	30	201	3
	10	60.6	73.3	279	36	76	150	138	54	35	35	35	243	2
	11	72.0	51.9	285	38	41	82	102	68	39	37	37	269	1
	12	78.0	0.0	286	38	39	41	59	74	59	41	39	277	12
			Half day totals		217	697	983	747	248	181	172	171	1100	
June 21	6	12.2	110.2	130	44	123	127	55	10	10	10	10	28	6
	7	24.3	103.4	209	46	176	201	108	19	19	19	19	88	5
	8	36.9	96.8	244	36	171	214	135	28	26	26	26	151	4
	9	49.6	89.4	263	34	136	193	139	35	32	32	32	203	3
	10	62.2	79.7	273	38	86	146	122	45	36	36	36	244	2
	11	74.2	60.9	278	40	46	81	88	56	40	38	38	268	1
	12	81.5	0.0	280	40	41	42	52	60	52	42	41	276	12
			Half day totals		252	744	972	672	222	186	180	180	1119	
July 21	6	10.7	107.7	113	33	105	112	53	8	8	8	8	22	6
	7	23.1	100.6	203	37	167	198	114	19	18	18	18	81	5
	8	35.7	93.6	241	31	163	216	145	28	26	26	26	145	4
	9	48.4	85.5	261	34	128	195	150	37	31	31	31	199	3
	10	60.9	74.3	271	37	78	148	134	53	35	35	35	240	2
	11	72.4	53.3	277	39	43	82	99	66	40	38	38	265	1
	12	78.6	0.0	278	40	40	42	58	71	58	42	40	273	12
			Half day totals		227	694	964	724	245	184	176	175	1089	

TABLE 3a (continued)

Solar Position and Intensity; Solar Heat Gain Factors for 32°N Latitude

Date	Time	Solar position		Direct normal irradiation,	Solar Heat Gain Factors, Btu/(hr)(sq ft)[a]									Time
	A.M.	Alt.	Azimuth	(Btu/hr ft²)	N	NE	E	SE	S	SW	W	NW	Hor.	P.M.
Aug 21	6	6.5	100.5	59	9	50	59	34	3	3	3	3	7	6
	7	19.1	92.8	189	18	140	189	127	16	15	15	15	61	5
	8	31.8	84.7	239	25	141	219	170	30	23	23	23	127	4
	9	44.3	75.0	263	30	104	200	180	56	29	29	29	185	3
	10	56.1	61.3	275	33	55	152	167	84	34	33	33	229	2
	11	66.0	38.4	281	36	38	83	131	104	41	36	36	256	1
	12	70.3	0.0	283	37	37	40	80	111	80	40	37	265	12
			Half day totals		169	552	929	858	349	180	159	158	999	
Sep 21	7	12.7	81.9	163	10	95	159	128	19	9	9	9	30	5
	8	25.1	73.0	240	18	103	215	199	60	18	18	18	96	4
	9	36.8	62.1	272	24	64	202	218	105	24	24	24	158	3
	10	47.3	47.5	287	29	32	154	208	141	30	29	29	204	2
	11	55.0	26.8	294	31	32	81	174	164	59	31	31	234	1
	12	58.0	0.0	296	32	32	34	120	171	120	34	32	244	12
			Half day totals		128	355	846	1004	575	194	128	127	844	
Oct 21	7	6.8	73.1	98	4	43	92	85	23	4	4	4	9	5
	8	18.7	64.0	229	13	63	195	205	90	13	13	13	62	4
	9	29.5	53.0	273	19	33	193	239	144	20	19	19	125	3
	10	38.7	39.1	292	24	25	147	234	183	32	24	24	173	2
	11	45.1	21.1	301	27	27	75	202	206	85	27	27	203	1
	12	47.5	0.0	304	28	28	30	151	214	151	30	28	213	12
			Half day totals		100	205	718	1044	750	225	101	100	677	
Nov 21	7	1.5	65.4	1	0	0	1	1	0	0	0	0	0	5
	8	12.7	56.6	196	9	29	156	183	100	9	9	9	32	4

	9	22.6	46.1	262	15	17	172	241	166	16	15	15	87	3
	10	30.8	33.2	288	20	20	134	244	209	45	20	20	135	2
	11	36.2	17.6	300	22	22	67	218	234	108	22	22	165	1
	12	38.2	0.0	303	23	23	25	171	243	171	25	23	175	12
			Half day totals		76	92	521	955	820	258	77	76	505	
Dec 21	8	10.3	53.8	176	7	18	135	166	96	7	7	7	22	4
	9	19.8	43.6	257	13	14	162	238	171	15	13	13	72	3
	10	27.6	31.2	287	18	18	127	246	217	52	18	18	119	2
	11	32.7	16.4	300	20	20	63	222	243	116	20	20	148	1
	12	34.6	0.0	304	21	21	23	177	252	177	23	21	158	12
			Half day totals		67	76	482	947	844	273	68	67	440	↑
					N	NW	W	SW	S	SE	E	NE	HOR.←	P.M.

[a] To convert from English to SI units: (Btu/hr)/ft^2 × 3.152 = W/m^2

Reprinted by permission from *ASHRAE Handbook of Fundamentals*, American Society of Heating, Refrigerating and Air Conditioning Engineers, New York, 1967, 471.

TABLE 3b

Solar Position and Intensity; Solar Heat Gain Factors for 40° N. Latitude

Date	Time A.M.	Solar position		Direct normal irradiation (Btu/hr ft^2)	Solar heat gain factors, Btu/(hr)(sq ft)[a]									Time P.M.
		Alt.	Azimuth		N	NE	E	SE	S	SW	W	NW	Hor.	
Jan 21	8	8.1	55.3	141	5	17	111	133	75	5	5	5	13	4
	9	16.8	44.0	238	11	12	154	224	160	13	11	11	54	3
	10	23.8	30.9	274	16	16	123	241	213	51	16	16	96	2
	11	28.4	16.0	289	18	18	61	222	244	118	18	18	123	1
	12	30.0	0.0	293	19	19	20	179	254	179	20	19	133	12
			Half day totals		59	68	449	903	815	271	59	59	353	

TABLE 3b (continued)

Solar Position and Intensity; Solar Heat Gain Factors for 40° N. Latitude

Date	Time A.M.	Solar position: Alt.	Solar position: Azimuth	Direct normal irradiation ($Btu/hr\ ft^2$)	Solar heat gain factors, Btu/(hr)(sq ft)[a]: N	NE	E	SE	S	SW	W	NW	Hor.	Time P.M.
Feb 21	7	4.3	72.1	55	1	22	50	47	13	1	1	1	3	5
	8	14.8	61.6	219	10	50	183	199	94	10	10	10	43	4
	9	24.3	49.7	271	16	22	186	245	157	17	16	16	98	3
	10	32.1	35.4	293	20	21	142	247	203	38	20	20	143	2
	11	37.3	18.6	303	23	23	71	219	231	103	23	23	171	1
	12	39.2	0.0	306	24	24	25	170	241	170	25	24	180	12
			Half day totals		81	144	634	1035	813	250	81	81	546	
Mar 21	7	11.4	80.2	171	8	93	163	135	21	8	8	8	26	5
	8	22.5	69.6	250	15	91	218	211	73	15	15	15	85	4
	9	32.8	57.3	281	21	46	203	236	128	21	21	21	143	3
	10	41.6	41.9	297	25	36	153	229	171	28	25	25	186	2
	11	47.7	22.6	304	28	28	78	198	197	77	28	28	213	1
	12	50.0	0.0	306	28	28	30	145	206	145	30	28	223	12
			Half day totals		112	310	849	1100	692	218	112	112	764	
Apr 21	6	7.4	98.9	89	11	72	88	52	5	4	4	4	11	6
	7	18.9	89.5	207	16	141	201	143	16	14	14	14	61	5
	8	30.3	79.3	253	22	128	225	189	41	21	21	21	124	4
	9	41.3	67.2	275	26	80	203	204	83	26	26	26	177	3
	10	51.2	51.4	286	30	37	153	194	121	32	30	30	218	2
	11	58.7	29.2	292	33	34	81	161	146	52	33	33	244	1
	12	61.6	0.0	294	33	33	36	108	155	108	36	33	253	12
			Half day totals		153	509	969	1003	489	196	146	145	962	
May 21	5	1.9	114.7	1	0	0	0	0	0	0	0	0	0	7
	6	12.7	105.6	143	35	128	141	71	10	10	10	10	30	6
	7	24.0	96.6	216	28	165	209	131	20	18	18	18	87	5

	8	35.4	87.2	249	27	149	220	164	29	25	25	25	146	4
	9	46.8	76.0	267	31	105	197	175	53	30	30	30	196	3
	10	57.5	60.9	277	34	54	148	163	83	35	34	34	234	2
	11	66.2	37.1	282	36	38	81	130	105	42	36	36	258	1
	12	70.0	0.0	284	37	37	40	82	112	82	40	37	265	12
			Half day totals		203	643	1002	874	356	194	171	170	1083	
June 21	5	4.2	117.3	21	10	21	6	1	1	1	1	2	7	
	6	14.8	108.4	154	47	142	151	70	12	12	12	12	39	6
	7	26.0	99.7	215	37	172	207	122	21	20	20	20	97	5
	8	37.4	90.7	246	29	156	215	152	29	26	26	26	153	4
	9	48.8	80.2	262	33	113	192	161	45	31	31	31	201	3
	10	59.8	65.8	272	35	62	145	148	69	36	35	35	237	2
	11	69.2	41.9	276	37	40	80	116	88	41	37	37	260	1
	12	73.5	0.0	278	38	38	41	71	95	71	41	38	267	12
			Half day totals		242	714	1019	810	311	197	181	180	1121	
July 21	5	2.3	115.2	2	0	2	1	0	0	0	0	0	0	7
	6	13.1	106.1	137	37	125	137	68	10	10	10	10	31	6
	7	24.3	97.2	208	30	163	204	127	20	19	19	19	88	5
	8	35.8	87.8	241	28	148	216	160	29	26	26	26	145	4
	9	47.2	76.7	259	32	106	194	170	52	31	31	31	194	3
	10	57.9	61.7	269	35	56	146	159	80	36	35	35	231	2
	11	66.7	37.9	274	37	39	81	127	102	42	37	37	255	1
	12	70.6	0.0	276	38	38	41	80	109	80	41	38	262	12
			Half day totals		211	645	986	850	347	197	177	176	1074	
Aug 21	6	7.9	99.5	80	12	67	82	48	5	5	5	5	11	6
	7	19.3	90.0	191	17	135	191	135	17	15	15	15	62	5
	8	30.7	79.9	236	23	126	216	180	40	22	22	22	122	4
	9	41.8	67.9	259	28	82	197	196	79	28	28	28	174	3
	10	51.7	52.1	271	32	40	149	187	116	34	32	32	213	2
	11	59.3	29.7	277	34	35	81	156	140	52	34	34	238	1

TABLE 3b (continued)

Solar Position and Intensity; Solar Heat Gain Factors for 40° N. Latitude

Date	Time A.M.	Solar position: Alt.	Solar position: Azimuth	Direct normal irradiation (Btu/hr ft²)	Solar heat gain factors, Btu/(hr)(sq ft)[a]: N	NE	E	SE	S	SW	W	NW	Hor.	Time P.M.
	12	62.3	0.0	279	35	35	38	105	149	105	38	35	247	12	
			Half day totals		161	503	936	961	471	202	154	153	945		
Sept 21	7	11.4	80.2	149	8	84	146	121	21	8	8	8	25	5	
	8	22.5	69.6	230	16	87	205	199	71	16	16	16	82	4	
	9	32.8	57.3	263	22	47	195	226	124	23	22	22	138	3	
	10	41.6	41.9	279	26	28	148	221	165	30	26	26	180	2	
	11	47.7	22.6	287	29	29	77	192	191	77	29	29	206	1	
	12	50.0	0.0	290	30	30	32	141	200	141	32	30	215	12	
			Half day totals		16	300	803	1045	672	221	117	116	738		
Oct 21	7	4.5	72.3	48	1	20	45	41	12	1	1	1	3	5	
	8	15.0	61.9	203	10	49	173	187	88	10	10	10	43	4	
	9	24.5	49.8	257	17	23	180	235	151	18	17	17	96	3	
	10	32.4	35.6	280	21	22	139	238	196	38	21	21	140	2	
	11	37.6	18.7	290	23	23	70	212	224	100	23	23	167	1	
	12	39.5	0.0	293	24	24	26	165	234	165	26	24	177	12	
			Half day totals		83	143	610	989	783	245	84	83	535		
Nov 21	8	8.2	55.4	136	5	17	107	128	72	5	5	5	14	4	
	9	17.0	44.1	232	12	13	151	219	156	13	12	12	54	3	
	10	24.0	31.0	267	16	16	122	237	209	50	16	16	96	2	
	11	28.6	16.1	283	19	19	61	218	240	116	19	19	123	1	
	12	30.2	0.0	287	19	19	21	176	250	176	21	19	132	12	
			Half day totals		61	71	442	884	798	267	62	61	353		

Date	Time A.M.	Alt.	Azimuth	Direct normal irradiation	N	NE	E	SE	S	SW	W	NW	Hor.	Time P.M.
Dec 21	8	5.5	53.0	88	2	7	67	83	49	3	2	2	6	4
	9	14.0	41.9	217	9	10	135	205	151	12	9	9	39	3
	10	20.7	29.4	261	14	14	113	232	210	55	14	14	77	2
	11	25.0	15.2	279	16	16	56	217	242	120	16	16	103	1
	12	26.6	0.0	284	17	17	18	177	253	177	18	17	113	12
			Half day totals		49	54	380	831	781	273	50	49	282	↑
					N	NW	W	SW	S	SE	E	NE	HOR.	←P.M.

[a] To convert from English to SI units: (Btu/hr/ft²) × 3.152 = W/m²

Reprinted by permission from *ASHRAE Handbook of Fundamentals*, American Society of Heating, Refrigerating, and Air Conditioning Engineers, New York, 1967, p. 472.

TABLE 3c

Solar Position and Intensity; Solar Heat Gain Factors for 48° N. Latitude

Date	Time A.M.	Solar position: Alt.	Azimuth	Direct normal irradiation, (Btu/hr ft²)	Solar heat gain factors, Btu/(hr) (sq ft)[a]: N	NE	E	SE	S	SW	W	NW	Hor.	Time P.M.
Jan 21	8	3.5	54.6	36	1	4	28	34	19	1	1	1	2	4
	9	11.0	42.6	185	7	8	117	176	128	9	7	7	25	3
	10	16.9	29.4	239	11	11	105	216	195	50	11	11	55	2
	11	20.7	15.1	260	14	14	52	208	233	115	14	14	77	1
	12	22.0	0.0	267	15	15	16	171	245	171	16	15	85	12
			Half day totals		41	44	319	735	705	256	41	41	202	
Feb 21	7	1.8	71.7	3	0	1	3	3	0	0	0	0	0	5
	8	10.9	60.0	180	7	36	149	166	82	7	7	7	24	4
	9	19.0	47.3	247	13	15	168	230	155	14	13	13	66	3
	10	25.5	33.0	275	17	17	131	242	207	43	17	17	105	2
	11	29.7	17.0	288	19	19	65	221	240	112	19	19	129	1
	12	31.2	0.0	291	20	20	21	176	251	176	21	20	138	12
			Half day totals		65	88	508	936	803	260	65	65	392	

TABLE 3c (continued)

Solar Position and Intensity; Solar Heat Gain Factors for 48° N. Latitude

Date	Time A.M.	Solar position		Direct normal irradiation, (Btu/hr ft²)	Solar heat gain factors, Btu/(hr) (sq ft)[a]									Time P.M.
		Alt.	Azimuth		N	NE	E	SE	S	SW	W	NW	Hor.	
Mar 21	7	10.0	78.7	152	7	80	145	123	22	7	7	7	20	5
	8	19.5	66.8	235	13	75	204	206	81	13	13	13	67	4
	9	28.2	53.4	270	19	33	193	239	142	19	19	19	117	3
	10	35.4	37.8	287	22	23	146	237	189	33	22	22	156	2
	11	40.3	19.8	295	24	24	73	210	218	94	24	24	180	1
	12	42.0	0.0	297	25	25	27	161	228	161	27	25	188	12
			Half day totals		98	256	790	1111	765	244	99	98	634	
Apr 21	6	8.6	97.8	108	12	86	107	64	6	6	6	6	14	6
	7	18.6	86.7	205	15	133	200	149	17	14	14	14	60	5
	8	28.5	74.9	247	20	111	219	197	55	20	20	20	114	4
	9	37.8	61.2	269	25	60	198	216	107	25	25	25	161	3
	10	45.8	44.6	281	28	30	148	210	150	30	28	28	197	2
	11	51.5	24.0	287	30	30	78	181	177	69	30	30	219	1
	12	53.6	0.0	289	31	31	33	132	187	132	33	31	227	12
			Half day totals		143	459	961	1086	604	225	139	138	879	
May 21	5	5.2	114.3	41	16	39	38	13	2	2	2	2	4	7
	6	14.7	103.7	162	35	140	160	84	12	12	12	12	39	6
	7	24.6	93.0	218	23	158	212	142	21	19	19	19	91	5
	8	34.6	81.6	248	26	132	218	178	38	24	24	24	142	4
	9	44.3	68.3	264	29	82	194	192	77	29	29	29	186	3
	10	53.0	51.3	274	32	39	145	184	116	34	32	32	219	2
	11	59.5	28.6	279	34	35	79	155	142	54	34	34	240	1
	12	62.0	0.0	280	35	35	38	106	150	106	38	35	247	12
			Half day totals		209	637	1058	1002	483	220	170	169	1043	

June 21	5	7.9	116.5	77	35	76	72	23	5	5	5	5	12	7
	6	17.2	106.2	172	46	154	169	84	14	14	14	14	51	6
	7	27.0	95.8	219	29	165	211	135	22	21	21	21	102	5
	8	37.1	84.6	245	28	139	215	167	34	26	26	26	152	4
	9	46.9	71.6	260	31	90	190	179	66	31	31	31	193	3
	10	55.8	54.8	269	34	45	142	171	101	36	34	34	225	2
	11	62.7	31.2	273	36	37	78	142	125	49	36	36	245	1
	12	65.4	0.0	275	36	36	39	96	134	96	39	36	252	12
			Half day totals		258	728	1098	952	434	224	186	185	1105	
July 21	5	5.7	114.7	42	18	42	40	14	2	2	2	2	5	7
	6	15.2	104.1	155	36	138	156	82	12	12	12	12	41	6
	7	25.1	93.5	211	24	156	207	138	21	20	20	20	92	5
	8	35.1	82.1	240	27	132	214	174	38	25	25	25	142	4
	9	44.8	68.8	256	30	83	191	187	75	30	30	30	184	3
	10	53.5	51.9	266	33	41	143	180	113	35	33	33	217	2
	11	60.1	29.0	271	35	37	79	151	138	53	35	35	238	1
	12	62.6	0.0	272	36	36	39	104	146	104	39	36	245	12
			Half day totals		218	643	1044	978	472	222	175	174	1040	
Aug 21	6	9.1	98.3	98	13	81	100	60	7	6	6	6	16	6
	7	19.1	87.2	189	16	127	189	140	18	15	15	15	61	5
	8	29.0	75.4	231	21	110	211	188	53	21	21	21	113	4
	9	38.4	61.8	253	26	62	192	208	102	26	26	26	159	3
	10	46.4	45.1	265	30	32	145	202	144	32	30	30	193	2
	11	52.2	24.3	271	32	32	78	175	171	68	32	32	215	1
	12	54.3	0.0	273	33	33	35	128	180	128	35	33	222	12
			Half day totals		152	454	927	1040	584	227	147	146	868	
Sep 21	7	10.0	78.7	131	7	71	128	108	21	7	7	7	19	5
	8	19.5	66.8	215	14	72	191	193	77	14	14	14	65	4
	9	28.2	53.4	251	20	33	184	227	136	20	20	20	113	3
	10	35.4	37.8	269	23	25	141	228	182	34	23	23	151	2
	11	40.3	19.8	277	26	26	73	203	211	92	26	26	174	1
	12	42.0	0.0	280	26	26	28	156	220	156	28	26	182	12
			Half day totals		104	246	744	1050	736	242	104	104	612	

TABLE 3c (continued)

Solar Position and Intensity; Solar Heat Gain Factors for 48° N. Latitude

Date	Time A.M.	Solar position: Alt.	Solar position: Azimuth	Direct normal irradiation, (Btu/hr ft²)	Solar heat gain factors, Btu/(hr) (sq ft)[a]: N	NE	E	SE	S	SW	W	NW	Hor.	Time P.M.
Oct 21	7	2.0	71.9	3	0	1	3	3	0	0	0	0	0	5
	8	11.2	60.2	165	8	35	139	155	76	8	8	8	24	4
	9	19.3	47.4	232	13	16	161	219	147	15	13	13	66	3
	10	25.7	33.1	261	17	18	127	233	199	43	17	17	103	2
	11	30.0	17.1	274	20	20	64	213	231	109	20	20	127	1
	12	31.5	0.0	278	20	20	22	170	242	170	22	20	136	12
			Half day totals		67	91	488	895	768	256	68	67	387	
Nov 21	8	3.6	54.7	36	1	4	28	34	19	1	1	1	2	4
	9	11.2	42.7	178	7	8	115	171	125	9	7	7	25	3
	10	17.1	29.5	232	12	12	104	211	191	49	12	12	55	2
	11	20.9	15.1	254	14	14	52	204	228	113	14	14	77	1
	12	22.2	0.0	260	15	15	16	168	240	168	16	15	85	12
			Half day totals		41	45	316	719	690	252	42	41	202	
Dec 21	9	8.0	40.9	140	5	5	86	133	100	7	5	5	13	3
	10	13.6	28.2	214	9	9	91	194	179	49	9	9	37	2
	11	17.3	14.4	242	12	12	46	195	220	111	12	12	57	1
	12	18.6	0.0	250	12	12	14	163	233	163	14	12	64	12
			Half day totals		32	32	241	621	623	244	33	32	139	↑
					N	NW	W	SW	S	SE	E	NE	HOR.	←P.M.

[a] To convert from English to SI units: (Btu/hr/ft²) × 3.152 = W/m².

Reprinted by permission from *ASHRAE Handbook of Fundamentals*, American Society of Heating, Refrigerating, and Air Conditioning Engineers, New York, 1967, p. 473.

TABLE 4

Shading Coefficients

Type of glass		No indoor shading	Venetian blinds: Medium	Venetian blinds: Light	Roller shades: Dark	Roller shades: White
Regular sheet	3mm	1.00	0.64	0.55	0.59	0.25
Regular plate	6mm	0.95	0.64	0.55	0.59	0.25
	12mm	0.88	0.64	0.55	0.59	0.25
Heat absorbing	6mm	0.70	0.54	0.52	0.45	0.30
	12mm	0.50	0.54	0.52	0.45	0.30
Insulating						
Regular sheet		0.90	0.57	0.51	0.60	0.25
Plate		0.83	0.57	0.51	0.60	0.25
Reflective		0.2—0.4	0.2—0.33			

Shading coefficient (applies to all shading columns)

Reprinted by permission from *ASHRAE Handbook of Fundamentals,* American Society of Heating, Refrigerating and Air Conditioning Engineers, New York, 1967, 482.

internal shading are presented in Table 4. If external surfaces shade the window, SHGF values for a north orientation are used. Thus, the instantaneous heat gain due to solar energy passing through a window can be expressed as:

$$q_{sg} = (\text{SHGF})\ (\text{SC})A$$

When the transmission component due to temperature difference is also included, the heat gain through a window can be expressed as:

$$q_g = [(\text{SHGF})\text{SC} + U\ (T_o - T_i)]\ A \tag{4}$$

Example 2 — Determine the peak solar heat gain for a southwest-facing window at 40°N. Latitude, July 21. The window is equipped with light color venetian blinds. From Table 3b solar heat gain factors are 159 Btu/hr-ft^2 at 2 p.m., 170 Btu/hr-ft^2 at 3 p.m., and 160 Btu/hr-ft^2 at 4 p.m. Therefore, the peak SHGF is 170 Btu/hr-ft^2 at 3 p.m. for a sunlit window. (Note: The p.m. hours appear at the right-hand side of Table 3b while the orientation for p.m. hours appears along the bottom.) From Table 4, q_{ws} = (SHGF)(SC); q_{ws} = (170)(0.55) = 93.5 Btu/hr-ft^2; = 93.5(3.15) = 294.5 watts/m^2. However, before proceeding, two other factors should be considered.

First, shading from overhangs or other projections can significantly reduce solar heat gain through a window. The depth of a shadow cast by a horizontal projection above a window can be calculated from the solar altitude angle β and the wall azimuth angle γ. The solar altitude β and the solar azimuth angle ϕ are given in Table 3 (a, b, and c). The wall azimuth

$$\gamma = \phi \pm \psi$$

Where ψ is the wall azimuth measured east or west from south. The depth of a shadow below a horizontal projection of width d is given by

$$y = d\ \frac{\tan \beta}{\cos \gamma}$$

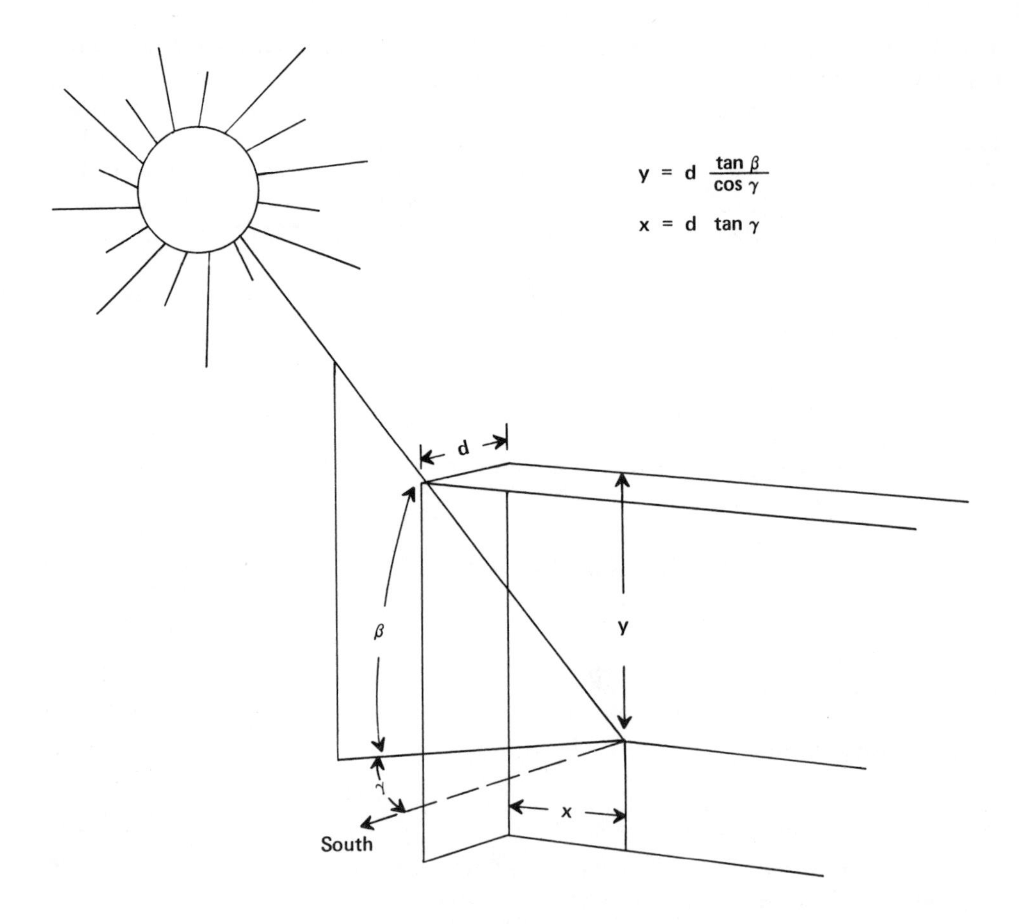

FIGURE 2. Calculation of shadow position.

The width of a shadow cast by a vertical projection of depth d is $x = d \tan \gamma$.

Figure 2 illustrates the calculation of the shadow position.

Example 3 — A 1.2m by 2.5m window is inset from the face of the wall 0.15m. Calculate the shading provided by the inset at 2 p.m. sun time if the window is facing south at 32° N. Latitude, August 21.

For a south-facing window, $\phi = 0$ and thus $\gamma = \phi$. From Table 3a

- $\beta = 56.1°$
- $\gamma = 61.3$
- $y = d \tan \beta / \cos \gamma$
  $= (0.15) \tan 56.1 / \cos 61.3$
  $= 0.46\text{m}$
- $x = d \tan \gamma$
  $= (0.15) \tan 61.3$
  $= 0.27\text{m}$

Sunlit area $= (2.5 - 0.27)(1.25 - 0.46)$
$= 1.76\text{m}^2$

Shaded area $= 1.36\ \text{m}^2$

The second factor which must be considered is that solar energy entering a space does not appear instantaneously as a load on the cooling system. The radiant energy is first absorbed by the surfaces in the space, then their temperature increases at a rate that is dependent on their mass. Thus, the solar energy absorbed is "delayed" before being transferred to the air in the space by convection. This process may involve a significant time delay; therefore, it is the usual practice to use an average value for the SHGF in Equation 4. The average is based on the SHGF over the preceding 3 to 5 hr depending on the mass of the space components. The shorter time period is used for averaging with light construction and the longer for spaces with more massive elements such as concrete floors.

The transmissivity, τ, of an opaque wall is zero and thus $\varrho + \alpha = 1$. Therefore, in this case Equation 3 reduces to

$$q_{sw} = (U_w \alpha / h_o) I_t A \qquad (5)$$

When the transmission due to temperature is added: $q_w = (U_w \alpha / h_o) I_t A + U_w A (T_o - T_i)$ which can be rearranged to give

$$q_w = U_w A [(T_o + \alpha I_t / h_o) - T_i] \qquad (6)$$

From this equation it is apparent that if the first term in the bracket is replaced by an equivalent temperature (called the sol-air temperature) Equation 6 takes the same form as in the case of thermal transmission without solar heat gain:

$$q_w = U_w A (T_e - T_i)$$

This procedure provides a convenient means of including solar loads. However, in the case of an opaque wall, the effects of mass can be quite pronounced, and using the temperature difference $(T_e - T_i)$ may significantly overestimate the heat gain. Again, to resolve this in the simplest way possible, an averaging process is used. Here the average takes the form of an equivalent temperature difference which is defined as

$$\text{TETD} = T_{ea} - T_i + \lambda(T_{e\delta} - T_{ea})$$

In this equation T_{ea} = the daily average sol-air temperature; $T_{e\delta}$ = the sol-air teoperature δ hours prior to the time for which the calculation is being made; λ = a weighting factor depending on the mass of the wall.

Representative values of λ and δ are shown in Table 5 (a and b) for several wall types. As the TETD is influenced by the absorptivity of the wall surface, values are given for "dark" surfaces ($\alpha/h_o = 0.3$) and "light" surfaces ($\alpha/h_o = 0.15$). Interpolation between dark and light should be used for medium-color walls. Table 5 provides TETD values for a number of typical walls. These tables presume a maximum outside temperature of 35°C and a daily temperature range of 11°C. (Note: as these tables are drawn from the *ASHRAE Handbook of Fundamentals, 1967,* they remain in English units. To convert to SI, divide the TETD in °F by 1.8.) Corrections as indicated in the table footnotes should be applied for other conditions. Thus the heat gain through the wall is given by:

$$q_w = U_w A(\text{TETD}) \qquad (7)$$

Example 4 — Determine the peak heat gain through a west-facing brick veneer wall (similar in cross section to that shown in Figure 1) July 21. The wall is a dark red

TABLE 5a

Total Equivalent Temperature Differentials (°F) for Calculating Heat Gain Through Sunlit and Shaded Walls

North latitude wall facing	Sun time: A.M. 8		10		12		P.M. 2		4		6		8		10		12		Amplitude decrement factor, λ time lag, δ hr		South latitude wall facing
	Exterior color of wall — D = dark, L = light																				
	D	L	D	L	D	L	D	L	D	L	D	L	D	L	D	L	D	L	λ	δ	
Frame																					
NE	37	18	46	25	30	19	26	19	28	22	26	21	19	16	13	13	9	8			SE
E	41	19	63	33	53	28	28	20	29	23	27	21	21	17	14	13	11	9			E
SE	24	12	50	26	55	31	43	28	30	24	27	21	20	17	13	11	11	8			NE
S	7	3	16	11	37	22	49	31	45	30	28	22	19	16	13	11	10	8	0.75	2	N
SW	9	4	14	8	22	14	44	28	64	40	63	39	37	25	14	11	11	8			NW
W	8	4	14	8	21	14	28	20	61	38	76	46	53	33	14	12	11	9			W
NW	7	4	13	8	20	14	26	19	39	27	59	37	49	31	13	11	10	8			SW
N	15	8	13	8	19	13	24	18	27	21	25	21	27	21	11	10	8	7			S
4 in. Brick or Stone Veneer + Frame																					
NE	7	5	34	18	42	23	29	18	25	18	27	21	25	20	19	16	14	12			SE
E	9	6	38	20	57	30	48	28	27	20	29	22	27	20	21	17	16	12			E
SE	9	5	24	13	46	24	50	29	40	26	29	21	27	20	21	16	15	12			NE
S	7	5	9	5	16	10	34	21	44	28	41	28	27	20	19	16	14	12	0.62	4	N
SW	9	5	10	6	16	9	22	14	40	26	57	35	56	35	34	23	15	12			NW
W	9	5	11	6	16	10	21	15	27	20	55	35	67	41	48	30	16	12			W

NW	7	5	9	5	14	9	20	14	25	17	35	25	52	33	44	27	14	12			SW
N	5	4	15	9	13	9	18	13	23	17	25	20	23	19	25	19	12	10			S
8 in. Hollow Tile or 8 in. Cinder Block																					
NE	10	7	10	7	37	21	33	20	23	16	25	19	25	19	22	17	16	13			SE
E	13	8	13	8	46	25	48	27	36	22	28	20	28	20	25	18	19	14			E
SE	12	8	12	7	34	18	44	25	42	25	30	21	27	20	24	18	18	14			NE
S	10	7	10	7	14	9	25	15	36	23	39	26	32	23	22	17	16	13	0.48	5	N
SW	12	8	12	7	16	9	20	13	27	18	45	28	51	32	43	27	18	14			NW
W	13	8	13	8	16	10	21	13	25	17	38	25	55	34	55	33	19	14			W
NW	10	7	10	7	14	9	18	12	22	16	26	19	40	20	46	29	16	13			SW
N	7	6	7	5	14	9	15	11	19	14	22	17	22	18	23	18	13	11			S
4 in. Brick or 4 in. Concrete																					
NE	25	13	38	20	34	20	32	21	31	23	29	22	22	18	15	13	10	9			SE
E	30	15	52	27	53	29	42	26	36	26	31	24	23	19	17	14	10	9			E
SE	19	9	41	21	52	29	48	29	39	27	32	24	24	19	16	14	10	9			NE
S	3	1	14	8	32	19	46	29	47	31	37	27	26	20	17	14	11	9	0.69	3	N
SW	3	1	9	5	18	12	39	25	59	37	64	40	45	30	26	19	15	12			NW
W	3	2	9	5	16	11	30	21	56	36	72	45	54	35	31	21	17	13			W
NW	3	1	9	5	16	11	24	17	40	27	56	36	45	30	26	19	15	12			SW
N	8	4	11	7	17	12	24	17	27	21	29	23	25	20	17	14	11	10			S
8 in. Brick or 8 in. Concrete or 12 in. Hollow Tile or 12 in. Cinder Block																					
NE	10	7	19	11	26	15	28	17	29	19	29	20	28	20	24	18	20	16			SE
E	11	7	23	13	36	19	39	22	38	23	36	23	33	23	28	20	22	17			E
SE	9	6	17	10	29	16	37	21	39	23	37	24	33	23	28	20	23	17			NE
S	7	5	7	5	13	8	23	14	32	20	36	23	34	23	28	21	23	17	0.39	5	N
SW	9	6	8	5	10	7	17	11	29	19	42	26	47	30	41	27	33	22			NW
W	10	7	9	6	11	7	15	10	25	17	40	26	51	32	46	29	36	24			W
NW	9	6	8	5	10	6	14	9	20	14	30	21	40	26	37	25	30	21			SW
N	7	5	8	5	11	7	14	10	18	13	22	16	24	19	23	18	19	15			S

TABLE 5a (continued)

Total Equivalent Temperature Differentials (°F) for Calculating Heat Gain Through Sunlit and Shaded Walls

North latitude wall facing	Sun time: A.M. 8		10		12		P.M. 2		4		6		8		10		12		Amplitude decrement factor, λ time lag, δ hr		South latitude wall facing
	Exterior color of wall — D = dark, L = light																				
	D	L	D	L	D	L	D	L	D	L	D	L	D	L	D	L	D	L	λ	δ	
12 in. Brick or 12 in. Concrete																					
NE	12	9	16	10	21	13	23	14	25	16	26	17	26	18	25	18	23	17			SE
E	14	10	19	12	26	15	31	18	33	20	33	21	32	21	30	20	27	19			E
SE	14	10	16	10	22	13	28	16	32	19	33	20	32	21	30	20	27	19			NE
S	13	9	11	8	12	8	17	11	23	15	28	18	29	20	28	19	25	18	0.24	7	N
SW	17	11	14	10	14	9	15	10	21	14	30	19	36	23	37	24	34	22			NW
W	18	12	16	10	15	9	15	10	19	13	28	18	37	23	39	25	37	24			W
NW	15	10	13	9	12	8	14	9	16	11	22	15	29	19	31	21	29	20			SW
N	10	8	10	8	11	8	12	9	15	11	17	13	20	15	21	16	19	15			S
4 in. Brick or 4 in. Concrete + Air Space + Finish																					
NE	14	8	30	15	33	18	31	20	31	22	30	22	26	20	20	16	14	12			SE
E	17	9	39	20	49	26	44	26	39	25	34	24	28	22	21	17	15	13			E
SE	10	6	28	15	43	23	47	27	42	27	36	25	29	22	22	17	15	13			NE
S	3	2	8	5	21	12	36	22	43	28	40	27	32	23	23	18	16	13	0.57	3	N

SW	4	2	6	4	13	8	27	17	46	29	58	36	52	33	36	24	23	17			NW
W	4	2	7	4	12	8	21	14	40	26	60	37	60	38	42	27	27	19			W
NW	4	2	6	4	12	8	18	13	29	21	45	30	48	32	34	24	23	17			SW
N	6	3	9	5	13	9	19	13	24	18	27	21	27	21	21	17	15	13			S
8 in. Brick or 8 in. Concrete + Air Space + Finish																					
NE	11	8	15	10	22	13	25	15	26	16	27	18	27	19	26	18	22	17			SE
E	12	9	18	11	28	16	34	19	36	21	35	22	33	22	30	21	26	19			E
SE	11	8	14	9	22	13	30	17	35	20	35	22	34	22	31	21	26	19			NE
S	10	8	9	6	11	7	17	11	25	16	31	20	32	21	30	21	26	19	0.31	6	N
SW	14	10	12	8	11	7	14	9	22	14	32	20	40	25	40	26	35	23			NW
W	16	10	13	8	12	8	14	9	19	13	30	19	41	26	44	28	39	25			W
NW	13	9	11	7	11	7	13	8	16	11	23	16	32	21	35	23	31	21			SW
N	9	7	9	6	10	7	12	8	15	11	19	14	21	16	22	17	20	16			S

Notes: Explanation: Total heat transmission from solar radiation and temperature difference between outside and room air, Btu per (hr) (sq ft wall area) = Equivalent temperature differential from above table × Heat transmission coefficient for wall, Btu per (hr) (sq ft) (°F)

1. *Application.* These values may be used for all normal air conditioning estimates, usually without corrections, when the load is calculated for the hottest weather. Correction for latitude (Note 2) is necessary only where extreme accuracy is required. There may be jobs where the indoor room temperature is considerably above or below 80° F, or where the outdoor design temperature is considerably above 95° F, in which case it may be desirable to make correction to the temperature differentials shown. The solar intensity on all walls other than east and west varies considerably with time of year.
2. *Corrections. Outdoor minus room temperature.* If the outdoor maximum design temperature minus room temperature is different from the base of 20°F, correct as follows: When the difference is greater (or less) than 20°F, add the excess to (or subtract the deficiency from) the above differentials.

   For Outdoor Daily Range of Temperature other than 20°F, correct Equivalent Temperature Difference as follows:

Outdoor daily range	Medium construction	Heavy construction
For each 1° difference *less* than 20°	*Add* ¼°	*Add* ½°
For each 1° difference *greater* than 20°	*Subtract* ¼°	*Subtract* ½°
Maximum Correction	3°	5°

TABLE 5a (continued)

Total Equivalent Temperature Differentials (°F) for Calculating Heat Gain Through Sunlit and Shaded Walls

For Light construction, apply no correction.

Color of exterior surface of wall. Use temperature differentials for light walls only where the permanence of the light wall is established by experience. For cream colors use the values for light walls. For medium colors interpolate half way between the dark and light values. Medium colors are medium blue, medium green, bright red, light brown, unpainted wood, natural color concrete, etc. Dark blue, red, brown, green, etc., are considered dark colors.

For latitudes other than 40° north; and in other months. These table values will be approximately correct for the east or west wall in any latitude (0° to 50° North or South) during the hottest weather. In the lower latitudes when the maximum solar altitude is approximately 80° to 90° (the maximum occurs at noon) the temperature differential for either a south or north wall will be approximately the same as a north or the total equivalent temperature differential (TETD) for any wall facing and for any latitude for any month may be estimated by Equation 30.

3. *For insulated walls,* use same temperature differentials as used for uninsulated walls.

This table is in °F. To convert to °C divide °F by 1.8.

From *ASHRAE Handbook of Fundamentals*, American Society of Heating, Refrigerating and Air Conditioning Engineers, New York, 1967, p. 492. With permission.

TABLE 5b

Total Equivalent Temperature Differentials for Calculating Heat Gain Through Roofs

	Sun time												Amplitude decrement factor, λ time lag, δ hr	
	A.M.						P.M.							
Description of roof construction[a]	2	4	6	8	10	12	2	4	6	8	10	12	λ	δ
Light construction roofs — Exposed to sun														
1'' Wood[b]	−5	−6	−6	16	52	81	94	84	56	20	5	−4	0.95	2

2'' Insulating board + 1'' wood[b]	2	−3	−6	−1	20	50	75	85	77	53	26	7	0.82	4
2'' Concrete or 2'' Plank	−2	−5	−6	11	41	69	85	83	63	32	13	1	0.83	3
2'' Wood[b]	1	−3	−6	24	51	77	85	75	48	23	6	1	0.82	3
Medium construction roofs — exposed to sun														
2'' Insulating board + 2'' concrete or 2'' gypsum plank	10	3	−1	−1	13	36	60	75	75	60	38	19	0.69	5
2'' Gypsum or 2'' concrete + 4'' rock wool + ½'' plaster	13	5	0	−1	9	30	54	71	75	64	43	22	0.66	5
2'' Wood[b] + 4'' rock wool + ½'' plaster	29	19	12	6	7	16	32	49	61	63	54	39	0.48	7
Heavy construction roofs — exposed to sun														
4'' Concrete	9	3	−1	5	23	46	65	74	68	49	31	15	0.64	4
6'' Concrete	20	12	7	6	16	32	49	61	63	54	40	26	0.48	5
2'' Insulating board + 4'' concrete	27	19	14	10	14	25	39	51	57	54	45	32	0.38	6
2'' Insulating board + 6'' concrete	32	27	22	18	17	23	31	41	47	48	44	35	0.26	7

Notes: Explanation: Total heat transmission from solar radiation and temperature difference between outdoor and room air. Btu per (hr) (sq ft) of roof area = Equivalent temperature differential from above table × Heat transmission coefficient for summer Btu per (hr) (sq ft) (°F)

1. *Application.* These values may be used for all normal air conditioning estimates; usually without correction, in latitude 0° to 50° north or south when the load is calculated for the hottest weather. Note 3 explains how to adjust the temperature differential for other room and outdoor temperatures.
2. *Attics.* If the ceiling is insulated and if a fan is used in the attic for positive ventilation, the total temperature differential for a roof exposed to the sun may be decreased 25%.
3. *Corrections. For temperature difference when outdoor maximum design minus room temperature is dfferent from 20° F.* If the outdoor design temperature minus room temperature is different from the base of 20° F, correct as follows: When the difference is greater (or less) than 20° F add the excess to (or subtract the deficiency from) the above differentials.

For Outdoor Daily Range of Temperature other than 20° F, correct Equivalent Temperature Difference as follows:

Outdoor daily range	Medium construction	Heavy construction
For each 1° difference less than 20°	Add ¼°	Add ½°
For each 1° difference greater than 20°	Subtract ¼°	Subtract ½°
Maximum Correction	3°	5°

TABLE 5b (continued)

Total Equivalent Temperature Differentials for Calculating Heat Gain Through Roofs

For Light construction, apply no correction.

Light Colors. Credit should not be taken for light colored roofs except where the permanence of the light color is established by experience, as in rural areas or where there is little smoke.

For solar transmission in latitudes other than 40° north, and in other months. The table values of temperature differentials will be approximately correct for a roof in the following months:

North Latitude		South Latitude	
Latitude °	Months	Latitude °	Months
0	All months	0	All months
10	All months	10	All months
20	All months except Nov, Dec, Jan	20	All months except May, June, July
30	Mar, Apr, May, June, July, Aug, Sept	30	Sept, Oct, Nov, Dec, Jan, Feb, Mar
40	April, May, June, July, Aug	40	Oct, Nov, Dec, Jan, Feb
50	May, June, July	50	Nov, Dec, Jan

For other months, light colored roofs, or both, use the total equivalent temperature differential from Equation 30. For room air temperature less than 75°F, add the difference between 75 and room air temperature; if greater than 75, subtract the difference.

[a] Includes ¾ in. felt roofing with or without slag. May also be used for shingle roof.

[b] Nominal thickness of the wood.

This table is in °F. To convert to °C, divide °F by 1.8.

Reprinted by permission from *ASHRAE Handbook of Fundamentals,* American Society of Heating, Refrigerating, and Air Conditioning Engineers, New York, 1967, 491.

color, the outside-inside temperature difference is 22°F, and the daily temperature range is 20°F.

From Table 5, $TETD_{max}$ = 67°F at 8 p.m. However, since the maximum outdoor temperature minus room temperature differs from 20°F this must be corrected.

- TETD = 67 + 2 = 69° = 38°C
- q_{max} = UA(TETD)
- q_{max}/A = (0.356)(38°C)
  = 13.53 watts/m²

3. Infiltration Loads

The infiltration of outside air into the space influences both the air temperature and the humidity level. Usually, a distinction is made between the two effects, referring to the temperature effect as "sensible" load and the humidity effect as "latent" load. Note that this terminology applies to the other load components as well. For example, transmission and solar loads are sensible as they affect only temperature, while internal loads, such as those arising from occupancy, have both a sensible and a latent component. Heat loss or heat gain due to infiltration is then expressed as

$$q_{is} = 1.23\,\dot{Q}\,(T_o - T_i)$$

$$q_{i\ell} = 3.00\,\dot{Q}\,(W_o - W_i) \tag{8}$$

where $\dot{Q}$ = volumetric flow rate of outside air into space ℓ/s; W = humidity ratio: g water per kg of air.

The volumetric flow rate is rather difficult to determine with any measure of precision. It will vary with the quality of construction, wind speed and direction, and temperature difference. One procedure that is often used in load calculations is to specify the infiltration in terms of the number of air changes per hour. (One air change per hour would be a volumetric flow rate numerically equal to the internal volume of the space.) The number of air changes per hour may be estimated as a function of wind velocity and temperature difference as

$$\text{Number of Air Changes} = a + bv + c\,|\Delta T|$$

where V = wind velocity; ΔT = difference between the inside and outside air temperature. Values for the coefficients a, b, and c are given in Table 6. Unfortunately, estimates of infiltration load must, at present, be primarily based on experience and judgment rather than on a satisfactory array of engineering data.

4. Internal Loads

The primary sources of internal heat gain are lights, occupants, and equipment operating within the space. Internal loads in residential space are generally small, but they are a major factor in most other types of buildings. The amount of space heat gain due to lighting depends on the wattage of the lamps and the type of fixture. When fluorescent lighting is used, the energy dissipated by the ballast must also be included in the internal load. (This additional energy may be estimated as 0.2 × bulb wattage). Table 7a indicates loads from occupants as a function of activity. The greatest uncertainty in estimating this load component is the number of occupants. If the number

TABLE 6

Infiltration
$I = a + bV + c\Delta T$

Construction	a	b	c
Tight	0.15	0.010	0.007
Average	0.20	0.015	0.014
Loose	0.25	0.020	0.022

Note: I = air changes per hour; V = wind velocity, km/h; ΔT = temperature difference, °C.

TABLE 7a

Heat Gain from Occupants

Activity	Heat gain per person (watts)
Sleeping	70
Seated, quiet	100
Standing	120
Walking, 3 km/h	200
Office work	120
Teaching	160
Retail shop	160
Industrial	300—600

TABLE 7b

Space per Occupant

Type	Occupancy
Residence	2—6 occupants per residence[a]
Office	10—15 m^2/occupant
Retail	3—5 m^2/occupant
School	2.5 m^2/occupant
Auditorium	1.5 m^2/occupant

[a] Maximum 2 occupants per bedroom.

of occupants is unknown, values such as those in Table 7b may be used. In the case of equipment it is necessary to estimate the power used along with the period and/or frequency of use. Here again it is necessary to exercise judgment and not to devote undue time and effort to the analysis.

5. Procedure For Calculating Heat Loss and Heat Gain

In calculating heat loss and heat gain for a building, it is important to use an organized, step-by-step procedure. The necessary steps may be outlined as follows:

1. Select design values for outdoor dry bulb temperature, temperature range, and coincident wet bulb temperature as needed for each season from Table 2.
2. Select indoor design temperatures that are appropriate to the activities to be carried out in the space and for the season.
3. Determine if any special conditions will exist, such as adjacent unconditioned spaces. Estimate temperatures in the adjacent spaces as necessary.
4. On the basis of building plans and specifications, calculate heat transfer coefficients and areas for the building components in each enclosing surface. Any surfaces connecting with spaces to be maintained at the same temperature may be omitted, i.e., interior walls.
5. On the basis of building components and design values of wind velocity and temperature difference, estimate the rate of infiltration of unconditioned outside air into the space and the associated heat loss or heat gain.
6. Determine the additional building characteristics such as location, orientation, external shading, color, and mass that will influence solar heat gain.

7. On the basis of season, building components, and design conditions, determine the appropriate temperature differences and solar heat gain factors to be used. (Note: Due to solar irradiation and the thermal capacitance of some building materials, the appropriate temperature differences are not always simply the difference between the inside and the outside design temperatures. For sunlit opaque roofs and walls, the TETD values should be used.)
8. On the basis of the heat transfer coefficients, areas, temperature differences, etc., determined above, utilize Equations 4, 7, and 8 to calculate the rate of heat loss or heat gain to the space.
9. For spaces with appreciable heat gain from internal sources (lights, equipment, or people), calculate the internal heat gain. In the case of a net heat loss, include the internal heat gain as a credit only if the rate of internal heat generation is reasonably constant. In the case of a net heat gain, include all coincident internal gains when determining capacity requirements.
10. Sum all of the pertinent load components to determine the maximum capacity required for heating and cooling. If the building is to be operated intermittently, additional capacity may be required.

The above procedure and the preceding discussion have been brief. The reader is directed to the most recent *ASHRAE Handbook and Product Directory, Fundamentals Volume* or similar sources for a more complete discussion of the details and for more extensive tabular data.

B. Estimating Annual Heating and Cooling Loads

1. Methods of Analysis

For the purposes of sizing heating and cooling equipment, the procedures outlined above have proved to be reasonably successful, particularly when the traditional margin of safety was added to assure adequate performance. However, heating and cooling systems have become more complex and expensive, and energy costs have risen to become a significant part of operating expenses. Owners, architects, and engineers are increasingly concerned with minimizing operating costs as well as first cost.

This concern suggests that some optimization of design be attempted. The optimization can be accomplished only if a reasonably accurate means is available to evaluate heating and cooling loads over the whole range of operating conditions. Such an analysis goes considerably beyond just estimating peak loads. It involves making similar calculations of load for some number of intermediate conditions. The number required may range from one or two to several thousand depending on the complexity of the building and its heating and cooling system. The way in which interaction between the external environment, building components and systems, and internal loads influences energy consumption is the basic criterion which must be examined to choose analysis procedures. For example, a single family residence with a single thermostat may be handled with relatively few calculations, while a hospital or multistory office building with a complex control system may require calculations on an hour-by-hour basis. When hour-by-hour calculations are required, a computerized analysis should be considered.

The "complexity" of a building should be judged by examining the characteristics of the loads and the control system. The factors which govern the heating or cooling load in a space can be grouped into two categories: those that are influenced by climatic variables, and those that are governed by use and schedule. The transmission, solar, and infiltration loads are in the first category. Internal loads are in the second, while ventilation loads are influenced by both. In those buildings, such as a residence, where loads are primarily transmission, solar, and infiltration, annual estimates of energy consumption based on a correlation with average daily temperature, degree-days, etc., are often quite good. For buildings where the loads are primarily internal, variations in occupancy schedule and hours of operation may dominate. However, climatic factors cannot be ignored because of their influence on ventilation. Often in large buildings the complexity of the interaction among load components prevents a simple correlation based on a single parameter such as temperature.

As the need for multiple-parameter energy analysis arose at a time when computers were beginning to find application in related fields, it seems quite natural that a number of computerized procedures for load calculation and energy estimates have been developed. Many of these computer procedures employ hour-by-hour calculations which consider the transient heat transfer behavior of buildings and the interactions which occur among the external environment, the building materials and systems, and the internal loads. These computer analysis procedures provide a much more complete look at a building's thermal performance than any hand calculation procedure could ever accomplish. Computerized building modeling and system simulation can be expected to play an increasingly important role in the evaluation of the thermal performance of buildings in the future. The time will soon arrive when it will be feasible to run a computer analysis on several design options for every major building project.

Computerized building modeling analysis services are available through a number of private consulting firms, some equipment manufacturers, and some electric and natural gas utility company associations. ASHRAE[2] has prepared a bibliography of energy analysis programs which gives a brief description of each program, the input requirements, and pertinent information as to program availability. In addition to the programs listed in Reference 2, two programs developed under federal government sponsorship are available on national computer service networks. There are, therefore, a number of means of access to computerized analysis. The difficulty with the computerized analysis procedures is that they generally require extensive, detailed input data, and in most cases some degree of expertise on the part of the user. However, computerized energy analysis is a rapidly changing field, and significant improvements in flexibility, cost, and availability can be expected in the future.

Several procedures that do not require computers are available for estimating the effect of design decisions on the thermal characteristics of buildings. Although these procedures do not provide the detail or precision of computerized analysis, they are of value particularly in the preliminary design stage.

2. Simplified Estimates of Thermal Performance

The heat loss/heat gain of a building is affected by variations in the external environment, building material characteristics, and internal load; therefore, each of these factors must be considered even in a simplified analysis. However, the calculational procedures become very extensive unless some assumptions are made. For example, the climatic factors which affect a building include temperature, humidity, solar irradiation, and wind velocity. Although the characteristic variations of these parameters in a given season are somewhat related, there are no correlations of one with another that can be used without reservation. Therefore, choosing a single climatic parameter as input to an analysis immediately reduces the accuracy of the result. It is necessary, however, to choose a single climatic parameter as input if extensive and often repetitive calculations are to be avoided. Similar simplifications must also be made with regard to internal loads.

The climatic variable most often chosen as input for energy calculations is the dry bulb temperature. The temperature input may be in the form of a mean daily temperature (or the deviation of the mean daily temperature from a reference value as in the degree-

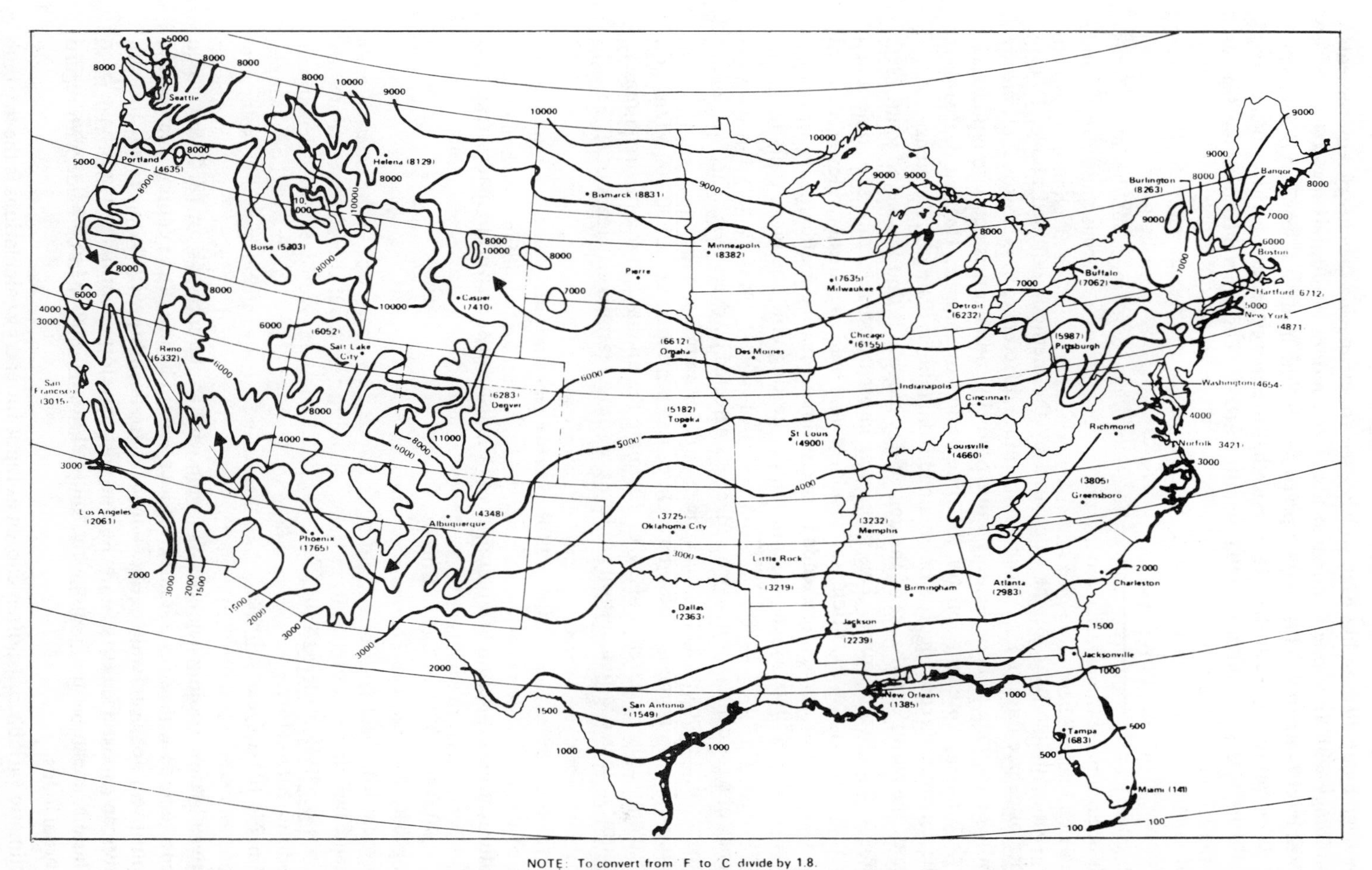

FIGURE 3. Annual heating degree-days (Base 65°F) (From Energy Conservation for Existing Buildings, ECM-1, Federal Energy Administration, June 1975, 57. With permission.

day procedure) or on a frequency-of-occurrence basis (where the number of hours that the temperature falls into a given range is used). In either case, the heat loss or heat gain is assumed to vary linearly with temperature.

The degree-day method has been widely used for estimating the energy requirements for buildings with envelope-dominated loads and indoor conditions maintained at a constant level 24 hr/day. This method presumes that no heating is required if the mean daily temperature is greater than 18°C and that the energy required with a mean daily temperature of 0°C would double that required at 9°C. The general form of the degree-day calculation is as follows:

$$E = \frac{q_{HD} \cdot DD \cdot 24}{\Delta T \cdot \eta_s \cdot Hv} \quad (9)$$

where q_{HD} = design heat loss as calculated by the methods outlined above; DD = number of degree-days; ΔT = design temperature difference; η_s = seasonal efficiency of heating equipment; Hv = heating value of fuel in units consistent with q and E; E = fuel or energy consumption for the time period of interest. Heating degree-days may be determined from Figure 3 or obtained from tabular listings in Reference 3.

One of the major drawbacks of the degree-day method is that it is keyed to an 18°C base mean daily temperature. This value was chosen as a reference at a time when residential buildings were often not insulated and had significantly lower internal loads. Therefore this procedure tends to overestimate heating requirements for newer, well-insulated buildings which may not require heating until the mean daily temperature drops considerably below 18°C. A second problem arises in that the seasonal efficiency of a residential unit may be adversely affected by oversizing. This fact should be considered in choosing a seasonal efficiency η_s for a unit.

Example 5 — A home in Columbus, Ohio, has peak heat loss of 70,000 kJ/hr at a design temperature difference of 34°C. Estimate the annual heating fuel requirement if the home is heated with a natural gas furnace with a seasonal efficiency of 60%.

$$E = (q \cdot DD \cdot 24)/(\Delta T \cdot \eta \cdot Hv)$$

Since the degree-day values of Figure 3 are in English units, convert to that system

- q = 70,000 (1.055) = 73,850 Btu/hr
- ΔT = 34(1.8) = 61°F
- Hv = 10^6 Btu/10^3 ft^3
- DD = 5,660
- E = $\dfrac{(73{,}850)(5{,}660)(24)}{(61)(0.6)(10^6)} \times 10^3$ ft^3
- E = 274 10^3 ft^3/year

In recent years, cooling degree-days (again with a reference base of 18°C) have been compiled and published for a number of locations. The use of the cooling degree-days has, however, not been widespread, nor is its applicability well documented. Instead, cooling energy requirements for buildings with envelope-dominated loads have been estimated by means of the "equivalent full load hours" method in the following form:

$$E = \frac{q_{CD} \cdot EFL}{COP} \quad (10)$$

TABLE 8a

Occurrences of Dry Bulb Temperature Below 75°F

Location	Outdoor temperature, °F																		
	72	67	62	57	52	47	42	37	32	27	22	17	12	7	2	−3	−8	−13	−18
Albany, N.Y.	588	733	740	708	652	625	647	769	793	574	404	278	184	110	63	32	10	5	4
Albuquerque, N.M.	767	831	719	651	687	734	741	689	552	346	154	66	21	4	1	1	—	—	—
Atlanta, Ga.	1185	926	823	784	735	676	598	468	271	112	44	19	8	2	—	—	—	—	—
Bakersfield, Calif.	831	898	966	977	908	746	541	247	77	7	—	—	—	—	—	—	—	—	—
Birmingham, Ala.	1138	908	805	742	668	614	528	433	292	143	69	17	6	3	—	—	—	—	—
Bismark, N.D.	454	566	614	606	563	520	518	604	653	550	474	371	338	292	278	208	131	77	80
Boise, Idaho	492	575	643	702	786	798	878	829	522	307	148	53	26	14	6	2	—	—	—
Boston, Ma	676	819	804	781	766	757	828	848	674	429	256	151	74	35	4	9	1	—	—
Buffalo, N.Y.	646	772	760	700	666	624	647	756	849	602	426	267	170	81	5	24	2	—	—
Burlington, Vt.	573	670	703	694	655	603	637	716	752	561	491	336	272	216	135	81	39	17	8
Casper, Wyo.	423	532	592	642	606	670	782	831	806	683	495	324	200	116	73	45	30	15	5
Charleston, S.C.	1267	1090	889	787	651	576	434	321	192	79	27	5	—	—	—	—	—	—	—
Charleston, W.Va.	912	949	767	689	661	667	607	633	630	356	252	135	73	22	7	1	—	—	—
Charlotte, N.C.	1115	908	839	752	730	684	634	515	360	166	64	23	5	2	—	—	—	—	—
Chattanooga, Tenn.	1021	895	775	713	679	642	553	414	228	113	45	4	4	2	—	—	—	—	
Chicago, Ill.	762	769	653	592	569	543	591	800	822	551	335	196	117	85	59	25	12	3	—
Cincinnati, Ohio	879	843	726	639	611	599	627	698	711	460	249	131	68	44	18	8	2	—	—
Cleveland, Ohio	763	831	732	641	638	607	620	754	806	578	355	201	111	47	22	11	2	—	—
Columbus, Ohio	774	820	720	648	622	603	658	730	772	502	280	169	94	40	20	10	4	1	—
Corpus Christi, Tex.	1175	1041	748	551	444	302	180	83	27	9	3	—	—	—	—	—	—	—	—
Dallas, Tex.	831	795	693	656	629	576	504	371	231	91	34	17	4	1	—	—	—	—	—
Denver, Colo.	549	684	783	731	678	704	692	717	721	553	359	216	119	78	36	22	6	1	1
Des Moines, Iowa	707	751	681	600	585	512	510	627	747	557	405	281	211	152	104	59	23	8	1
Detroit, Mich.	721	783	695	633	592	566	595	808	884	618	377	248	131	61	17	4	1	—	—
El Paso, Tex.	933	839	749	760	687	611	494	369	233	104	34	10	2	—	—	—	—	—	—
Ft. Wayne, Ind.	728	777	699	608	569	552	601	725	905	596	381	205	124	69	40	19	6	1	—
Fresno, Calif.	709	803	921	1006	1036	952	673	426	168	34	—	—	—	—	—	—	—	—	—
Grand Rapids, Mich.	634	739	712	647	571	565	554	742	938	690	469	293	172	78	31	10	1	1	—
Great Falls, Mont.,	407	520	636	754	822	830	832	813	698	533	355	218	167	136	118	101	68	51	62

TABLE 8a (continued)

Occurrences of Dry Bulb Temperature Below 75°F

Location	Outdoor temperature, °F																		
	72	67	62	57	52	47	42	37	32	27	22	17	12	7	2	−3	−8	−13	−18
Harrisburg, Pa.	807	824	737	692	635	659	722	888	749	427	222	125	52	18	4	1	—	—	—
Hartford, Conn.	617	755	751	752	649	575	683	807	825	552	370	233	153	77	33	11	3	2	—
Houston, Tex.	1172	980	772	681	570	452	291	141	64	18	4	2	—	—	—	—	—	—	—
Indianapolis, Ind.	821	315	722	585	586	579	605	712	791	551	293	152	97	60	35	13	3	2	—
Jackson, Miss.	1169	922	790	677	618	605	484	367	224	103	41	6	2	2	1	—	—	—	—
Jacksonville, Fla.	1334	975	879	692	530	355	288	154	83	24	2	—	—	—	—	—	—	—	—
Kansas City, Mo.	761	723	601	572	553	562	628	625	591	407	265	175	99	51	21	4	—	—	—
Knoxville, Tenn.	1056	889	746	675	672	689	648	590	456	217	101	41	21	7	2	—	—	—	—
Las Vegas, Nev.	651	644	699	786	769	716	591	396	194	44	7	1	—	—	—	—	—	—	—
Little Rock, Ark.	94	803	725	672	638	669	605	509	363	172	50	25	5	1	—	—	—	—	—
Los Angeles, Calif.	881	1654	2193	1904	1054	428	107	10	—	—	—	—	—	—	—	—	—	—	—
Louisville, Ky.	869	758	693	654	619	634	649	703	631	332	169	97	45	25	8	3	1	—	—
Lubbock, Tex.	833	829	688	700	642	618	620	546	490	346	180	86	33	7	5	1	—	—	—
Memphis, Tenn.	977	798	715	690	618	633	614	532	374	196	74	25	10	4	—	—	—	—	—
Miami, Fla.	1705	810	452	277	147	71	26	4	—	—	—	—	—	—	—	—	—	—	—
Milwaukee, Wis.	597	753	749	634	585	591	611	774	913	659	421	285	176	116	83	47	18	4	3
Minneapolis, Minn.	621	690	695	602	588	482	500	560	632	609	514	383	311	246	186	119	62	31	16
Mobile, Ala.	1411	1038	882	698	609	506	377	214	109	49	7	3	—	—	—	—	—	—	—
Nashville, Tenn.	933	838	738	697	637	619	627	565	463	263	132	67	28	9	3	1	1	—	—
New Orleans, La.	1189	987	850	692	621	449	282	128	47	9	2	—	—	—	—	—	—	—	—
New York, N.Y.	926	877	754	745	722	796	838	858	603	330	188	2	26	10	1	—	—	—	—
Oklahoma City, Okla.	881	769	717	643	645	611	641	570	468	287	173	77	36	12	3	1	—	—	—
Omaha, Neb.	726	721	606	558	539	543	543	655	663	511	390	287	189	135	93	40	15	1	—
Philadelphia, Pa.	863	809	735	710	663	701	758	818	654	335	189	100	32	9	—	—	—	—	—
Phoenix, Ariz.	762	776	767	769	659	540	391	182	57	8	—	—	—	—		—	—	—	—
Pittsburgh, Pa.	722	910	799	678	637	587	631	688	774	569	360	233	159	60	30	7	1	—	—
Portland, Maine	407	627	780	808	760	748	772	839	820	599	408	293	190	109	60	29	15	5	1
Portland, Ore.	373	581	1001	1316	1274	1271	1238	772	343	123	40	10	4	1	—	—	—	—	—
Raleigh, N.C.	1087	937	848	762	707	672	638	527	410	236	103	38	11	1	—	—	—	—	—

Reno, Nev.	418	477	572	690	845	909	890	829	733	530	387	227	101	37	15	4	1	—	—
Richmond, Va.	953	850	784	745	690	673	699	632	478	285	138	67	19	2	1	—	—	—	—
Sacramento, Calif.	630	773	1071	1329	1298	1049	701	355	93	8	—	—	—	—	—	—	—	—	—
Salt Lake City, Utah	569	615	614	635	682	685	755	831	798	564	328	158	80	41	16	2	—	—	—
San Antonio, Tex.	1086	943	789	669	569	445	387	190	94	31	11	4	1	1	—	—	—	—	—
San Francisco, Calif.	285	665	1264	2341	2341	1153	449	99	10	—	—	—	—	—	—	—	—	—	—
Seattle, Wash.	258	448	750	1272	1462	1445	1408	914	427	104	39	20	3	—	—	—	—	—	—
Shreveport, La.	1063	886	772	679	619	609	516	361	200	72	23	6	2	—	—	—	—	—	—
Sioux Falls, S.D.	566	684	669	605	522	498	501	625	712	585	520	448	293	208	152	102	59	43	18
St. Louis, Mo.	823	728	646	575	585	578	620	671	650	411	219	134	77	40	15	7	1	—	—
Syracuse, N.Y.	627	735	723	717	656	641	651	720	830	547	392	282	190	102	55	23	5	2	2
Tampa, Fla.	1387	1187	877	570	345	216	137	48	10	1	—	—	—	—	—	—	—	—	—
Waco, Tex.	909	830	701	622	651	558	501	354	216	84	24	3	1	—	—	—	—	—	—
Washington, D.C.	960	766	740	673	690	684	790	744	542	254	138	54	17	2	—	—	—	—	—
Wichita, Kan.	758	709	641	603	589	592	611	584	607	426	273	161	85	45	14	3	1	—	—

Reprinted by permission from *ASHRAE Handbook and Product Directory, 1976 Systems,* American Society of Heating, Refrigerating, and Air Conditioning Engineers, New York, 1976, 43.11.

TABLE 8b

Occurrence of Dry Bulb Temperatures Above 75°F and Median Coincident Wet Bulb Temperatures

	77°	wb	82°	wb	87°	wb	92°	wb	97°	wb	102°	wb	107°	wb	112°	wb	Hr above 75°F
Albuquerque, N.M.	634	56°	503	58°	365	60°	217	61°	62	61°	3	62°	—	—	—	—	1784
Atlanta, Ga.	875	70°	603	71°	375	73°	167	74°	28	74°	2	74°	—	—	—	—	2048
Birmingham, Ala.	924	71°	671	72°	489	73°	255	74°	54	74°	2	74°	—	—	—	—	2395
Bismarck, N.D.	557	63°	247	65°	130	68°	69	69°	20	70°	4	70°	—	—	—	—	828
Boise, Idaho	382	58°	324	60°	234	62°	127	64°	36	65°	6	66°	—	—	—	—	1017
Boston, Ma	432	67°	241	69°	123	71°	42	73°	12	74°	—	—	—	—	—	—	432
Charlotte, N.C.	744	69°	558	71°	382	73°	191	75°	52	76°	4	76°	—	—	—	—	1932
Chicago, Ill.	554	67°	363	69°	211	71°	96	73°	26	75°	3	76°	—	—	—	—	1253
Columbus, Ohio	535	67°	395	69°	222	72°	88	74°	15	75°	1	76°	—	—	—	—	1251
Dallas, Tex.	929	70°	896	73°	667	74°	466	75°	285	76°	93	75°	11	75°	—	—	3347
Denver, Colo.	396	56°	328	57°	223	58°	81	59°	7	59°	—	—	—	—	—	—	1034
Des Moines, Iowa	567	67°	378	70°	214	73°	92	75°	28	75°	4	76°	—	—	—	—	1281
Detroit, Mich.	509	66°	307	69°	143	71°	48	73°	10	74°	—	—	—	—	—	—	1017
El Paso, Tex.	909	58°	756	60°	591	62°	418	63°	200	64°	38	64°	2	65°	—	—	2914
Fort Wayne, Ind.	507	67°	350	69°	180	72°	72	73°	13	75°	1	76°	—	—	—	—	1123

TABLE 8b (continued)

Occurrence of Dry Bulb Temperatures Above 75°F and Median Coincident Wet Bulb Temperatures

	77°	wb	82°	wb	87°	wb	92°	wb	97°	wb	102°	wb	107°	wb	112°	wb	Hr above 75°F
Great Falls, Mont.	295	56°	182	58°	102	59°	38	60°	4	65°	—	—	—	—	—	—	621
Houston, Tex.	1630	74°	956	75°	681	76°	324	77°	56	77°	1	78°	—	—	—	—	3649
Jackson, Miss.	1052	72°	686	73°	546	74°	343	76°	99	76°	14	77°	—	—	—	—	2741
Kansas City, Mo.	728	68°	594	70°	375	72°	218	74°	96	75°	24	74°	4	73°	—	—	2040
Las Vegas, Nev.	702	53°	664	55°	571	58°	467	59°	391	61°	267	63°	90	65°	10	66°	3160
Little Rock, Ark.	951	71°	687	73°	488	75°	294	76°	99	76°	23	76°	1	79°	—	—	2521
Los Angeles, Calif.	301	65°	89	65°	26	63°	7	63°	3	65°	1	70°	—	—	—	—	426
Louisville, Ky.	668	69°	538	71°	345	73°	159	75°	34	75°	5	75°	—	—	—	—	1748
Memphis, Tenn.	902	71°	666	72°	473	74°	292	76°	98	77°	18	78°	—	—	—	—	2449
Miami, Fla.	2717	72°	1977	75°	1006	76°	146	77°	2	78°	—	—	—	—	—	—	5847
Milwaukee, Wis.	391	68°	226	70°	98	72°	30	74°	6	75°	—	—	—	—	—	—	752
Minneapolis, Minn.	483	65°	307	68°	152	71°	28	74°	8	76°	—	—	—	—	—	—	977
Nashville, Tenn.	801	70°	580	71°	418	73°	216	75°	64	75°	11	75°	1	75°	—	—	2092
New Orleans, La.	1696	73°	983	75°	619	76°	231	77°	11	78°	—	—	—	—	—	—	3540
New York City, N.Y.	661	68°	378	70°	182	72°	59	74°	17	74°	2	75°	—	—	—	—	1298
Norfolk, Va.	899	71°	487	73°	283	75°	128	76°	17	77°	1	78°	—	—	—	—	1816
Oklahoma City, Okla.	798	69°	589	71°	419	72°	278	73°	116	73°	28	73°	2	72°	—	—	2230
Omaha, Neb.	525	68°	402	70°	257	72°	133	75°	43	76°	13	75°	1	75°	—	—	1374
Philadelphia, Pa.	646	69°	411	70°	228	72°	80	75°	20	76°	1	76°	—	—	—	—	1386
Phoenix, Ariz.	812	58°	792	62°	677	65°	597	67°	508	69°	320	70°	124	71°	18	71°	3858
Pittsburgh, Pa.	496	67°	316	69°	132	71°	31	73°	3	73°	—	—	—	—	—	—	978
Portland, Maine	268	66°	127	69°	60	72°	16	75°	1	75°	—	—	—	—	—	—	473
Portland, Ore.	214	63°	132	65°	53	67°	20	68°	4	70°	1	73°	—	—	—	—	424
Sacramento, Calif.	477	62°	368	64°	275	66°	182	68°	88	69°	29	70°	5	71°	—	—	1393
Salt Lake City, Utah	441	57°	367	58°	309	60°	179	61°	48	63°	3	64°	—	—	—	—	1374
San Francisco, Calif.	85	62°	38	63°	12	65°	4	67°	—	—	—	—	—	—	—	—	136
Seattle, Wash.	113	62°	58	64°	17	66°	5	66°	1	67°	—	—	—	—	—	—	194
St. Louis, Mo.	728	68°	563	70°	361	73°	189	75°	13	75°	1	75°	1	76°	—	—	1916
Tampa, Fla.	1894	73°	1320	74°	817	75°	241	76°	5	77°	—	—	—	—	—	—	4277
Washington, D.C.	723	69°	488	71°	283	73°	106	74°	22	75°	1	75°	—	—	—	—	1622
Wichita, Kan.	712	68°	543	70°	373	71°	254	72°	135	72°	—	—	—	—	—	—	2067

Adapted from Crow, Loren W., Weather data related to evaporative cooling, Table IIA, *ASHRAE Journal*, 66, June 1972. With permission.

q_{CD} = design cooling load; EFL = equivalent full load hours; COP = the cycle coefficient of performance. This procedure provides a quick reference value but should not be expected to provide anything more than a rough estimate.

An estimate of equivalent full load hours can be obtained by using the number of degree-hours above a chosen reference temperature. Degree-hours are determined from the frequency of occurrence of a given temperature over a cooling season. For example, if a reference temperature of 25°C were chosen, ten degree-hours would be accumulated for every hour the outdoor temperature was at 35°C, and so forth. The equivalent full load hour can thus be obtained by dividing degree-hours by the design temperature difference

$$EFL = \frac{\text{Degree-Hours above } 25^\circ C}{\text{Design Temperature Difference}} \quad (11)$$

Table 8 (a and b) provides frequency of occurrence of dry bulb temperature at selected locations in the continental United States.

Example 6 — Estimate the equivalent full load hours for a residential air conditioning system in Dallas, Texas. Assume the inside temperature is maintained at 77°F for the cooling season.

From Table 8b

	82°	87°	92°	97°	102°	107°
Dallas	896 hr	667	466	285	93	11

Thus the number of degree-hours above 77°F is

$$\begin{aligned} D\,Hrs &= 896(5) + 667(10) + 466(15) + 285(20) \\ &\quad + 93(25) + 11(30) \\ D\,Hrs &= 26{,}495 \end{aligned}$$

The design temperature for Dallas is 99°F. Thus $\Delta T = 99 - 77 = 22°$; then EFL = 26,495/22 = 1,204 hr.

The frequency of occurrence of dry bulb temperature data is also the basis of another, slightly more complicated procedure. This procedure is commonly referred to as the "bin" method because the temperature versus hours-of-occurrence data are grouped in 5°F increments or bins. The bin method should be used if the building is not occupied nor operated continuously. In such cases the bin hours are divided into those which correspond to operating hours and nonoperating hours. Reference 4 contains data that have been compiled for 0200 to 0900 hr, 1000 to 1700 hr, and 1800 to 0100 hr, which can be used as a basis for such analysis. The application of this procedure is outlined in Reference 3, and is illustrated in Volume II, Chapter 9.

3. Evaluation of Building Thermal Characteristics

A modification of the bin method can be used here to illustrate the impact of variations in the thermal characteristics of a building envelope on energy requirements. In this modified bin procedure, the load components discussed earlier are computed on a daily basis rather than for peak load hours. In addition, the solar loads are modified to account for possible cloud cover. The daily load is then divided by 24 and plotted at the mean daily temperature for the design day. This procedure is illustrated in Figure 4, where values for both heating and cooling are plotted. The two design day load

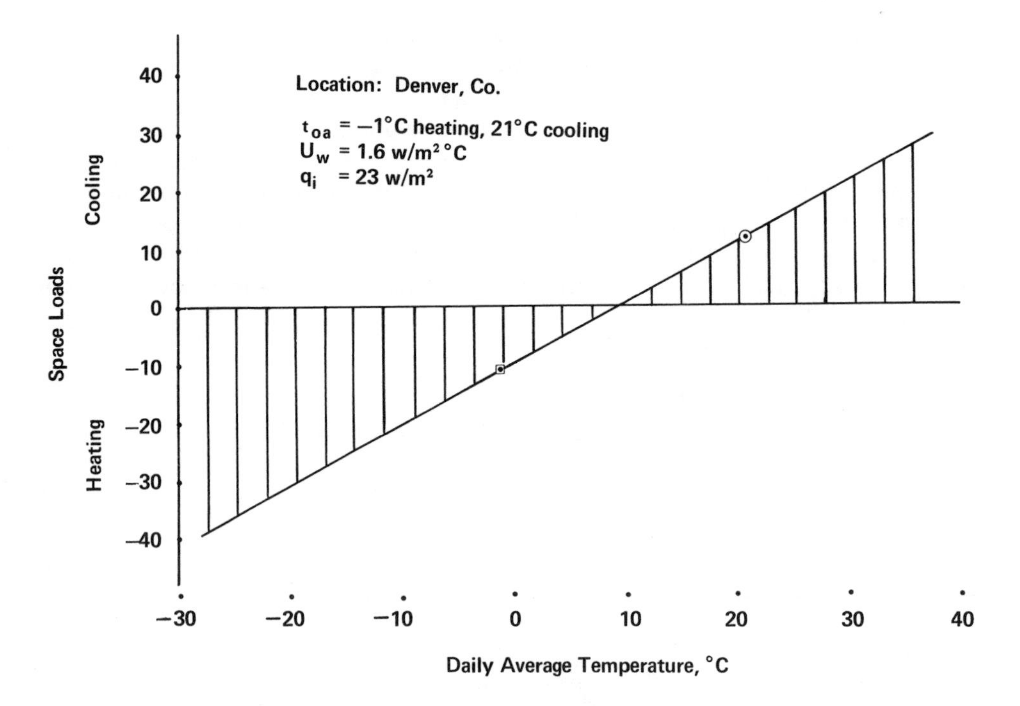

FIGURE 4. Estimating seasonal annual heating and cooling loads.

points are then connected with a straight line as indicated in the figure. This line is used to determine values of the average load versus temperature, which with the frequency of occurrence data is used to estimate envelope loads on a seasonal basis.

The daily loads are calculated as indicated below. This procedure is intended to apply to only those spaces in a building which have an outside wall or roof. The loads in interior spaces in large buildings are primarily internal loads and are governed by operating schedule; they are essentially independent of climate. Ventilation and infiltration loads are not included in this calculation.

In addition to those terms defined previously: q_D = daily load; ETD = equivalent temperature difference; I_{DT} = total daily irradiation on a surface; %S = percent possible sunshine; T_{oa} = mean daily temperature.

The load components are calculated as follows:

Loads for walls and roofs:

$$\text{with ETD} = (T_{oa} - T_i)\,24 + \alpha/h_o\,(I_{DT})\%S$$

$$q_w = U_w A_w(\text{ETD}) \tag{12}$$

Loads for windows:

$$q_g = [I^*_{DT}\,(SC)\%S + U\,(T_{oa} - T_i)24]\,A_g \tag{13}$$

where I^*_{DT} is the value of the solar radiation incident on the surface modified to account for external shading. Table 9 gives the depth of a shadow per foot of overhang averaged over a 5 hr period for August first.

TABLE 9
Shadow Factors

Direction window faces	Latitude, degrees 30	35	40	45	50
E	0.8	0.8	0.8	0.8	0.8
SE	1.6	1.4	1.3	1.1	1.0
S	5.4	3.6	2.6	2.0	1.7
SW	1.6	1.4	1.3	1.1	1.0
W	0.8	0.8	0.8	0.8	0.8

Note: The distance the shadow line falls below the edge of an overhang equals the width of the overhang multiplied by the shadow factor. Values are averages for the five hours of greatest solar intensity August 1.

4. Internal Loads

In this procedure, only the internal loads for those spaces which have at least one outside exposure wall or roof should be included. Lights q_L = (W/m²)(Area)(hours of operation); equipment q_E = (W)(hours of operation); people q_P = (125 W)(number of occupants)(hours of occupancy).

$$q_I = q_L + q_E + q_P \quad (14)$$

Total daily load:

$$q_{DT} = q_w + q_g + q_I \quad (15)$$

The interior zones are excluded as it is possible that cooling will be required in such a space at the same time that heating is required in the perimeter space. If, however, both the interior and perimeter spaces are served by the same system and controlled with a single thermostat, the interior space loads should also be included.

The total daily load q_{DT} is then divided by 24 and plotted against the average temperature for the day. In effect this procedure provides an average hourly load for the design day conditions. When average loads for the summer and winter design days are plotted as in Figure 4 and connected with a straight line, the result provides a load vs. temperature distribution for the building. With the frequency of occurrence temperature data this in turn can be used to estimate seasonal loads.

Example 7 — Estimate the annual heating and cooling loads for a single story office building located in Houston, Texas, given the data shown below.

- Construction: slab on grade, rectangular shape 15m × 25m
  flat roof, wood deck and built-up roofing U = 0.2 W/m²°C
  brick veneer wall, 3.5m high with U = 0.35 W/m²°C
  glazing, single glass 0.75m × 1.5m
  9 each east and west
  15 each north and south interval shading SC = 0.55
- Orientation: major axis running east-west
- Internal Loads: 16 W/m² 7 a.m. to 7 p.m., 3W/m² other
- Miscellaneous: Assume light color walls α/η_o = 0.026 m²°C/W
- Design Conditions: 29°N Latitude

TABLE 9a

Example 7

		Area m^2	U	$I_{DTc} W/m^2$	$I_{DTH} W/m^2$	ETD_c	ETD_H	qD_c	qD_H
Walls	N	70.6	0.35	1,225[a]	545[a]	70	−209	−1,730	−5,164
	E	42.4	0.35	3,941	2,189	120	−188	1,781	−2,790
	S	70.6	0.35	2,529	6,042	94	−138	2,323	−3,910
	W	42.4	0.35	3,941	2,189	120	−188	1,781	−2,790
Roof		375	0.20	7,210	3,689	179	−168	13,425	−12,600
Floor		375	0.00[b]	—	—	—	—	—	−1,913
Glass	N	16.9	6.25	1,065	472	—	—	11,999	−20,618
	E	10.1	6.25	3,427	1,902	—	—	16,352	−8,353
	S	16.9	6.25	2,170	5,254	—	—	19,182	1,606
	W	10.1	6.25	3,427	1,902	—	—	16,352	− 8,353
Transmission + Solar								84,925	−64,385
Internal								85,500	85,500
Hourly Average (Transmission + Solar + Internal)								7.1kW	0.9kW

[a] To obtain values of I_{DT} for walls, multiply SHGF of Table 3 by 1.15 and then by 3.15 to convert to W/ m.²

[b] For slab floor use "U" = 2.65 W/m and multiply by perimeter rather than by area.

	summer	winter
% sun	70%	50%
T_{oa}	27°C	11°C
T_i	25°C	20°C

Solution: Using the procedures of Equations 12 through 15 and the data of Tables 2, 3, and 8 the annual heating and cooling loads may be estimated. Note that the ventilation loads will be estimated separately. Moreover, as the data in the tables are in English units, appropriate conversion factors must be used.

These calculated average hourly loads are then plotted versus T_{oa} and a load-vs-temperature curve laid out. The load at the midpoint of each temperature "bin" may then be determined. These loads are then multiplied by the number of hours in each bin to estimate the annual heating and cooling loads as indicated in the table below.

Thus for this building and location the annual heating and cooling loads, exclusive of infiltration and ventilation loads, are heating = 1,127 kWh/year, and cooling = 33,602 kWh/year.

With this simple format it is possible easily to recalculate the design day loads for various design options and to determine the seasonal impact of the changes considered. A change in the thermal characteristics may move the load line up or down or change its slope. Some changes may reduce heating requirements but increase cooling requirements, and others may be beneficial in both seasons. This procedure provides a means of presenting seasonal load information in a simple graphic way.

Example 8 — If the building of Example 7 were located in Denver, Colorado, estimate the effect of the following thermal characteristics on the annual heating and cooling requirements.

- Walls U = 1.14 W/m²°C
  roof U = 1.14 W/m²°C
  glass, single sheet U = 6.25 W/m²°C
  internal load 23 W/m²

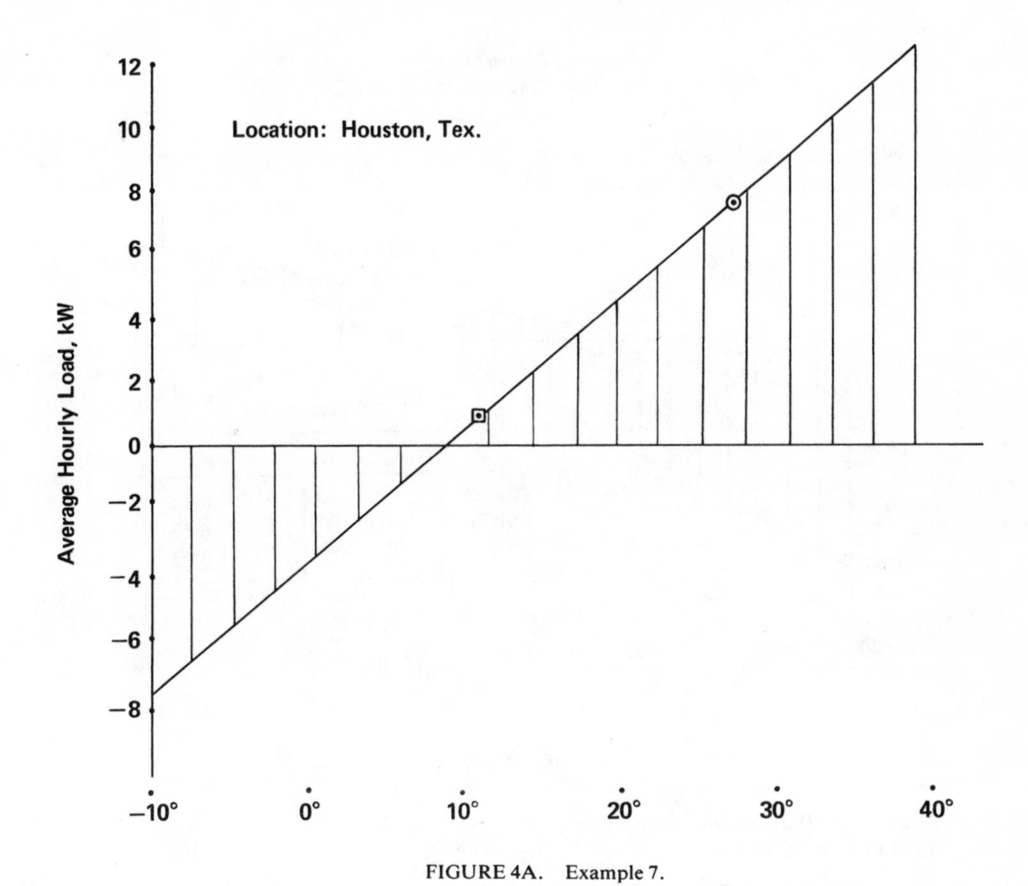

FIGURE 4A. Example 7.

TABLE 9b

Example 7

Bin °F	hr	Hourly load kW	Cooling load kWh	Bin °F	hr	Hourly load kW	Heating load kWh
102	1	11.6	11.6	47	452	−0.2	−90.8
97	56	10.5	588.0	42	291	−1.3	−378.3
92	324	9.4	3,045.6	37	141	−2.3	−324.3
87	681	8.4	5,720.4	32	64	−3.4	−217.6
82	956	7.3	6,978.8	27	18	−4.5	−81.0
77	1630	6.2	10,106.0	22	4	−5.6	−22.4
72	1172	5.2	6,094.4	17	2	−6.6	−13.2
67	980	4.1	4,018.0	12	0		
62	772	3.0	2,316.0				
57	681	1.9	1,293.9				
52	570	0.8	456.0				
Total	7800		33,602 kWh	Total	960		−1,128 kWh

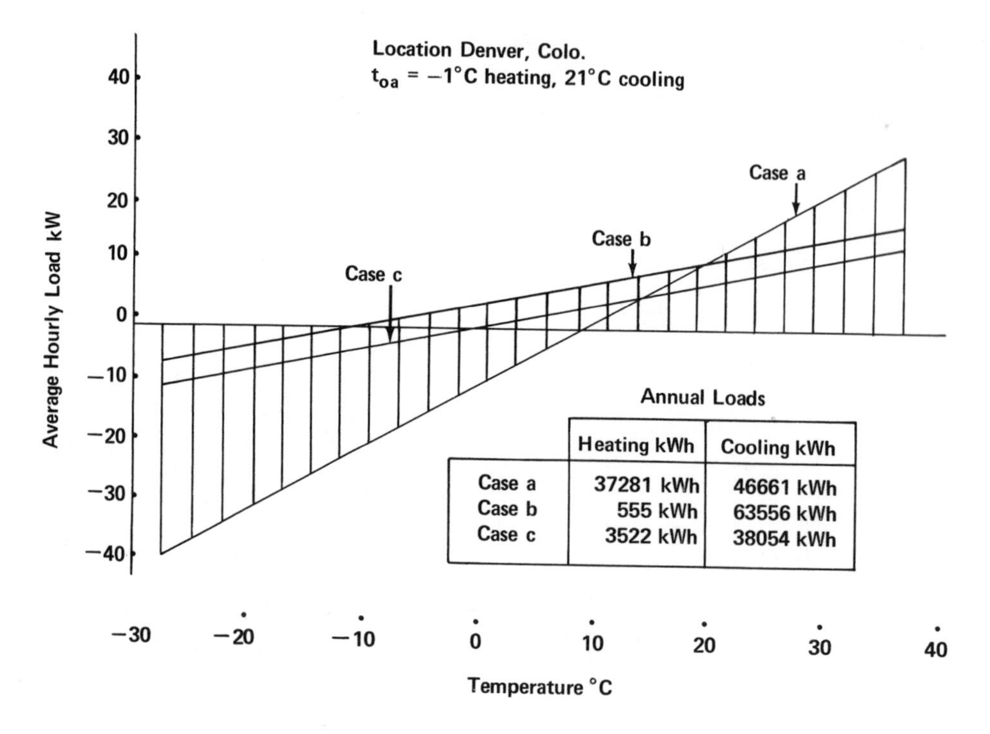

FIGURE 4B. Example 8.

- Walls U = 0.28 W/m²°C
  roof U = 0.28 W/m²°C
  glass, double U = 3.75 W/m²°C
  internal load 23 W/m²
- same as b except internal load = 15 W/m²

Utilizing the procedure described above, the following results are obtained for average daily loads at mean daily temperatures of −1°C (January) and 21°C (July).

	T_{oa}	Q_{wall} kWh/d	Q_{roof} kWh/d	Q_{glass} kWh/d	Q_{floor} kWh/d	Q_{int} kWh/d	Q_{DT} kWh/d
Case a	21°	5.97	32.87	29.56	—	225.25	293.65
	−1°	−119.04	−198.42	−158.22	−1.75	225.25	−252.96
Case b	21°	1.49	8.22	38.97	—	225.25	273.93
	−1°	−29.76	−49.60	−75.53	−1.75	225.25	68.61
Case c	21°	1.49	8.22	38.97	—	146.90	195.58
	−1°	−29.76	−49.60	−75.53	−1.75	146.90	−9.74

These data may be plotted as before to obtain the loads as a function of temperature.

These results indicate several points of interest. First, although increasing the thermal resistance of the envelope reduces the total of the heating and cooling loads by 24%, the cooling loads increase. This increase could, however, be offset by the use of outside air for cooling at temperatures below 15°C.

Second, the reduction in internal loads between cases b and c significantly decreases the cooling load but increases the heating load. Again, however, the total loads (heating and cooling) are lower by 35%, indicating significant savings.

For internal zones with no outside exposure, management of the internal loads is the critical factor, not the thermal characteristics of the space. The reader is directed

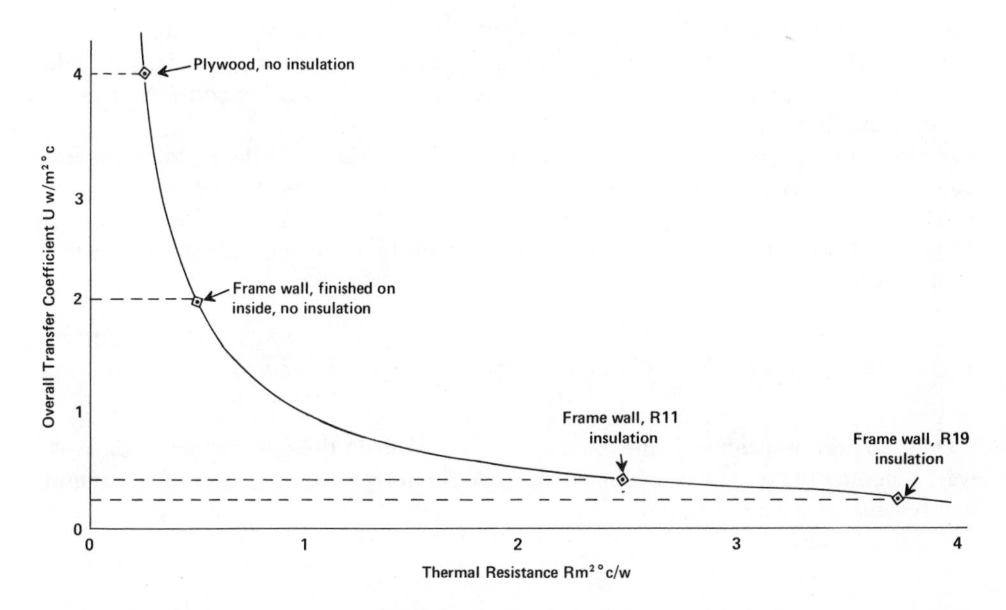

FIGURE 5. Overall heat transfer coefficient as a function of thermal resistance.

to Volume III, Chapter 2 for information on options for conservation associated with lighting, equipment, and system control.

C. Factors That Influence Thermal Performance

As noted in previous sections, the heat loss or heat gain through a building envelope is a function of the materials used; of geometric factors such as size, shape, and orientation; of internal loads; and of climatic factors. Procedures for estimating heat loss or heat gain on an hourly and on a seasonal basis have been presented. Utilizing these procedures it is now possible to consider the effects of variations of these factors on building heat loss and heat gain.

1. Materials

The two properties of building materials that are of particular importance in evaluating thermal performance are thermal resistance and thermal capacitance. The basic characteristics of each of these properties have been presented previously. It is useful at this point, however, to consider several brief examples of the analysis of these properties as related to energy conservation and solar energy application projects.

As indicated by Volume I, Chapter 2, Equation 36, the overall heat transfer coefficient U is inversely proportional to thermal resistance

$$U = \frac{1}{\Sigma R} \tag{16}$$

Plotting U versus R as in Figure 5 clearly indicates that the incremental benefit (that is, a reduction in the value of U and thus, in the heat transfer) decreases as the value of R increases. The implication is that there is an optimum point beyond which an increase in resistance can no longer be justified. This optimum must be determined by balancing increased first cost against reduced operating costs.

Reductions in operating costs may be based on estimates of reduction in the heat gain and heat loss calculated by means of the methods discussed above. Estimates of the cost must include all costs associated with the installation of materials which will increase thermal resistance. For example, if a thicker wall section is required for additional insulation, the incremental cost of the thicker wall must be considered as well as the insulation material costs.

A second aspect of thermal resistance arises from the fact that the building envelope contains a variety of components. Consider for example, a 100 m^2 wall which is 75% opaque with a total resistance of 2.5m^2°C/W and 25% glass with a resistance of 0.15 m^2°C/W. From Equations 1 and 15 it can be seen that the heat transfer rate will be proportional to

$$A_T/R_T = A_w/R_w + A_g/R_g = 75/2.5 + 25/0.15 = 30 + 167 = 197$$

Doubling the resistance of the opaque wall would drop the ratio to only 182. However, doubling the resistance of the window would drop the ratio to 114. This example simply illustrates the fact that if

$$A_w/R_w << A_g/R_g$$

there will be considerably less justification from an energy point of view in decreasing A_w/R_w than there is in decreasing A_g/R_g. Of course, the cost of accomplishing one versus the other must also be considered. However, all too frequently simple comparisons such as the above are overlooked, and the available financial resources misapplied.

Thermal capacitance, or the ability of material to store energy with a change in temperature, must also be examined. There is a direct link between the mass of a material and its thermal capacitance as the specific heats of most building materials are about the same. When discussing thermal capacitance, we are interested in transient behavior. As a result of capacitance, the rate of heat transfer through a massive wall is quite different from that through a light wall.

This characteristic difference is illustrated in Figure 6. In this figure the outside temperature variation is also shown. The inside temperature is assumed to be constant. Several observations should be made. First, the peak heat transfer rate is much lower for the more massive wall, and second, the peak is delayed. Both have a significant impact on capacity requirements. Finally, one should also note that the total heat transfer through the two walls for the daily cycle is approximately equal under the conditions assumed. This fact indicates that thermal mass alone will not reduce daily heat transfer.

There are, however, ways that thermal capacitance may be used to advantage for energy conservation under some climatic conditions. For example, if there is sufficient swing in the summer daily temperature range, the delayed heat transfer through massive walls can be offset by increasing the circulation of the cool night air through the space. In the winter, a combination of the massive walls and appropriately placed windows can be effective in minimizing energy requirements.

2. Size, Shape, and Orientation

Size, shape, and orientation have a direct influence on the thermal performance of a building. Size and shape together determine the ratio of surface area to useful internal space. This ratio may have a substantial impact on the energy required to maintain interior conditions at the desired level. A single large square building, for example, will have less surface area than several smaller buildings enclosing the same total

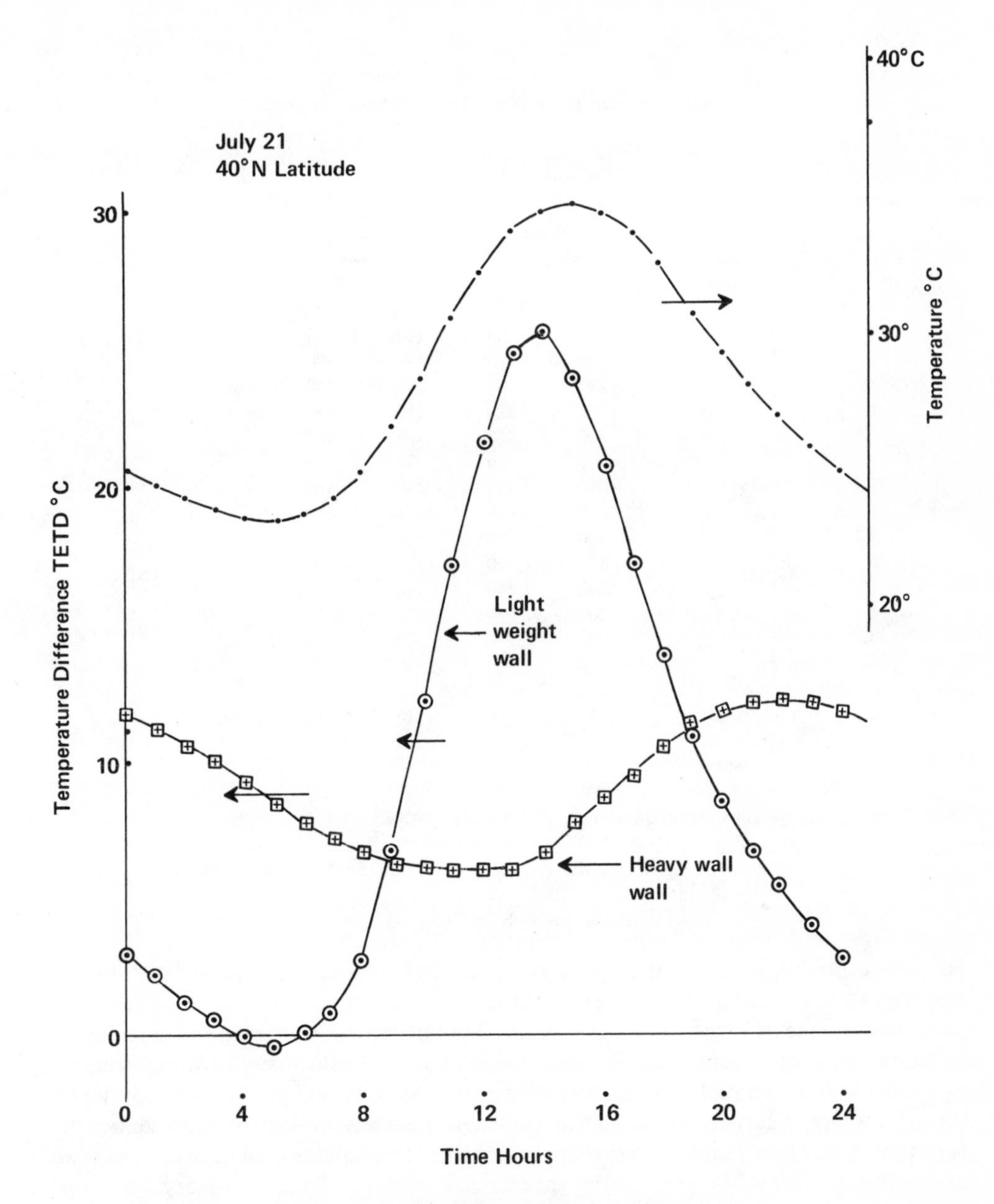

FIGURE 6. Effects of wall mass on TETD on a sunny day.

amount of space. Thus, the larger building would have less heat gain or heat loss than the smaller buildings under the same environmental conditions and might be considered more nearly "optimum" from a thermal performance point of view. The same conclusion can be drawn when comparing a relatively square building with one that has an extended floor plan and consequently a large amount of surface area for the enclosed volume.

Unfortunately, design decisions cannot be based solely on the thermal characteristics of size and shape. Other factors such as energy required for lighting, ventilation, and perhaps vertical transportation must be considered. In addition, criteria such as the limitations of the site, the intended function of the building, or aesthetic considerations, may play the significant role in determining size and shape. Therefore, it is not

TABLE 10

Solar Heat Gain Through a Window 750mm × 1250mm

40° N Latitude July 21
Btu/hr

Sun Time	Unshaded S	Unshaded W	Shaded S[a]	Shaded W	
6	100	100	100	100	
7	200	190	190	190	
8	290	260	260	260	
9	520	310	310	310	
10	800	350	350	350	
11	1,020	370	370	370	
Noon	1,090	410	380	410	
1	1,020	810	370	420	
2	800	1,460	350	560	
3	520	1,940	310	1,190	
4	290	2,160	260	1,850	
5	200	2,040	190	2,040	
6	100	1,370	100	1,370	
7	0	10	0	10	
July total[b]	6,950	11,780	3,540	9,810	Btu/day
June total	6,220	12,000	3,620	10,400	Btu/day
August total	9,420	10,900	3,080	9,620	Btu/day

% sunshine for June, July, and August = 70%

[a] A 0.91m overhang will completely shade a 2.5m high south wall for the entire day.
[b] To convert to kJ/day multiply by 1.055.

possible to arrive at a specific "rule of thumb" for design purposes. Any options relative to size and shape should be evaluated in the design process.

As an example of some of the questions which may arise in the case of a residential building, consider the trade-off between a one- and a two-story single family residence of the same floor area. The two-story house may have 10 to 15% less outside surface area. However, a larger percentage of the outside area is wall surface. In view of the fact that it is easier, and perhaps less expensive, to insulate ceiling areas than wall areas, the thermal advantage of the smaller overall area of the two-story house may be lost. The most significant difference between the two houses may be the amount of window area. If the two-story house has an increased window area, all the advantage of the more compact shape will be lost.

Building orientation is another feature which has energy-related implications but which may be primarily controlled by other factors, particularly the site. However, the orientation of windows with respect to solar and wind patterns may be of considerable importance. The ability to maintain comfortable conditions in perimeter spaces or to utilize day-lighting or natural ventilation is dependent upon the orientation of window areas. For example, exterior shading is approximately twice as effective in limiting solar heat gain as interior shading. In many instances, however, it is difficult to provide effective exterior shading on east- and west-facing windows. South-facing windows, on the other hand, can be shaded relatively easily during the summer months. In terms of utilizing daylight, windows with external shade are best as the shading eliminates any problems with glare.

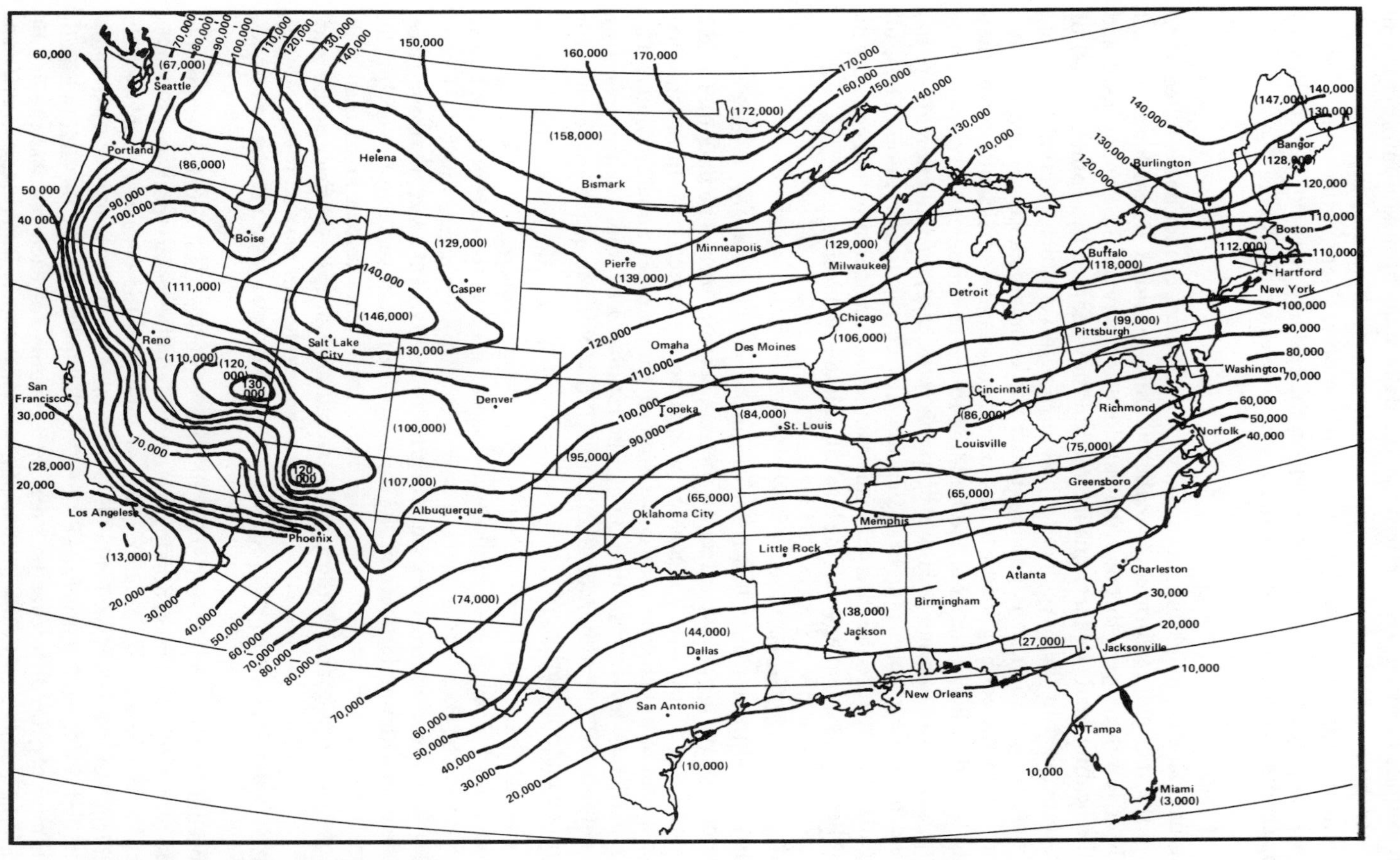

FIGURE 7. Annual wet bulb degree-hours below 54°F wet bulb and 68°F dry bulb (From Energy Conservation for Existing Buildings, ECM-1, Federal Energy Administration, June 1975, 61. With permission.)

Looking at the solar heat gain factors of Table 3 gives an indication of the impact of orientation on solar heat gain through a window during June, July, and August. Consider, for example, a comparison of a south-facing and a west-facing window at a 40° north latitude location. On the basis of the daily solar heat gain information in Table 10, percent sunshine, and the number of days in each month, the west window has a 53% increase in load over that of a south window. Shading the south window provides a 55% reduction. Effective shading on the east and west can be provided by trees, drop shades, louvers, or an overhang as extensive as a patio cover.

3. Internal Loads

The internal loads from lighting, occupancy, or equipment cannot be neglected when considering the thermal performance of the building envelope. These loads interact with the heat loss/heat gain through the envelope to increase the cooling load and decrease the heating load. In fact, this interaction may bear on the optimum level of insulation to be used in the envelope. In buildings with very high internal loads and 16- to 24 hr/day operation, it is possible under some circumstances to overinsulate. The reason is that the cooling load predominates and may be sufficiently larger than the heating load that heat loss from the building to the environment actually reduces the building's energy requirement. However, if this is the case, the reduction of internal loads would be a most important first step towards energy conservation. Thus, all options available to reduce the internal loads should be considered before considering the effects these loads might have on the level of wall thermal resistance.

4. Infiltration and Ventilation

As infiltration is the uncontrolled entry of outside air into a space, it may always be considered to have a negative effect on thermal performance. The data of Table 8 (a and b) and Figure 7 provide a means of estimating the effects of infiltration. First an average seasonal air change rate is determined for summer and for winter. The effect of a particular level of infiltration can be estimated by applying the following equation: defining $AC_w \equiv$ winter air change rate, VS ≡ Volume of the space, then for heating with no humidification:

$$q_H = AC_w(VS)\ 1.23\ DBDH_w \qquad (17)$$

where $DBDH_w \equiv$ the number of dry bulb degree-hours (°C) below 20°C (from Table 8a) or for heating with humidification:

$$q_H = AC_w(VS)\ 3.0\ WBDH_w \qquad (18)$$

where $WBDH_w \equiv$ the number of wet bulb degree-hours (°C) below 12°C (from Figure 7).

The procedure for cooling is somewhat more complicated as it involves estimating the outside air energy content (enthalpy). However, the necessary data can be determined from a standard psychrometric chart or table and the occurrence of dry bulb and coincident wet bulb data of Table 8b. To accomplish the calculation, first determine the enthalpy for each temperature "bin" in Table 8b and for the indoor condition. Then multiply the difference between the enthalpy for each bin and the indoor enthalpy by the number of hours in that bin, arriving at a seasonal energy rate, kJ-hr/kg.

$$SDH = \Sigma_i\ (h_{bin_i} - h_{room})\ hrs_{bin_i}$$

1. Enter your natural gas Total Annual Energy Use from Line 13 of your Energy Record ________(1)

2. a. Enter the number of therms you used during each month you did not use your furnace.

May ________

June ________

July ________

August ________

September ________

October ________

Total ________ (2a)

b. Divide the total by the number of months months you did not use your furnace to get your average monthly energy use for the months you did no space heating.

________ ÷ ________ (Number of months) = ________(2b)

c. Multiply Line 2b by 12.

________ (2b) X 12 = ________(2c)

3. Subtract Line 2c from Line 1. Annual gas consumption for heating. ________(3)

FIGURE 8. Energy use for space heating with natural gas (no summer space heating) (From *Solar for Your Present Home,* California Energy Resources Conservation and Development Commission, Sacramento, 1978. With permission.)

The seasonal cooling required to offset infiltration can then be estimated as

$$q_C = AC_S(VS)\ 1.16(SDH) \quad (19)$$

Example 9 — Estimate the annual heating and cooling loads resulting from infiltration if the air change rate in a building of the same size as that in Example 7 is 0.4 air changes per hour. Assume the building is located in Columbus, Ohio.

Solution: From Example 7 the building volume is VS = 15 × 25 × 3.5 = 1,312 5m^3, then AC(VS) = 525 m^3/hr. From Figure 8 $WBDH_w$ = 100,000°F hr; thus $WBDH_w$ = 55,555°C hr; and Q_H = (525)(55,555)(3.0) = 87.499 · 10^6 kJ; or Q_H = 24,305 kWh; assuming 20°C and 40% RH indoor conditions. For the summer, assuming 25°C and 60% RH indoor conditions: h_i = 31.5 Btu/lb = 73.4 kJ/kg and from Table 8b

	Natural Gas	No. of days in billing period	Therms per day	Electricity	No of days	kWh per day	Propane	No. of days	Gal. per day	Fuel Oil	No. of days	Gal. per day
1. January	______ therms*	______	______	______ KWH	______	______	______ gal.	______	______	______ gal.	______	______
2. February	______	______	______	______	______	______	______	______	______	______	______	______
3. March	______	______	______	______	______	______	______	______	______	______	______	______
4. April	______	______	______	______	______	______	______	______	______	______	______	______
5. May	______	______	______	______	______	______	______	______	______	______	______	______
6. June	______	______	______	______	______	______	______	______	______	______	______	______
7. July	______	______	______	______	______	______	______	______	______	______	______	______
8. August	______	______	______	______	______	______	______	______	______	______	______	______
9. September	______	______	______	______	______	______	______	______	______	______	______	______
10. October	______	______	______	______	______	______	______	______	______	______	______	______
11. November	______	______	______	______	______	______	______	______	______	______	______	______
12. December	______	______	______	______	______	______	______	______	______	______	______	______
13. Total	______ therms	______	______	______ KWH	______	______	______ gal.	______	______	______ gal.	______	______

*NOTE: 1 mcf = 10 therms. Your bill may be in mcf rather than therms.

FIGURE 9. Energy record. From *Solar for Your Present Home,* California Energy Resources Conservation and Development Commission, Sacramento, 1978. With permission.)

t_{db}	82°	87°	92°	97°	102°	°F
t_{wb}	69°	72°	74°	75°	76°	°F

thus

h	32.2	35.6	37.5	38.4	39.3	Btu/lb

and finally

Δh	1.63	9.55	13.98	16.08	18.17	kJ/kg
hr	395	222	88	15	1	hr
SDH	643.8	2120.1	1230.2	241.2	18.2	kJ-hr/kg

This indicates a seasonal cooling load due to infiltration of q_c = (525)(643.8 + 2120.1 + 1230.2 + 241.2 + 18.2)(1.16) or q_c = 722 kWh for the Columbus, Ohio, location.

These formulas may also be applied in estimating the seasonal impact of various levels of outside air being brought into a building for ventilation purposes. In the case of ventilation, however, the product AC(VS) is replaced by the number of ℓ/sec of outside air divided by 3.59.

III. ANALYSIS OF EXISTING BUILDINGS

In the discussion to this point the analysis presented has been applicable to both new and existing buildings, although the load calculation procedures described are most often associated with new building design. In considering the modification of an existing building, data should be gathered on existing energy use patterns. To assure their accuracy, calculated seasonal loads should then be compared with these data for the existing building. Actual consumption data provide a distinct advantage in analyzing the effect modifications will have on the building's energy use characteristics.

Electric and fuel consumption data should be tabulated by month for a period of at least a year. As billing periods often vary significantly, divide the monthly bill by the

ENERGY USE FOR ELECTRICAL APPLIANCES AND SPACE COOLING WITH NATURAL GAS FOR SPACE AND HOT WATER HEATING (NO WINTER SPACE COOLING)

1. Enter your Total Annual kWh from line 13 of your Energy Record (Figure 7.9). _____ kWh (1)

2. a. Enter the number of kWh you used for each month you did not use your air conditioner.

Oct ____________________

Nov ____________________

Dec ____________________

Jan ____________________

Feb ____________________

Mar ____________________

Total (2a) ____________________ kWh

b. Divide the total by the number of months you did not use your air conditioner.

__________ ÷ __________ (Number of months) = __________ (2b)

c. Multiply Line 2b by 12.

__________ (2b) X 12 = _____ (2c)

3. Subtract Line 2c from Line 1. _____ kWh
Annual kWh for cooling.

FIGURE 10. Energy use for electrical appliances and space cooling with natural gas for space and hot water (From *Solar for Your Present Home,* California Energy Resources Conservation and Development Commission, Sacramento, 1978. With permission.)

number of days in the billing period. Comparisons can then be made on an average day basis as in the previous analysis. The energy record form shown in Figure 9 can be used to compile this data. Since electricity and fuel may be used for functions other than heating and cooling, it is necessary to distribute consumption of each energy source among the appropriate categories. Table 11 lists typical consumption figures for major household appliances.

If fuel is used for heating and electricity for cooling, the procedure is somewhat simpler than if either electricity or fuel is used for both functions. For convenience the forms in Figures 8 and 10 may be used. The energy consumption estimates for lighting, appliances, and water heating should be checked against a monthly bill when heating or cooling is not included. For example, if a residence has electric air conditioning and uses natural gas for heating and water heating, the appliance and lighting estimates can be checked against electric billing for the winter months. Conversely, the water heating and gas appliance estimate may be checked against gas consumption

TABLE 11

Summary of Appliance Energy Data

I. Electric	kWh/month	Comments
Refrigerator	140	Frost-free may vary from 100 to 150 kWh/month depending on size and model. Manual will run 80 to 100 kWh/month if defrosted frequently enough to prevent extensive buildup on cooling coils
TV	40	Color TV runs roughly twice the power consumption of black and white; solid-state is approximately one half tube type for both color and black and white; thus, a solid-state color would use about the same power as a black and white tube type
Miscellaneous	50	This includes small appliances such as:
		kWh/month
		Hand iron 12
		Coffee maker 10
		Electric skillet 12
		Vacuum cleaner 4
		Hair dryer 3
		Stereo/radio 3
		Toaster 3
		Disposal 2
Freezer	120	About one fifth of residences nationwide have home food freezers
Dishwasher	30	Electric energy, exclusive of hot water; the dishwasher will use approximately 0.4 kWh/load for pumps, controls, etc., and 0.35 kWh/load for drying; various models range from 0.5 to 1.0 kWh/load
Clothes washer	10	Exclusive of hot water. Electric energy use will range from 0.2 to 0.3 kWh/load for various models; with a hot wash/cold rinse cycle, hot water consumption will be approximately 25 gal/load, while for warm wash/cold rinse it will be approximately 12 gal/load
Lighting	250	Depending on the wattage of the bulbs and the number of hours per month the lights are used, this could range from a low of 150 kWh/month to a high of 350 kWh/month

II. Electric/Gas	kWh/month	ft^3/month	Comments
Range/Oven	100	690	Based on approximately 6 hr/week oven baking, 1/2 hr/week broiling, and 5 hr/week operation of two surface elements on the range
Dryer	100	470 (+ 7 kWh electric)	Electric approximately 3 kWh/load; gas 14.5 ft^3/load plus 0.2 kWh/load electric
Yard light	18	1520	Gas yard lights burn continuously, using from 46 to 54 ft^3 of gas per day; the electric yard light is 60 W and is assumed to burn 10 hr/day, 365 days/yr
Water heater	505	2600	70 gal/day at 150°F
	690	3400	100 gal/day at 150°F; considerable savings could be achieved by reducing the hot water temperature to 135°F; for 70 gal/day use this would reduce the consumption to 415 kWh/month electric or 2100 ft^3/month gas

Figure 8 must be completed first.

1. Enter result from Line 2c of Figure 8 in therms. ——— (1)

2. Adjustments for other uses of gas:

   a. Cooking:
      Enter 70 for gas stove with pilot lights
      40 for gas stove without pilot lights
      0 for non-gas stove ——— (2a)

   b. Clothes drying:
      If you have a gas dryer, calculate your energy use as follows: 0.15 X ———— X 52 = ——— (2b)
      Number of loads per week

   c. Other uses:
      Estimate your gas consumption for other uses such as barbecues or outdoor lighting. Your utility can help you with this. Enter the total therms per year.
      See Table 10. ——— (2c)

   d. Add Lines 2a, 2b, and 2c. ——— (2d)

3. Substract Line 2d from Line 1. ——— (3)

4. Furnace pilot light correction. Your furnace pilot light consumes about 7 therms each month month that it burns. Enter the number of months per year your pilot light is on (some people turn it off during the summer) and multiply by 7. ———— X 7 = ——— (4)
   Number of months

5. Subtract Line 4 from Line 3. Annual gas consumption for hot water = ——— (5)

FIGURE 11. Energy use for water heating with natural gas (no summer space heating) (From *Solar for Your Present Home*, California Energy Resources Conservation and Development Commission, Sacramento, 1978. With permission.)

for a summer month when no heating occurs. In the case of estimating the water heating as a separate item, additional calculations will be required if gas is used for cooking, clothes drying, or for a yard light or grill. If the furnace has a standing pilot light, as most older units do, the gas consumed for this function must also be considered. Figure 11 provides a form for estimating water heating use. Table 11 provides information on gas appliances.

With the data obtained from the forms of Figures 8 and 10, the savings which can be expected from improved thermal performance can be estimated by calculating seasonal heating and cooling loads for the building as is and following any modifications made. It should be noted that the energy consumption derived from the billed consumption reflects the efficiency of the heating and cooling equipment as well as the heating and cooling loads.

A similar procedure is followed for a commercial building. However, in this case domestic hot water may represent only a small portion of the energy use, while lighting, air moving equipment, and office equipment are major items. Lighting energy use may be estimated by counting the number of fixtures and the number and wattage of bulbs in each fixture. Then, based on the number of hours of operation, an estimate of consumption may be made. If the fixtures are fluorescent, the bulb wattage must be multiplied by 1.2 to obtain the correct consumption figures. Office equipment is somewhat harder to estimate, in terms of both determining the power consumption and

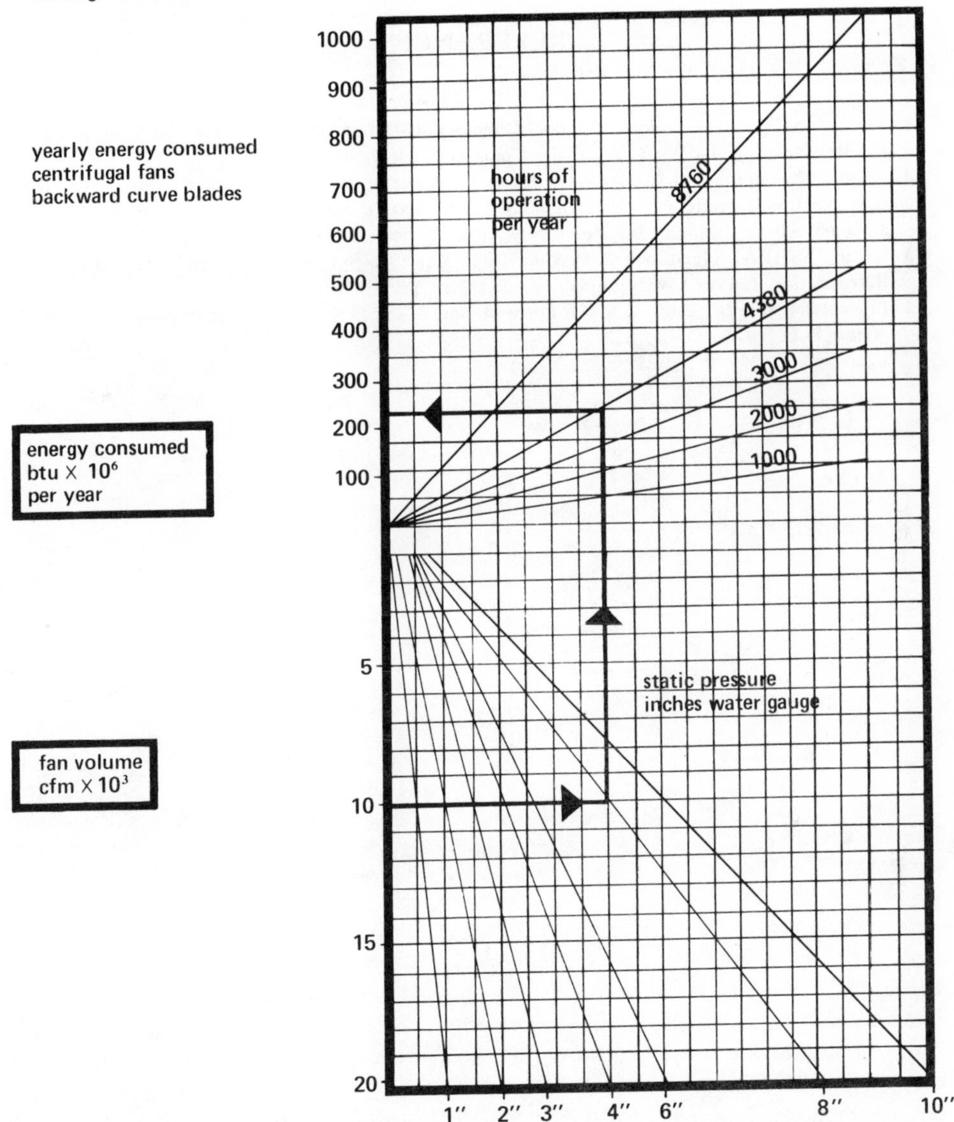

FIGURE 12. Estimating energy requirements for air-handling equipment (From Energy Conservation for Existing Buildings, ECM-2, Figure 24, Federal Energy Administration 1975, 201. With permission.)

estimating periods of operation. However, major equipment items should be included in estimates. Figure 12 provides data which may be used to estimate energy requirements for operating air-handling equipment based on the volumetric flow rate and hours of operation.

It is important to note that options for thermal conservation may not provide significant savings for large commercial buildings. In these large buildings the major cooling loads are attributable to internal heat sources. The major means of conserving energy is through reducing these internal loads and through modification of the heating and air conditioning system operation and controls, subjects which are treated elsewhere in this volume. However, in smaller commercial buildings, elementary schools, and similar types of structures, the principles of thermal conservation discussed here can be expected to provide savings.

REFERENCES

1. **Anon.**, ASHRAE Handbook & Product Directory, 1977 Fundamentals, American Society of Heating, Refrigerating, and Air-Conditioning Engineers, New York, 1977.
2. **Anon.**, Bibliography on Available Computer Programs in the General Area of Heating, Refrigeration, Air Conditioning and Ventilating, (NSF-RA-760002), American Society of Heating, Refrigerating and Air-Conditioning Engineers, New York, October 1975.
3. **Anon.**, ASHRAE Handbook & Product Directory, 1976 Systems, American Society of Heating, Refrigerating, and Air-Conditioning Engineers, New York, 1976.
4. Engineering Weather Data, U.S. Air Force Manual 88-29, U.S. Government Printing Office, Washington, D.C., 1978.

Chapter 2

ELECTRICAL ENERGY MANAGEMENT IN BUILDINGS

Craig B. Smith

TABLE OF CONTENTS

I. PRINCIPAL ELECTRICITY USES IN BUILDINGS

A. Introduction: Building Types

There are hundreds of building types, and buildings can be categorized in many ways — by use, type of construction, size, or thermal characteristics, to name a few. For simplicity, two designations will be used here: residential and nonresidential.

The residential category includes features common to single family dwellings, apartments, and hotels. The nonresidential category includes a major emphasis on office buildings, as well as a less detailed discussion of features common to retail stores, hospitals, restaurants, and laundries. Industrial facilities are not included here, but are discussed in Volume III, Chapters 3 and 4. The extension to other types is either obvious or can be pursued by referring to the literature listed in the references.

The approach taken in this chapter is to list two categories of specific strategies which are cost-effective methods for conserving electricity. The first category includes those measures which can be implemented at low capital cost using existing facilities and equipment in an essentially unmodified state. The second category includes technologies which require retrofitting, modification of existing equipment, or new equipment or processes. Generally, moderate to substantial capital investments are also required.

B. Residential Uses

Energy use by the residential/commercial sector in 1976 is shown in Table 1.[1] The most significant end uses are heating, ventilating, and air conditioning (HVAC) systems, lighting, and water heating, followed by refrigeration, cooking, and appliances.

HVAC energy use accounts for approximately 64% of total energy use. Of this total, about half is electricity. Electricity is used in space heating and cooling to drive fans and compressors, as a direct source of heat (resistance heating), or as an indirect heat source (heat pumps). Heat pumps are discussed elsewhere in Volume II, Chapter 9.

Lighting, which accounts for about 20% of U.S. electricity use (5% of total U.S. energy use in 1977), amounts to 10% of the electricity used in the residential sector.[2] In a typical office building, lighting consumes approximately 25% of the total energy used in the building, and approximately 50% of the electricity.[3] The bulk of residential lighting is incandescent, and offers substantial opportunities for improved efficiency.

Water heating amounts to about 11% of residential/commercial use, with gas being the predominant fuel. Electricity use for this purpose currently occurs in regions where there is cheap hydroelectricity or where alternative fuels are not available. The use of electricity can be expected to increase in the future as alternative fuels become more expensive or have their use curtailed. Solar water heating (discussed in Volume II, Chapters 5 and 6) is another alternative which will become increasingly attractive.

Refrigerators are another important energy use in the residential sector, accounting for about 6% of total energy and about 20% of electricity. The vast majority of refrigerators in use today are electric.

Cooking and other appliances account for the balance of energy use in the residential sector (14% of all energy, 26% of electricity).

C. Nonresidential Uses

In nonresidential buildings HVAC is again one of the major energy users. There are exceptions of course — in energy-intensive facilities such as laundries, the process energy will be most important.

Electricity is used to run fans, pumps, chillers, and cooling towers. Other uses include electric resistance heating (for example, in terminal reheat systems) or electric boilers.

TABLE 1

Energy Use in the Residential/Commercial Sector — 1976

Sector	Used as fuel[a] (10^{15} Btu)	Electricity conversion losses[b] (10^{15} Btu)	Used as electricity (10^{15} Btu)	Totals (10^{15} Btu)	Totals (10^{9} GJ)	Totals (%)
Space heating	12.46	1.40	0.70	14.56	15.38	56
Air conditioning	0.02	1.34	0.67	2.03	2.14	8
Water heating (residential)	0.96	1.34	0.67	2.97	3.14	11
Lighting	—	1.70	0.85	2.55	2.69	10
Refrigeration	—	0.58	0.29	0.87	0.92	3
Cooking	0.23	0.28	0.14	0.65	0.69	2
Other	1.10	0.94	0.47	2.51	2.65	10
Totals	14.77	8.28	3.79	26.14	27.61	100

[a] Multiply 10^{15} Btu by 1.05 to get 10^{9} GJ.
[b] Estimated from data in Reference 1 using an average heat rate of 10,200 Btu/kWh.

From Electric Power Research Institute, *EPRI Journal*, Vol. 2(10), 40, 1977. With permission.

Nonresidential lighting is generally next in importance to HVAC for total use of electricity, except in those nonresidential facilities with energy-intensive processes. Lighting is generally fluorescent (80%) with a small fraction (20%) incandescent. Incandescent lamps are largely used in older buildings or for decorative or esthetic applications. Recently, there has been a trend toward development and use of more efficient, high-intensity discharge (e.g., high pressure sodium) lamps in nonresidential facilities.

Water heating is another important energy use in nonresidential buildings, but here circulating systems (using a heater, storage tank, and pump) are more common. Many possibilities exist for using heat recovery as a source of hot water.

Refrigeration is an important use of energy in supermarkets and several other types of nonresidential facilities. It is now common practice to include heat recovery units on commercial refrigeration systems.

Computers are an increasingly common feature of office buildings and many nonresidential facilities. For the larger units, specially designed rooms with temperature and humidity control are required. As a rule of thumb, the electricity used by the computer must be at least doubled, since cooling must be provided to remove the heat from the equipment, lights, operating, and personnel.

In nonresidential facilities, the balance of the electricity use is for elevators, escalators, and office equipment such as duplicating machines and typewriters.

II. STRATEGIES FOR ELECTRICITY END USE MANAGEMENT

A. Setting up an Energy Management Program

The general procedure for establishing an energy management program in buildings involves five steps:

- Review historical energy use.
- Perform energy audits.
- Identify energy management opportunities.
- Implement changes to save energy.
- Monitor the energy management program, set goals, review progress.

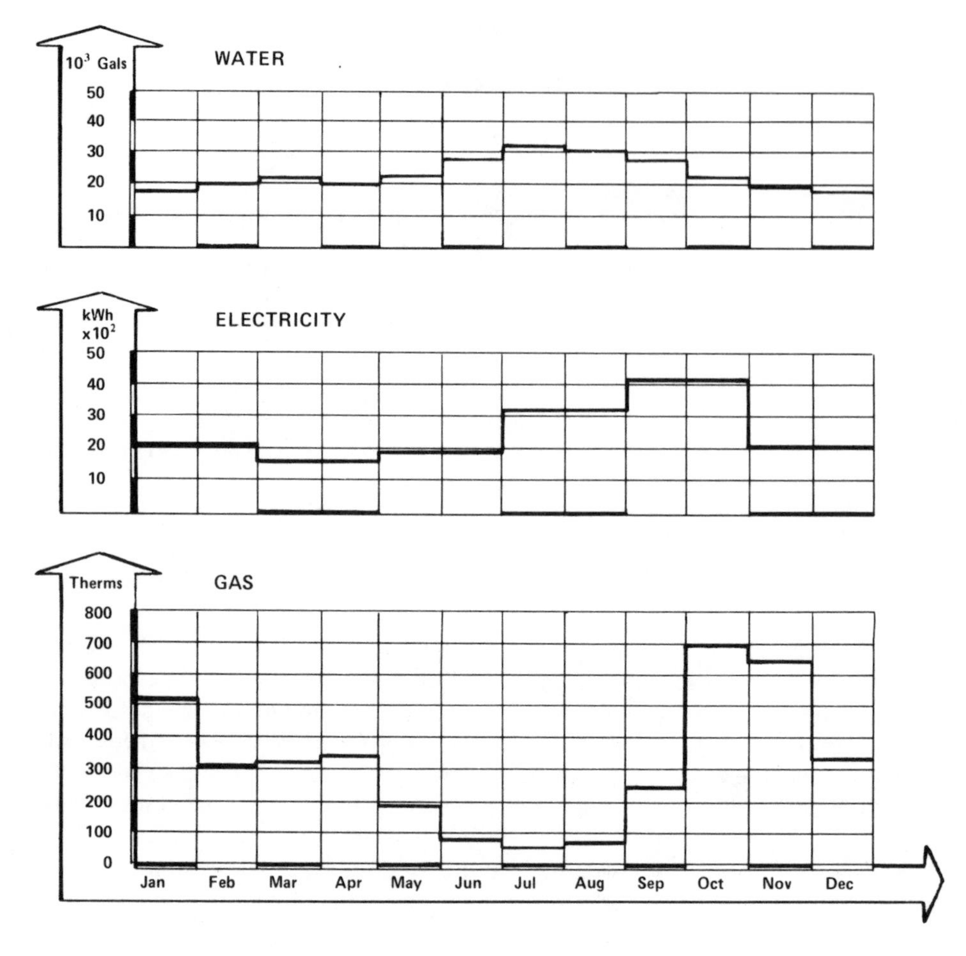

FIGURE 1. Historical energy use in a small manufacturing facility. (From Smith, Craig B., **Ed.**, *Efficient Electricity Use,* 2nd ed., Pergamon Press, New York, 1978. With permission.)

Each step will be described briefly.

1. Review of Historical Energy Use

Utility records can be compiled to establish electricity use for a recent 12-month period. These should be graphed on a form (see Figure 1) so annual variations and trends can be evaluated. By placing several years (e.g., last year, this year, and next year projected) on the form, past trends can be reviewed and future electricity use can be compared with goals. Alternatively, several energy forms can be compared, or energy use vs. production determined (meals served, for a restaurant, kilograms of laundry washed, for a laundry, etc.).

2. Perform Energy Audits

Figures 2 and 3 are forms for performing an energy audit of a building.[2] The Building Energy Survey Form, Figure 2, provides a gross indication of how energy is used in the building in meeting the particular purpose for which it was designed. This form would not be applicable to single family residences, but could be used with apartments. It is primarily intended for commercial buildings.

Figure 3 is a form used to gather information concerning energy used by each piece of equipment in the building. When totaled, the audit results can be compared with

Building Description

heating degree days ___

- Name: Age: ___ years
- Location:
- No. of floors___ Gross floor area___m^2 (ft^2) Net floor area___m^2 (ft^2)
- Percentage of surface area which is glazed___% cooling degree days___
- Type of air conditioning system; heating only___ evaporative___
  dual duct___ other (describe)___________________________
- Percentage breakdown of lighting equipment: Incandescent___%
  fluorescent___% high intensity discharge (type)___% other__________%

Building Mission

- What is facility used for:
- Full time occupancy (employees) ___ persons
- Transient occupancy (visitors or public) ___ persons
- Hours of operation per year ___
- Units of production per year ___ Unit is__________

Installed Capacity

- Total installed capacity for lighting ___kW
- Total installed capacity of electric drives greater than 7.5 kW (10 hp)
  (motors, pumps, fans, elevators, chillers, etc.) ___hp × 0.746 = ___kW
- Total steam requirements ___lbs/day or ___kg/day
- Total gas requirements ___ft^3/day or Btu/hr or ___m^3/day
- Total other fuel requirements ___________________________

Annual Energy End Use

Energy Form		×	Conversion		kBtu/yr	Metric Units		Conversion		MJ/yr
• Electricity	___kWh/yr	×	3.41	=	______	___kWh/yr	×	3.6	=	______
• Steam	___lb/yr	×	1.00	=	______	___kg/yr	×	2.32	=	______
• Natural gas	___cf/yr	×	1.03	=	______	___m^3/yr	×	38.4	=	______
• Oil	___gals/yr	×	{#2 139; #6 150}	=	______	___ℓ/yr	×	{#2 38.88; #6 41.8}	=	______
• Coal	___tons/yr	×	24,000	=	______	___kg/yr	×	28.0	=	______
• Other	___	×			______	___	×	___	=	______
	Totals				______					______

Energy Use Performance Factors (EUPF's) for Building

- EUPF 1 = MJ/yr(kBtu/yr) ÷ Net floor area =___MJ/m^2 yr(kBtu/ft^2 yr)
- EUPF 2 = MJ/yr(kBtu/yr) ÷ Average annual occupancy =___MJ/person · yr (kBtu/person · yr)
- EUPF 3 = MJ/yr(kBtu/yr) ÷ Annual units of production =___MJ/unit · yr (kBtu/unit · yr)

FIGURE 2. Building energy survey form. (From Smith, Craig B., Ed., *Efficient Electricity Use,* 2nd ed., Pergamon Press, New York, 1978. With permission.)

Symbols: k = 10^3
k = 10^6

Conversion Factors

Multiply	by	to get
kWh	3.6	MJ
Btu/hr	0.000293	kW
hp	0.746	kW

Plant Name ———— By —— Date ———— Sheet — of —
Location ———— Period of Survey: 1 day 1 wk 1 mo 1 yr
Department ———— Notes ————

Type Fuel	Equip. ID No	Equipment Description	Power: Name Plate Rating (Btu/hr, kW, hp, etc.)	Power: Conv. Factor to kW	Power: kW	Est. % Load (100%, 50%, etc.)	Est. Hrs. Use Per Period	kW	Conv. Factor	Total Energy Use Per Period (MJ)

FIGURE 3. Energy audit data sheet. (From Smith, Craig B., Ed., *Efficient Electricity Use*, 2nd ed., Pergamon Press, New York, 1978. With permission.)

the historical energy use records plotted on Figure 1. The energy audit results show a detailed breakdown and permit identification of major energy using equipment items.

Figure 4 shows partial results of an audit on a building housing an electronics manufacturing operation. The results of the audit have been processed by a computer which makes the calculations and prints a summary of the results.

3. Identify Energy Management Opportunities

An overall estimate should be made of how effectively the facility uses its energy resources. This is difficult to do in many cases, because so many operations are unique. An idea can be obtained, however, by comparing similar buildings having similar climates. Tables 2 and 3 show representative values and indicate the range in performance factors which is possible.

Next, areas or equipment which use the greatest amounts of electricity should be examined. Each item should be reviewed and these questions asked:

- Is this actually needed?
- How can the same equipment be used more efficiently?
- How can the same purpose be accomplished with less energy?
- Can the equipment be modified to use less energy?
- Would new, more efficient equipment be cost effective?

4. Implement Changes

Once certain actions to save energy have been identified, an economic analysis will be necessary to establish the economic benefits and to determine if the cost of the

Roof Equip ID	Description	Name Plate Rating	Units	Conversion Factor	Power (KW)	Est. % Load	Est. Hours	Energy (KWH)	Energy (GJE)	Energy (MBTU)
11.001	AIR CONDITIONER	15.000	KW	1.0000000	1.500000E 1	50.000	264.00	1.979999E 3	2.138399E 1	
11.002	AIR CONDITIONER	10.000	KW	1.0000000	1.000000E 1	50.000	264.00	1.320000E 3	1.425599E 1	
11.003	ELEC STRIP HEAT	13.000	KW	1.0000000	1.300000E 1	50.000	264.00	1.716000E 3	1.853279E 1	
11.004	ELEC STRIP HEAT	15.000	KW	1.0000000	1.500000E 1	50.000	264.00	1.979999E 3	2.138399E 1	
11.005	HEATER	17000.000	BTU/HR	0.0002930	4.980998E 0	50.000	264.00	6.574917E 2	7.100908E 0	
11.006	1-1/2 TON NATL GAS	6800.000	BTU/HR	0.0002930	1.992399E 0	50.000	264.00	2.629966E 2	2.840363E 0	
11.007	EXHAUST FANS (5)	0.400	KW	1.0000000	4.000000E −1	100.000	264.00	1.056000E 2	1.140479E 0	
11.008	A/C UNITS (2)	0.800	KW	1.0000000	8.000000E −1	50.000	264.00			3.801600E −1
11.009	EXHAUST FANS (4)	1.600	KW	1.0000000	1.600000E 0	100.000	264.00	4.223999E 2	4.561917E 0	
11.010	8 TON UNIT	10.000	KW	1.0000000	1.000000E 1	50.000	264.00	1.320000E 3	1.425599E 1	
11.011	INTAKE FILTERS (6)	3.000	KW	1.0000000	3.000000E 0	75.000	264.00	5.939997E 2	6.415195E 0	
11.012	CENTRIFUGAL EXHAUST	0.900	KW	1.0000000	9.000000E −1	75.000	264.00	1.782000E 2	1.924559E 0	
					7.667340E 1			1.053669E 4	1.137962E 2	3.801600E −1

Lunch Room Equip. ID	Description	Name Plate Rating	Units	Conversion Factor	Power (KW)	Est. % Load	Est. Hours	Energy (KWH)	Energy (GJE)	Energy (MBTU)
12.001	GAS WATER HEATER	*1500000.0	BTU/HR	0.0002930	4.394997E 2	10.000	720.00			1.139183E 2
12.002	BOILER	560000.000	BTU/HR	0.0002930	1.640799E 2	15.000	0.00			0.000000E 0
12.003	HOT WATER CIRC PUMP	0.750	HP	0.7460000	5.595000E −1	50.000	264.00	7.385397E 1	7.976227E −1	
12.004	SUPPLY FAN MOTOR	7.500	HP	9.7460000	5.595000E 0	50.000	264.00	7.385398E 2	7.976227E 0	
12.005	LIGHTS (11)	2.200	KW	1.0000000	2.200001E 0	100.000	360.00	7.920000E 2	8.553597E 0	
12.006	VENDING MACHINES	2.800	KW	1.0000000	2.800000E 0	100.000	360.00	1.008000E 3	1.088639E 1	
					6.147341E 2			2.612394E 3	2.821385E 1	1.139183E 2

Mechanics Equip. ID	Description	Name Plate Rating	Units	Conversion Factor	Power (KW)	Est. % Load	Est. Hours	Energy (KWH)	Energy (GJE)	Energy (MBTU)
13.001	COMPRESSOR MOTOR	60.000	HP	0.7460000	4.475999E 1	50.000	264.00	5.908316E 3	6.380980E 1	
13.002	SUPPLY FAN	20.000	HP	0.7460000	1.492000E 1	100.000	264.00	3.938879E 3	4.253989E 1	
13.003	HEATER (BOILER)	500000.000	BTU/HR	0.0002930	1.464999E 2	15.000	0.00			0.000000E 0
13.004	INCAND. LIGHTS (2)	0.400	KW	1.0000000	4.000000E −1	100.000	360.00	1.440000E 2	1.555200E 0	
					2.065799E 2			9.991195E 3	1.079049E 2	0.000000E 0

FIGURE 4. Sample energy audit results. Note that the symbol E followed by a number represents 10 raised to that power, e.g., E 2 = 10^2.

action is justified. (Refer to Volume I, Chapters 3, 4, and 6 for guidance.) Those changes which satisfy the economic criteria of the building owner (or occupant) will then be implemented. Economic criteria might include a minimum return on investment (e.g., 50%), a minimum payback period (e.g., 2 years) or a minimum benefit-cost ratio (e.g., 2.0).

5. Monitor the Program, Establish Goals

This is the final — and perhaps most important — step in the program. A continuing monitoring program is necessary to ensure that energy savings do not gradually disappear as personnel return to their old ways of operation, equipment gets out of calibration, needed maintenance is neglected, etc. Also, setting goals (they should be realistic) provides energy management personnel with targets against which they can gauge their performance and the success of their programs.

6. Summary of Energy Management Programs

The foregoing has been outlined in two tables to provide a step-by-step procedure for electrical energy management in buildings. Table 4 is directed at the homeowner or apartment manager, while Table 5 has been prepared for the commercial building owner or operator. Industrial facilities are treated separately; refer to Volume III, Chapter 4, Electrical Power Management in Industry.

One problem in performing the energy audit is determining the energy used by each item of equipment. In many cases published data are available — as in Table 6 for residential appliances. In other cases, engineering judgments must be made, the manufacturer consulted, or instrumentation provided to actually measure energy use.

TABLE 2

Sample Energy Use Performance Factors (EUPF's)[a,b]

Facility description[b]	Net area (m^2)	EUPF #1[c] kBtu/ $ft^2 \cdot$year	EUPF #1[c] MJ/ $m^2 \cdot$year	EUPF #2 MBtu/ person·year	EUPF #2 GJ/ person·year
Small engineering office*	279	43.6	496	11.6	12.2
Engineering office and laboratory*	929	28.7	326	14.4	15.2
Elementary school	4,000	72.7	826	4.86	5.13
Hospital laundry	4,200	687	7,810	404	427
Elementary school	7,000	61.4	698	3.84	4.06
City administration building*	16,250	119	1,350	16.1	17.0
Office building with private clubs	16,700	175	1,990	52.7	55.6
Police administration building*	41,500	202	2,300	—	—
Office building	46,500	106	1,210	—	—
Office building	48,000	139	1,580	—	—
Los Angeles city hall complex*	138,600	162	1,840	51.5	54.4

[a] Buildings are located in California, Wisconsin, Louisiana, Virginia, and Pennsylvania. Data are for period 1973—1976.

[b] Buildings marked with an asterisk have had energy management programs which have reduced energy use by 20 to 30% compared to pre-1973 values.

[c] Multiply kBtu/ft^2 by 11.37 to get MJ/m^2.

From Smith, C. B., **Ed.**, *Efficient Electricity Use*, 2nd ed., Pergamon Press, New York, 1978. With permission.

TABLE 3

Representative Electricity Use in Buildings

Facility description	Net area (m^2)	Net area (ft^2)	Total electricity use kWh/ $m^2 \cdot$year	Total electricity use kWh/ $ft^2 \cdot$year	Total electricity use kWh/ Occupant·year	Lighting installed capacity W/m^2	Lighting installed capacity W/ft^2
Small engineering office	279	3,000	67	6.3	1,670	—	—
Engineering office and laboratory	929	10,000	22	2.0	750	16	1.5
Elementary school	4,000	43,500	229	21.3	1,430	31	2.9
Hospital laundry	4,200	45,000	242	22.6	11,950	26	2.4
Elementary school	7,000	75,000	194	18.0	1,130	30	2.8
City administration building	16,250	175,000	234	21.7	2,920	—	2.0—4.0
Office building with private clubs	16,700	180,000	345	32.1	9,620	33	3.1
Police administration building	41,500	447,000	194	18.0	—	—	—
Office building	46,500	500,000	127	11.8	—	—	—
Office building	48,000	517,000	243	22.6	—	—	—
Los Angeles city hall complex	138,600	1,492,000	202	18.7	4,940	—	—

From Smith, C. B., **Ed.**, *Efficient Electricity Use*, 2nd ed., Pergamon Press, New York, 1978. With permission.

TABLE 4

An Energy Management Plan for the Homeowner or Apartment Manager[a]

First Step: Review Historical Data

1. Collect utility bills for a recent 12-month period.
2. Add up the bills and calculate total kWh, total $, average kWh (divide total by 12), average $, and note the months with the lowest and highest kWh.
3. Calculate a seasonal variation factor (svf) by dividing the kWh for the greatest month by the kWh for the lowest month.

Second Step: Perform Energy Audits

4. Identify all electrical loads greater than 1 kW (1000 W). Refer to Table 6 for assistance. Most electrical appliances have labels indicating the wattage. If not, use the relation W = V × A.
5. Estimate the number of hours per month each appliance is used.
6. Estimate the percentage of full load (pfl) by each device under normal use. For a lamp, it is 100%; for water heaters and refrigerators, which cycle on and off, about 30%, for a range, about 25% (only rarely are *all* burners *and* the oven used), etc.
7. For each device, calculate kWh by multiplying: kW × hours/month × pfl = kWh/month.
8. Add up all kWh calculated by this method. The total should be smaller than the average monthly kWh calculated in (2.)
9. Note: if the svf is greated than 1.5, the load shows strong seasonal variations, e.g., summer air conditioning, winter heating, etc. If this is the case, make two sets of calculations, one for the lowest month (when the fewest loads are operating) and one for the highest month.
10. Make a table listing the wattage of each lamp and the estimated numbers of hours of use per month for each lamp. Multiply watts times hours for each, sum, and divide by 1000. This gives kWh for the lighting loads. Add this to the total shown.
11. Add the refrigerator, television, and all other appliances or tools which use 5 kWh per month or more.
12. By this process you should now have identified 80 to 90% of electricity using loads. Other small appliances which are used infrequently can be ignored. The test is to now compare with the average month (high or low month if svf is greater than 1.5). If your total is too high, you have over estimated the pfl or the hours of use.
13. Now rank each appliance in descending order of kWh used per month. Your list should read approximately like this:

First:	Heating (in cold climates) ; air conditioning would be first in hot climates.
Second:	Waterheating
Third:	Lighting
Fourth:	Refrigeration
Fifth:	Cooking
Sixth:	Television
Seventh to last:	All others

Third Step: Apply Energy Management Principles

14. Attack the highest priority loads first. There are three general things which can be done: (1) reduce kW (smaller lamps, more efficient appliances); (2) reduce pfl ("oven cooked" meals, change thermostats, etc.); (3) reduce hours of use (turn lights off, etc.). Refer to the text for detailed suggestions.

Fourth Step: Monitor Program, Calculate Savings

15. After the energy management program has been initiated, examine subsequent utility bills to determine if you are succeeding.
16. Calculate savings by comparing utility bills. Note: since utility rates are rising, your utility bills may not be any lower. In this case it is informative to calculate what your bill would have been without the energy management program.

[a] Although the discussion is limited to electricity, a similar approach can be used with each fuel or energy form.

TABLE 5

An Energy Management Program for the Commercial Building Operator

First Step: Review Historical Data

1. Collect utility bills for a recent 12-month period.
2. Add up the bills and calculate total kWh, total $, average kWh (divide total by 12), average $, and note the months with the lowest and highest kWh.
3. Calculate a seasonal variation factor (svf) by dividing the kWh for the greatest month by the kWh for the lowest month.
4. Prepare a graph of historical energy use (see Figure 1).

Second Step: Perform Energy Audits

5. Evaluate major loads. In commercial buildings loads can be divided into four categories:
   a. HVAC (fans, pumps, chillers, heaters, cooling towers)
   b. Lighting
   c. Office equipment and appliances (elevators, typewriters, cash registers, copy machines, hot water heaters, etc.)
   d. Process equipment (as in laundries, restaurants, bakeries, shops, etc.)

Items a, b, and c are common to all commercial operations and will be discussed here. Item d overlaps with industry and the reader should also refer to Volume III, Chapter 4. Generally items a, b, and d account for the greatest use of electricity and should be examined in that order.

6. In carrying out the energy audit, focus on major loads. Items which together comprise less than 1% of the total connected load in kW can often be ignored with little sacrifice in accuracy.
7. Use the methodology described above and in Volume III, Chapter 4, "Electrical Power Management in Industry", for making the audit.
8. Compare audit results with historical energy use. If 80 to 90% of the total (according to the historical records) has been identified, this is generally adequate.

Third Step: Formulate the Energy Management Plan

9. Secure management commitment. The need for this varies with the size and complexity of the operation. However, any formal program will cost something, in terms of salary for the energy coordinator as well as (possibly) an investment in building modifications and new equipment. At this stage it is very important to project current energy usage and costs ahead for the next 3 to 5 years, make a preliminary estimate of potential savings (typically 10 to 50% per year), and establish the potential payback or return on investment in the program.
10. Develop a list of energy management opportunities (EMO's) (e.g., install heat recovery equipment in building exhaust air), estimate the cost of each EMO, and also the payback. Methods for economic analysis are given in Volume I, Chapter 3. For ideas and approaches useful for identifying EMO's, refer to the text.
11. Communicate the plan to employees, department heads, equipment operators, etc. Spell out who will do what, why there is a need, what the potential benefits and savings are. Make the point (if appropriate) that "the energy you save may save your job". If employees are informed, understand the purpose, and realize that the plan applies to everyone, including the President, cooperation is increased.
12. Set goals for department managers, building engineers, equipment operators, etc., and provide monthly reports so they can measure their performance.
13. Enlist the assistance of all personnel in: (1) better "housekeeping and operations", e.g., turning off lights, keeping doors closed; (2) locating obvious wastes of electricity; e.g., equipment operating needlessly, better methods of doing jobs.

Fourth Step: Implement the Plan

14. Implementation should be done in two parts. First, carry out operational and housekeeping improvements with a goal of, say, 10% reduction in electricity use at essentially no cost and no reduction in quality of service or quantity of production. Second, carry out those modifications (retrofitting of buildings, new equipment, process changes) which have been shown to be economically attractive.
15. As changes are made it is important to continue to monitor electricity usage to determine if goals are being realized. Additional energy audits may be justified.

Fifth Step: Evaluate Progress, Management Report

16. Compare actual performnce to the goals established in Item 12. Make corrections for weather variations, increases or decreases in production or number of employees, addition of new buildings, etc.
17. Provide a summary report of energy quantities and dollars saved, and prepare new plans for the future.

TABLE 6

Residential Energy Usage — Typical Appliances (Electricity and Gas)[a]

Electric Appliances	Power (watts)	Typical use (kWh/year)	Typical use (GJe/year)
Food Preparation			
Blender	300	1	0.0036
Broiler	1,140	85	0.31
Carving Knife	92	8	0.03
Coffee maker	1,200	140	0.50
Deep fryer	1,448	83	0.30
Dishwasher	1,201	363	1.31
Egg cooker	516	14	0.05
Frying pan	1,196	100	0.36
Hot plate	1,200	90	0.32
Mixer	127	2	0.0072
Oven, microwave (only)	1,450	190	0.68
Range			
With oven	12,200	700	2.52
With self-cleaning oven	12,200	730	2.63
Roaster	1,333	60	0.22
Sandwich grill	1,161	33	0.12
Toaster	1,146	39	0.14
Trash compactor	400	50	0.18
Waffle iron	1,200	20	0.07
Waste dispenser	445	7	0.03
Food preservation			
Freezer			
Manual defrost — 16 cu. ft.	—	1,190	4.28
Automatic defrost — 16.5 cu. ft.	—	1,820	6.55
Refrigerators/freezers			
Manual defrost, 12 cu. ft.	—	1,500	5.40
Automatic defrost, 17.5 cu. ft.	—	2,250	8.10
Laundry			
Clothes dryer	4,856	993	3.57
Iron (hand)	1,100	60	0.22
Washing machine (automatic)	512	103	0.37
Washing machine (non-automatic)	286	76	0.27
Water heater	2,475	4,219	15.19
(quick recovery)	4,474	4,811	17.32
Housewares			
Clock	2	17	0.06
Floor polisher	305	15	0.05
Sewing machine	75	11	0.04
Vacuum cleaner	630	46	0.17
Comfort conditioning			
Air cleaner	50	216	0.78
Air conditioner (room)	860	860	3.10
Bed covering	177	147	0.53
Dehumidifier	257	377	1.36
Fan (attic)	370	291	1.05
Fan (circulating)	88	43	0.15
Fan (rollaway)	171	138	0.50
Fan (window)	200	170	0.61
Heater (portable)	1,322	176	0.63
Heating pad	65	10	0.04
Humidifier	177	163	0.59

TABLE 6 (continued)

Residential Energy Usage — Typical Appliances (Electricity and Gas)[a]

Electric Appliances	Power (watts)	Typical use (kWh/year)	Typical use (GJe/year)
Health and beauty			
Germicidal lamp	20	141	0.51
Hair dryer	600	25	0.09
Heat lamp (infrared)	250	13	0.05
Shaver	15	0.5	0.0013
Sun lamp	279	16	0.06
Tooth brush	1.1	1.0	0.0036
Vibrator	40	2	0.0072
Home entertainment			
Radio	71	86	0.31
Radio/record player	109	109	0.39
Television			
Black and white	100	220	0.79
Tube type			
Solid state	45	100	0.36
Color	240	528	1.90
Tube type			
Solid state	145	320	1.15

Gas appliances

	Typical use (kft^3/yr)	Typical use (GJ/yr)
Clothes dryer	5.0	5.3
Furnace	65	69
Gas light	18	19
Pool heater	50—150	53—158
Range	10	11
Water heater	30	32

Note: The letter e, e.g., GJe, denotes the electrical energy used.

[a] From Smith, C. B., **Ed.,** *Efficient Electricity Use,* 2nd ed., Pergamon Press, New York, 1978. With permission.

B. Electricity-Saving Techniques by Category of End Use

In the discussion which follows, strategies for saving energy which can be implemented in a short time at zero or low capital cost are discussed. Retrofit and new design strategies are then described. The ordering of topics corresponds approximately to their importance in terms of building energy use. Energy use specifically for a process (e.g., heating) is excluded except as it relates to buildings and their occupants. Some discussion of process energy use is included in Volume III, Chapter 4, however.

1. Residential HVAC

Residential HVAC units using electricity are generally heat pumps, refrigeration systems, and electrical resistance heaters. Heaters range from electric furnace types, small radiant heaters, duct heaters, and strip or baseboard heaters, to embedded floor or ceiling heating systems. Efficiency for heating is usually high since there are no stack or flue losses and the heater transfers heat directly into the living space.

Cooling systems range from window air conditioning to central refrigeration or heat pump systems. Evaporative coolers are also used in some climates.

Principal operational and maintenance strategies for existing equipment include:

- System maintenance and cleanup
- Thermostat calibration and setback
- Time clocks, night cool down
- Improved controls and operating procedures
- Heated or cooled volume reduction
- Reduction of infiltration and exfiltration losses

System maintenance is an obvious but often neglected energy-saving tool. Dirty heat transfer surfaces decrease in efficiency. Clogged filters increase pressure drops and pumping power. Inoperable or malfunctioning dampers can waste energy and prevent proper operation of the system.

In residential systems heating and cooling is generally controlled by the room thermostat. As a first step the calibration of the thermostat should be checked, since these low cost devices can be inaccurate by as much as ±5°C. Several manufacturers now offer replacement thermostats with night setback controllers. Thus two setpoints are provided; one for day time use when heating to approximately 18°C (65°F) is recommended, the other for night time when temperatures are permitted to drop to 10 to 13°C (50 to 55°F). Depending on temperature conditions, hours of use, etc., correct temperature settings and night time temperature setback can save 5 to 10% of the annual heating bill.

A similar approach can be used with air conditioners. Thermostats should be set to 24°C (75°F) or higher.* Time clocks are sometimes useful for heating or cooling control, for example, for shutting systems off at night or for starting heating or cooling systems early in the morning.

Sometimes simple changes in controls or operating procedures will save energy. In cooling, use night air for summer cool-down. When the outside air temperature is cool, turn off the refrigeration unit and circulate straight outside air. If fan units have more than one speed, use the lowest speed which provides satisfactory operation. Check the balance of the system and the operation of dampers and vents, to insure that heating and cooling is provided in the correct quantities where needed.

Energy savings can be achieved by reducing the volume of the heated or cooled space. This can be accomplished by closing vents, doors, or other appropriate means. Usually it is not necessary to heat or cool an entire residence; the spare bedroom is rarely used, halls can be closed off, etc.

A major cause of energy wastage is infiltration and exfiltration. In a poorly "sealed" residence, infiltration of cold or hot air will increase heating or cooling energy use. In many residences this is the major loss of heated or cooled air. If ventilation rates are excessive due to high fan speeds, the dwelling can be overpressurized, increasing the leakage of heated or cooled air from the structure.

Check for open doors and windows, open fireplace dampers, inadequate weather stripping around windows and doors, and any other openings which can be sealed.

In retrofit or new designs, two general strategies should be observed:

- Reduce heating and cooling needs to the minimum level practical by proper selec-

* Caution: Some systems provide auxiliary heating. Be careful that increasing the thermostat setting does not increase energy use.

tion of site, dwelling orientation on the site, and dwelling design and construction.

- Provide the most efficient heating/cooling system possible. Generally, this will require provisions for heat storage and heat recovery.

These two considerations are particularly important for installations where solar heating is being considered.

In retrofit or new design projects the following techniques will save energy:

- Site selection and building orientation
- Building envelope design
- Selection of efficient heating/cooling equipment

Site selection and building orientation is not always under the control of the owner/occupant. Where possible, though, select a site sheltered from temperature extremes and wind. Orient the building (in cold climates) with a maximum southerly exposure to take advantage of direct solar heating in winter. Use earth berms to reduce heat losses on northerly exposed parts of the building. Deciduous trees provide summer shading but permit winter solar heating. See Volume II, Chapter 4 for a design discussion.

Building envelope design can improve heat absorption and retention in winter and summer coolness. The first requirement is to design a well-insulated, thermally tight structure. Guidelines for accomplishing this depend on climate. Consult the literature, local contractors, or your utility representative to determine recommendations for your area.

In general, the most efficient electric heating and cooling system is the heat pump. Commercially available equipment demonstrates a wide range of efficiency. Heating performance is measured in terms of a coefficient of performance (COP) which describes how many units of heat are supplied for each unit of electricity used. Typical COP's range from 2 to 5. Cooling performance of residential air conditioners is measured in terms of an energy efficiency ratio (EER) which describes the ratio of cooling capacity to electrical power input. In purchasing new equipment, selection of equipment with the highest COP and EER should be considered. See Volume II, Chapter 9 for further discussion of heat pumps.

Sizing of equipment is important, since the most efficient operation generally occurs at or near full load. Selection of oversized equipment is thus initially more expensive, and will also lead to greater operating costs.

The efficiency of heat pumps declines as the temperature difference between the heat source and heat sink decreases. Since outside air is generally the heat source, heat is most difficult to get when it is most needed. For this reason heat pumps often have electrical backup heaters for extremely cold weather.

An alternate approach is to design the system using a heat source other than outside air. Examples include heated air, such as is exhausted from a building, a deep well, providing water at a constant year-round temperature, or a solar heat source. There are a great many variations on solar heating and heat pump combinations. Refer to Volume II, Chapters 7, 8, and 9 for a more detailed discussion.

2. Nonresidential HVAC[2]

HVAC systems in nonresidential installations generally involve a central plant. Although the basic principles are similar to those discussed above in connection with residential systems, the equipment is larger and control more complex.

Efficiency of many existing HVAC systems can be improved. Modifications can reduce energy use by 10 to 15%, often with building occupants unaware that changes have been made.

The basic function of HVAC systems is to heat, cool, dehumidify, humidify, and provide air mixing and ventilation. The energy required to carry out these functions depends on the building design, its duty cycle (e.g., 24 hr per day use as in a hospital vs. 10 hr per day in an office), the type of occupancy, the occupants' use patterns and training in using the HVAC system, the type of HVAC equipment installed, and finally, daily and seasonal temperature and weather conditions to which the building is exposed.

A complete discussion of psychrometrics, HVAC system design, and commercially available equipment types is beyond the scope of this chapter.

Energy management strategies will be described in three parts:

- Equipment modifications (control, retrofit, and new designs)
  - —Fans —Chillers —Ducts and dampers
  - —Pumps
- Economizer systems and enthalpy controllers
- Heat recovery techniques

a. Equipment Modifications (Control, Retrofit, and New Designs)

i. Fans

All HVAC systems involve some motion of air. The energy needed for this motion can make up a large portion of the total system energy used. This is especially true in moderate weather when the heating or cooling load drops off, but the distribution systems often operate at the same level.

Control — Simple control changes can save electrical energy in the operation of fans. Examples include turning off large fan systems when relatively few people are in the building or stopping ventilation a half hour before the building closes. The types of changes which can be made will depend upon the specific facility. Some changes involve more sophisticated controls which may already be available in the HVAC system.

Retrofit — The capacity of the building ventilation system is usually determined by the maximum cooling or heating load in the building. This load has been changing due to reduced outside air requirements, lower lighting levels, and wider acceptable comfort ranges. As a result it is now feasible to decrease air flow in many existing commercial buildings.

The volume rate of air flow through the fan, Q, varies directly with the speed of the impeller's rotation. This is expressed as follows for a fan whose speed is changed from N_1 to N_2.

$$Q_2 = \left(\frac{N_2}{N_1}\right) Q_1$$

The pressure developed by the fan, P, (either static or total) varies as the square of the impeller speed.

$$P_2 = \left(\frac{N_2}{N_1}\right)^2 P_1$$

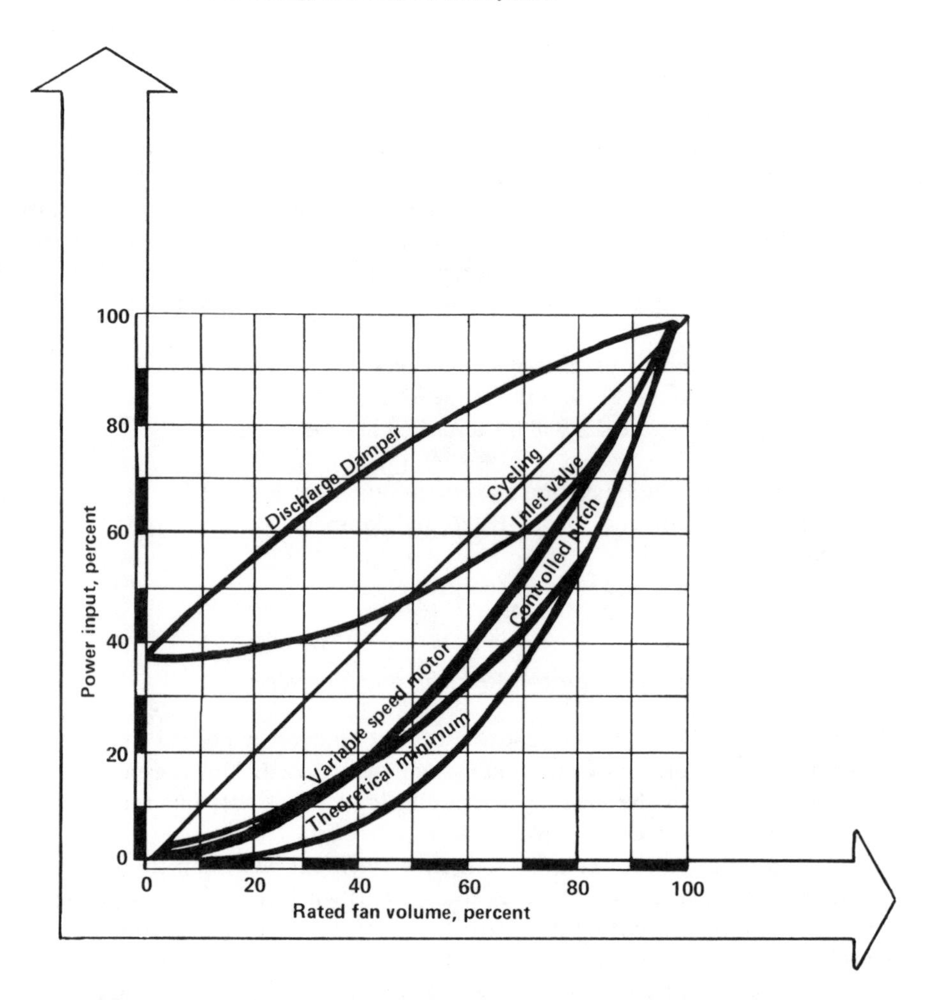

FIGURE 5. Fan power consumption for various types of part load controls. (From Smith, Craig B., Ed., *Efficient Electricity Use,* 2nd ed., Pergamon Press, New York, 1978. With permission.)

The power needed to drive the fan, H, varies as the cube of the impeller speed.

$$H_2 = \left(\frac{N_2}{N_1}\right)^3 H_1$$

The result of these laws is that for a given air distribution system (specified ducts, dampers, etc.), if the air flow is to be doubled, eight (2^3) times the power is needed. Conversely, if the air flow is to be cut in half, one eighth ($\frac{1}{2}^3$) of the power is required. This is useful in HVAC systems because even a small reduction in air flow (say 10%) can result in significant energy savings (27%).

The manner in which the air flow is reduced is critical in realizing these savings. Maximum savings are achieved by sizing the motor exactly to the requirements. Simply changing pulleys to provide the desired speed will also result in energy reductions according to the cubic law. The efficiency of existing fan motors tends to drop off below the half load range.

If variable volume air delivery is required, it may be achieved through inlet vane control, outlet dampers, variable speed motors, controlled pitch fans, or cycling. The relative efficiency of these approaches is shown in Figure 5. Energy efficiency in a retrofit design is best obtainable with variable speed motors or controlled pitch fans.

New design — The parameters for new design are similar to those for fan retrofit. It is desirable, when possible, to use a varying ventilation rate which will decrease as the load decreases. A system such as Variable Air Volume incorporates this in the interior zones of a building. In some cases there will be a trade-off between power saved by running the fan slower and the additional power needed to generate colder air. The choices should be determined on a case-by-case basis.

ii. Pumps

Pumps are found in a variety of HVAC applications such as chilled water, heating hot water, and condenser water loops. They are another piece of peripheral equipment which can use a large portion of HVAC energy, especially at low system loads.

Control — The control of pumps is often neglected in medium and large HVAC systems where it could significantly reduce the demand. A typical system would be a three chiller installation where only one chiller is needed much of the year. Two chilled water pumps in parallel are designed to handle the maximum load through all three chillers. Even when only one chiller is on, both pumps are used. By manual adjustments two chillers could be by-passed and one pump turned off. All systems should be reviewed in this manner to ensure that only the necessary pumps operate under normal load conditions.

Retrofit — Pumps follow laws similar to fan laws, the key being the cubic relationship of power to the volume pumped through a given system. Small decreases in flow rate can save significant portions of energy.

In systems in which cooling or heating requirements have been permanently decreased, flow rates may be reduced also. A simple way to do this is by trimming the pump impeller. The pump curve must be checked first, however, because pump efficiency is a function of the impeller diameter, flow rate, and pressure rise. It should be ensured that after trimming, the pump will still be operating in an efficient region. This is roughly the equivalent of changing fan pulleys in that the savings follow the cubic law of power reduction.

Another common method for decreasing flow rates is to use a pressure reducing valve. The result is equivalent to that of the discharge damper shown in Figure 5. The valve creates an artificial use of energy which can be responsible for much of the work performed by the pumps.

New design — In a variable load situation, common to most HVAC systems, more efficient systems with new designs are available than the standard constant volume pump. (These may also apply to some retrofit situations.)

One option is the use of several pumps of different capacity so that a smaller pump can be used when it can handle the load and a larger pump used the rest of the time. This can be a retrofit modification as well when a back-up pump provides redundancy. Its impeller would be trimmed to provide the lower flow rate.

Another option is to use variable speed pumps. While their initial cost is greater, they offer an improvement in efficiency over the standard pumps. The economic desirability of this or any similar change can be determined by estimating the number of hours the system will operate under various loads.

iii. Chillers

Chillers are often the largest single energy user in the HVAC system. The chiller cools the water used to extract heat from the building and outside air. By optimizing chiller operation the performance of the whole system is improved.

Two basic types of chillers are often found in commercial and industrial applications: absorptive and mechanical chillers. Absorptive units boil water, the refrigerant, at a low pressure through absorption into a high concentration lithium bromide solution. Mechanical chillers cool through evaporation of a refrigerant, such as freon, at a low pressure after it has been compressed, cooled, and passed through an expansion valve.

There are three common types of mechanical chillers. They have similar thermodynamic properties, but use different types of compressors. Reciprocating and screw-type compressors are both positive displacement units. The centrifugal chiller uses a rapidly rotating impeller to pressurize the refrigerant.

All of these chillers must reject heat to a sink outside the building. Some use air-cooled condensers, but most large units operate with evaporative cooling towers. Cooling towers have the advantage of rejecting heat to a lower temperature heat sink because the water approaches the ambient wet-bulb temperature while air-cooled units are limited to the dry-bulb temperature. This results in a higher condensing temperature, which lowers the efficiency of the chiller. Air-cooled condensers are used because they require much less maintenance than cooling towers.

Mechanical cooling can also be performed by Direct Expansion (DX) units. These are very similar to chillers except that they cool the air directly instead of using the chilled water as a heat transfer medium. They eliminate the need for chilled water pumps and also reduce efficiency losses associated with the transfer of the heat to and from the water. DX units must be located close (∼30 m) to the ducts they are cooling so they are typically limited in size to the cooling required for a single air handler. A single large chiller can serve a number of distributed air handlers. Where the air handlers are located close together, it can be more efficient to use a DX unit.

Controls — Mechanical chillers operate on a principle similar to the heat pump. The objective is to remove heat from a low temperature building and deposit it in a higher temperature atmosphere. The lower the temperature rise which the chiller has to face, the more efficiently it will operate. It is useful, therefore, to maintain as warm a chilled water loop and as cool a condenser water loop as possible.

Energy can be saved by using lower temperature water from the cooling tower to reject the heat. However, as the condenser temperature drops, the pressure differential across the expansion valve drops, starving the evaporator of refrigerant. Many units with expansion valves, therefore, operate at a constant condensing temperature, usually 41°C (105°F), even when more cooling is available from the cooling tower. Field experience has shown that in many systems, if the chiller is not fully loaded, it can be operated with a lower cooling tower temperature.

Retrofit — Where a heat load exists and the wet-bulb temperature is low, cooling can be done directly with the cooling tower. If proper filtering is available, the cooling tower water can be piped directly into the chilled water loop. Often a direct heat exchanger between the two loops is preferred to protect the coils from fouling. Another technique is to turn off the chiller but use its refrigerant to transfer heat between the two loops. This "thermocycle" uses the same principle as heat pipes, and only works on chillers with the proper configuration.

A low wet-bulb temperature during the night can also be utilized. It requires a chiller which handles low condensing temperatures and a cold storage tank. This technique may become even more desirable as time-of-day or demand pricing for electricity increase.

New design — In the purchase of a new chiller, an important consideration should be the load control feature. Since the chiller will be operating at partial load most of the time, it is important that it can do so efficiently.

In addition to control of single units, it is sometimes desirable to use multiple compressor reciprocating chillers. This allows some units to be shut down at partial load. The remaining compressors operate near full load, usually more efficiently.

Commonly, in commercial and industrial buildings, a convenient source of heat for a heat pump is the building exhaust air. This is a constant source of warm air available throughout the heating season. A typical heat pump design could generate hot water for space heating from this source at around 32 to 35°C (90 to 95°F). Heat pumps designed specifically to use building exhaust air can reach 66°C (150°F).

Another application of the heat pump is a continuous loop of water traveling throughout the building with small heat pumps located in each zone. Each small pump can both heat and cool, depending upon the needs of the zone. This system can be used to transfer heat from the warm side of a building to the cool side. A supplemental cooling tower and boiler are included in the loop to compensate for net heating or cooling loads.

A double bundle condenser can be used as a retrofit design for a centralized system. This creates the option of pumping the heat either to the cooling tower or into the heating system hot duct. Some chillers can be retrofitted to act as heat pumps. Centrifugal chillers will work much more effectively with a heat source warmer than outside air (exhaust air, for example). The compression efficiency of the centrifugal chiller falls off as the evaporator temperature drops.

Because they are positive displacement machines, reciprocating and screw-type compressors operate more effectively at lower evaporator temperatures. They can be used to transfer heat across a larger temperature differential. Multistage compressors increase this capacity even further.

iv. Ducting-Dampers

Controls — In HVAC systems using dual ducts, static pressure dampers are often placed near the start of the hot or cold plenum run. They control the pressure throughout the entire distribution system and can be indicators of system operation. Often in over-designed systems, the static pressure dampers may never open more than 25%. Fan pulleys can be changed to slow the fan and open the dampers fully, eliminating the previous pressure drop. The same volume of air is delivered with a significant drop in fan power.

Retrofit — Other HVAC systems use constant volume mixing boxes for balancing which create their own pressure drops as the static pressure increases. An entire system of these boxes could be overpressurized by several inches of water without affecting the air flow, but the required fan power would increase. (One inch of water pressure is about 250 N/m^2 or 250 Pa.) These systems should be monitored to ensure that static pressure is controlled at the lowest required value. It may also be desirable to replace the constant volume mixing boxes with boxes without volume control to eliminate their minimum pressure drop of approximately 1 in. of water. In this case, static pressure dampers will be necessary in the ducting.

Leakage in any dampers can cause a loss of hot or cold air. Neoprene seals can be added to blades to slow leakage considerably. If a damper leaks more than 10% it can be less costly to replace the entire damper assembly with effective positive-closing damper blades rather than to tolerate the loss of energy.

New design — In the past small ducts were installed because of their low initial cost despite the fact that the additional fan power required offset the initial cost on a life cycle basis. ASHRAE 90-75 provides a guideline to reduce this additional use of electricity by defining an air transport factor which may not be less than 4.0.

$$\text{Air Transport Factor} = \frac{\text{Space Sensible Heat Removal*}}{\text{(Supply + Return Fan(s) Power Input)*}}$$

This sets a maximum limit on the fan power than can be used for a given cooling capacity. As a result the air system pressure drop must be low enough to permit the desired air flow. In small buildings this pressure drop is often largest across filters, coils, and registers. In large buildings the duct runs may be responsible for a significant fraction of the total static pressure drop, particularly in high velocity systems.

v. Systems

The use of efficient equipment is only the first step in the optimum operation of a building. Equal emphasis should be placed upon the combination of elements in a system and the control of those elements.

Control — Many systems use a combination of hot and cold to achieve moderate temperatures. Included are dual duct, multizone, and terminal reheat systems, and some induction, variable air volume and fan coil units. Whenever combined heating and cooling occurs, the temperatures of the hot and cold ducts or water loops should be brought as close together as possible, while still maintaining building comfort.

This can be accomplished in a number of ways. Hot and cold duct temperatures are often reset on the basis of the temperature of the outside air or the return air. A more complex approach is to monitor the demand for heating and cooling in each zone. For example, in a multizone building, the demand of each zone is transferred back to the supply unit by electric or pneumatic signals. At the supply end hot and cold air are mixed in proportion to this demand. The cold air temperature should be just low enough to cool the zone calling for the most cooling. If the cold air were any colder, it would be mixed with hot air to achieve the right temperature. This creates an overlap in heating and cooling not only for that zone but for all the zones because they would all be mixing in the colder air.

If no zone calls for total cooling, then the cold air temperature can be icreased gradually until the first zone requires cooling. At this point, the minimum cooling necessary for that multizone configuration is performed. The same operation can be performed with the hot air temperature until the first zone is calling for heating only.

Note that simultaneous heating and cooling is still occurring in the rest of the zones. This is not an ideal system but it is a first step in improving operating efficiency.

The technique for resetting hot and cold duct temperatures can be extended to the other systems which have been mentioned. It may be performed automatically with pneumatic or electric controls, or manually. In some buildings it will require the installation of more monitoring equipment (usually only in the zones of greatest demand) but the expense should be relatively small and the payback period short.

Nighttime temperature setback is another control option which can save energy without significantly affecting the comfort level. Energy is saved by shutting off or cycling fans. Building heat loss may also be reduced because the building is cooler and no longer pressurized.

In moderate climates complete night shutdown can be used with a morning warm-up period. In colder areas where the average night temperature is below 4°C (40°F), it is usually necessary to provide some heat during the night. Building setback temperature is partially dictated by the capacity of the heating system to warm the building in the morning. In some cases it may be the mean radiant temperature of the building rather than air temperature which determines occupant comfort.

* Expressed in either Btu/hr or W.

Some warm-up designs use "free" heating from people and lights to help attain the last few degrees of heat. This also provides a transition period for the occupants to adjust from the colder outdoor temperatures.

In some locations during the summer, it is desirable to use night air for a cool-down period. This "free cooling" can decrease the temperature of the building mass which accumulates heat during the day. In some buildings with high heat content (such as libraries or buildings that have thick walls), a long period of night cooling may decrease the building mass temperature by a degree or two. This represents a large amount of cooling that the chiller will not have to perform the following day.

Retrofit — Retrofitting HVAC systems may be an easy or difficult task depending upon the possibility of using existing equipment in a more efficient manner. Often retrofitting involves control or ducting changes which appear relatively minor but will greatly increase the efficiency of the system. Some of these common changes, such as decreasing air flow, are discussed elsewhere in this chapter. This section will describe a few changes appropriate to particular systems.

Both dual duct and multizone systems mix hot and cold air to achieve the proper degree of heating or cooling. In most large buildings the need for heating interior areas is essentially nonexistent, due to internal heat generation. A modification which adjusts for this is simply shutting off air to the hot duct. The mixing box then acts as a variable air volume box, modulating cold air according to room demand as relayed by the existing thermostat. (It should be confirmed that the low volume from a particular box meets minimum air requirements.)

Savings from this modification come mostly from the elimination of simultaneous heating and cooling. Since fans in these systems are likely to be controlled by static pressure dampers in the duct after the fan, they do not unload very efficiently and represent only a small portion of the savings.

b. Economizer Systems and Enthalpy Controllers

The economizer cycle is a technique for introducing varying amounts of outside air to the mixed air duct. Basically it permits mixing warm return air at 24°C (75°F) with cold outside air to maintain a preset temperature in the mixed air plenum (typically 10 to 15°C, 50 to 60°F). When the outside temperature is slightly above this set point, 100% outside air is used to provide as much of the cooling as possible. During very hot outside weather, minimum outside air will be added to the system.

A major downfall of economizer systems is poor maintenance. The failure of the motor or dampers may not cause a noticeable comfort change in the building, because the system is often capable of handling the additional load. Since the problem is not readily apparent, corrective maintenance may be put off indefinitely. In the meantime the HVAC system will be working harder than necessary, wasting energy and money. A continual maintenance program is necessary for any economizer installation.

Typically, economizers are controlled by the dry-bulb temperature of the outside air rather than its enthalpy (actual heat content). This is adequate most of the time, but can lead to unnecessary cooling of air. When enthalpy controls are used to measure wet-bulb temperatures, this cooling can be reduced. However, enthalpy controllers are more extensive and less reliable.

The rules which govern the more complex enthalpy controls for cooling-only applications are as follows:

- When outside air enthalpy is greater than that of the return air or when outside air dry-bulb temperature is greater than that of the return air, use minimum outside air.

- When the outside air enthalpy is below the return air enthalpy and the outside dry-bulb temperature is below the return air dry-bulb temperature but above the cooling coil control point, use 100% outside air.
- When outside air enthalpy is below the return air enthalpy and the outside air dry-bulb temperature is below the return air dry-bulb temperature and below the cooling coil controller setting, the return and outside air are mixed by modulating dampers according to the cooling set point.

These points are valid for the majority of cases. When mixed air is to be used for heating and cooling, a more intricate optimization plan will be necessary, based on the value of the fuels used for heating and cooling.

c. Heat Recovery

Heat recovery is often practiced in industrial processes which involve high temperatures. It can also be employed in HVAC systems.

Systems are available which operate with direct heat transfer from the inlet air to the exhaust air. These are most reasonable when there is a large volume of exhaust air, for example, in once-through systems, and when weather conditions are not moderate.

Common heat recovery systems are broken down into two types, regenerative and recuperative. Regenerative units use alternating air flow from the hot and cold stream over the same heat storage/transfer medium. This flow may be reversed by dampers or the whole heat exchanger may rotate between streams. Recuperative units involve continuous flow; the emphasis is upon heat transfer through a medium with little storage.

The rotary regenerative unit, or heat wheel, is one of the most common heat recovery devices. It contains a corrugated or woven heat storage material which gains heat in the hot stream. This material is then rotated into the cold stream where the heat is given off again. The wheels can be impregnated with a dessicant to transfer latent as well as sensible heat. Purge sections for HVAC applications can reduce carry-over from the exhaust stream to acceptable limits for most installations.

The heat transfer efficiency of heat wheels generally ranges from 60 to 85% depending upon the installation, type of media, and air velocity. For easiest installation the intake and exhaust ducts should be located near each other.

Another system which can be employed with convenient duct location is a plate type air-to-air heat exchanger. This system is usually lighter though more voluminous than heat wheels. Heat transfer efficiency is typically in the 60 to 75% range. Individual units range from 1,000 to 11,000 SCFM and can be grouped together for greater capacity. Almost all designs employ counterflow heat transfer for maximum efficiency.

Another option to consider for nearly contiguous ducts is the heat pipe. This is a unit which uses a boiling refrigerant within a closed pipe to transfer heat. Since the heat of vaporization is utilized, a great deal of heat transfer can take place in a small space.

Heat pipes are often used in double wide coils which look very much like two steam coils fastened together. The amount of heat transferred can be varied by tilting the tubes to increase or decrease the flow of liquid through the capillary action. Heat pipes cannot be "turned off" so by-pass ducting is often desirable. The efficiency of heat transfer ranges from 55 to 75% depending upon the number of pipes, fins per inch, air face velocity, etc.

Runaround systems are also popular for HVAC applications, particularly when the supply and exhaust plenums are not physically close. Runaround systems involve two

TABLE 7

Energy Saving Modifications for Water Heaters

	Percent reduction in annual fuel use	
	Gas	Electric
Operating strategies		
Lower hot water temperature (per 10°F reduction)	5.8%	4.5%
Modifications		
Increased tank insulation		
10 cm (4 in), justified at 1¢/kWh	—	8.2%
18 cm (7 in), justified at 4¢/kWh	—	11.0%
8 cm (3 in), justified at 10¢/therm	21.6%	—
13 cm (5 in), justified at 40¢/therm	25.0%	—
Hot water plumbing insulation		
(7.5 m (25 ft) of exposed pipe)	1.6%	1.6%
Solar preheat tank	20—60%	gas/electric
Design		
Heat recovery from air conditioning	20—80%	gas/electric
Low excess air	8.0%	gas/electric
Electric ignition with damper	13.0%	gas/electric
Reduced flue temperatures	10.0%	gas/electric
Reduced burner rate	2.0%	gas/electric
Instantaneous heater with small tank	8.0%	gas/electric
Heat pump water heater	50.0%	gas/electric
Programmed off periods	10.0%	gas/electric
Indirect heater with electric pilot	17.0%	gas/electric
Ambient preheat tank	8.0%	gas/electric
Condensation of flue gases	9.0%	gas/electric

From Smith, C. B., **Ed.**, *Efficient Electricity Use,* 2nd ed., Pergamon Press, New York, 1978. With permission.

coils (air-to-water heat exchangers) connected by a piping loop of water or glycol solution and a small pump. The glycol solution is necessary if the air temperatures in the inlet coils are below freezing.

Standard air-conditioning coils can be used for the runaround system and some equipment manufacturers supply computer programs for size optimization. Precaution should be used when the exhaust air temperature drops below 0°C (32°F) which would cause freezing of the condensed water on its fins. A three-way by-pass valve will maintain the temperature of the solution entering the coil at just above 0°C (32°F). The heat transfer efficiency of this system ranges from 60 to 75% depending upon the installation.

Another system similar to the runaround in layout is the dessicant spray system. Instead of using coils in the air plenums, it uses spray towers. The heat transfer fluid is a dessicant (lithium chloride) which transfers both latent and sensible heat—desirable in many applications. Tower capacities range from 7,700 to 92,000 SCFM; multiple units can be used in large installations. The enthalpy recovery efficiency is in the range of 60 to 65%.

3. Water Heating

Residential water heaters typically range in size from 76 ℓ (20 gal) to 303 ℓ (80 gal). Electric units generally have one or two immersion heaters, each rated at 2 to 6 kW depending on tank size. Energy input for water heating depends on the temperature at which water is delivered, the supply water temperature, and standby losses from the water heater, storage tanks, and piping.

In single tank residential systems, major savings can be obtained by:

- Thermostat temperature setback to 60°C (140°F)
- Time clock control
- Supplementary tank insulation
- Hot water piping insulation

Refer to Table 7 for typical savings.

The major source of heat loss from electric water heaters is standby losses through the tank walls and from piping, since there are no flame or stack losses in electric units. The heat loss is proportional to the temperature difference between the tank and its surroundings. Thus lowering the temperature to 60°C will result in two savings: (1) a reduction of the energy needed to heat water and (2) a reduction in the amount of heat lost. Residential hot water uses do not require temperatures in excess of 60°C; for any special use which does, it would be advantageous to provide a booster heater to meet this requirement when needed, rather than maintain 100 to 200 ℓ of water continuously at this temperature with associated losses.

When the tank is charged with cold water, both heating elements operate until the temperature reaches a set point. After this initial rise, one heating element thermostatically cycles on and off to maintain the temperature, replacing heat which is removed by withdrawing hot water or which is lost by conduction and convection during standby operation.

Experiments indicate that the heating elements may be energized only 10 to 20% of the time, depending on the ambient temperature, demand for hot water, water supply temperature, etc. By carefully scheduling hot water usage, this time can be greatly reduced. In one case a residential water heater was operated for 1 hr in the morning and 1 hr in the evening. The morning cycle provided sufficient hot water for clothes washing, dishes, and other needs. Throughout the day the weter in the tank, although gradually cooling, still was sufficiently hot for incidental needs. The evening heating cycle provided sufficient water for cooking, washing dishes, and bathing. Standby losses were eliminated during the night and much of the day. Electricity use was cut to a fraction of the normal amount. This method requires the installation of a time clock to control the water heater. A manual override can be provided to meet special needs.

Supplementary tank insulation can be installed at a low cost to reduce standby losses. The economic benefit depends on the price of electricity and the type of insulation installed. However, paybacks of a few months up to a year are typical. Hot water piping should also be insulated, particularly when hot water tanks are located outside or when there are long piping runs. If copper pipe is used, it is particularly important to insulate the pipe for the first 3 to 5 m where it joins the tank, since it can provide an efficient heat conduction path.

Since the energy input depends on the water flow rate and the temperature difference between the supply water temperature and the hot water discharge temperature, energy use is reduced by reducing either of these two quantities. Hot water demand can be reduced by cold water clothes washing, and by providing hot water at or near the use temperature, to avoid the need for dilution with cold water. Supply water should be provided at the warmest temperature possible. Since reservoirs and underground piping systems are generally warmer than the air temperature on a winter day in a cold climate, supply piping should be buried, insulated, or otherwise kept above the ambient temperature.

Solar systems are available today for heating hot water. Simple inexpensive systems can preheat the water, reducing the amount of electricity needed to reach the final temperature. Alternatively, solar heaters (some with electric backup heaters) are also available, although current costs* correspond to 5- to 10-year paybacks.

Heat recovery is another technique for preheating or heating water, although opportunities in residences are limited. This is discussed in more detail under commercial water heating.

Apartments and larger buildings use a combined water heater/storage tank, a circulation loop, and a circulating pump. Cold water is supplied to the tank, which thermostatically maintains a preset temperature, typically 71°C (160°F). The circulating pump maintains a flow of water through the circulating loop, so hot water is always available instantaneously upon demand to any user. This method is also used in hotels, office buildings, etc.

Adequate piping and tank insulation is even more important here, since the systems are larger and operate at higher temperature. The circulating hot water line should be insulated, since it will dissipate heat continuously otherwise.

Commercial/industrial hot water systems offer many opportunities for employing heat recovery. Examples of possible sources of heat include air compressors, chillers, heat pumps, refrigeration systems, and water-cooled equipment. Heat recovery permits a double energy savings in many cases. First, recovery of heat for hot water or space heating reduces the direct energy input needed for heating. The secondary benefit comes from reducing the energy used to dissipate waste heat to a heat sink (usually the atmosphere). This includes pumping energy and energy expended to operate cooling towers and heat exchangers. Solar hot water systems are also finding increasing use. Interestingly enough, the prerequisites for solar hot water systems also permit heat recovery. Once the hot water storage capacity and backup heating capability has been provided for the solar hot water system, it is economical to tie in other sources of waste heat, e.g., water jackets on air compressors.

4. Lighting

There are seven techniques for improving the efficiency of lighting systems:

- Delamping
- Relamping
- Improved controls
- More efficient lamps and devices
- Task-oriented lighting
- Increased use of daylight
- Room color changes, lamp maintenance

The first two techniques and possibly the third are low cost and may be considered operational changes. The last four items generally involve retrofit or new designs.

The first step in reviewing lighting electricity use is to perform a lighting survey. An inexpensive hand-held lightmeter can be used as a first approximation; however distinction must be made between raw intensities (lux or footcandles) recorded in this way and *Equivalent Sphere Illumination* (ESI) values.

Many variables can affect the "correct" lighting values for a particular task: task complexity, age of employee, glare, etc. For reliable results consult a lighting specialist or refer to the literature.[4] Fundamentals of lighting are discussed in Reference 2 also and will not be repeated here.

* 1977 cost is approximately $1000.

TABLE 8

Relamping Opportunities

Change office lamps (2700 hr per year)		Energy savings/cost savings			
		kWh	GJ	3¢/ kWh	5¢/ kWh
From	To	To save annually			
1 300-W incandescent	1 100-W mercury vapor	486	5.25	$14.58	$24.30
2 100-W incandescent	1 40-W fluorescent	400	4.32	12.00	20.00
7 150-W incandescent	1 150-W sodium vapor	2360	25.5	70.80	118.00
Change industrial lamps (3000 hr per year)					
1 300-W incandescent	2 40-W fluorescent	623	6.73	18.69	31.15
1 1000-W incandescent	2 215-W fluorescent	1617	17.5	48.51	80.85
3 300-W incandescent	1 250-W sodium vapor	1806	19.5	54.18	90.30
Change store lamps (3300 hr per year)					
1 300-W incandescent	2 40-W fluorescent	685	7.40	20.55	34.25
1 200-W incandescent	1 100-W mercury vapor	264	2.85	7.92	13.20
2 200-W incandescent	1 175-W mercury vapor	670	7.24	20.10	33.50

From Smith, C. B., **Ed.**, *Efficient Energy Use,* 2nd ed., Pergamon Press, New York, 1978. With permission.

The lighting survey indicates those areas of the building where lighting is potentially inadequate or excessive. Deviations from illumination levels which are adequate can occur for several reasons: overdesign; building changes; change of occupancy; modified layout of equipment or personnel, more efficient lamps, improper use of equipment, dirt buildup, etc.

Once the building manager has identified areas with potentially excessive illumination levels, he or she can apply one or more of the seven techniques listed above. Each of these will be described briefly.

Delamping refers to the removal of lamps to reduce illumination to acceptable levels. With incandescent lamps, bulbs are removed. With fluorescent or high intensity discharge lamps, ballasts account for 10 to 20% of total energy use and should be disconnected after lamps are removed.

Fluorescent lamps often are installed in luminaires in groups of two or more lamps where it is impossible to remove only one lamp. In such cases an artificial load (called a "phantom tube") can be installed in place of the lamp which has been removed.

Relamping refers to the replacement of existing lamps by lamps of lower wattage or increased efficiency. Low wattage fluorescent tubes are available which require 15 to 20% less wattage (but produce 10 to 15% less light). In some types of high intensity discharge lamps, a more efficient lamp can be substituted directly. However, in most cases, ballasts must also be changed. (See Table 8.).

Improved controls permit lamps to be used only when and where needed. For example, certain office buildings have all lights for one floor on a single contactor. These lamps will be switched on at 6 a.m. before work begins, and are not turned off until 10 p.m. when maintenance personnel finish their clean-up duties.

Energy usage can be cut by as much as 50% by installing individual switches for each office or work area, installing time clocks, using photocell controls, or instructing custodial crews to turn lights on as needed and turn them off when work is complete.

TABLE 9

Properties of Incandescent and Electrical Discharge Systems for Illumination Purposes

Lamp type	Power lamp efficacy			Ballast power	Ballast + lamp efficacy		Fitting loss	Overall efficacy	
	(We)	lm/We	%	(We)	lm/We	%		lm/We	%
Incandescent	100	15	5	—	—	—	10—50 (indoor)	7—13	2.5—4.5
Incandescent halogen	100	20—30	13	—	—	—	10—50 (indoor)	15—27	6.5—12
Fluorescent	40	80	22	13.5	60	16.5	10—50 (indoor)	30—54	8.5—15
	65	85	—	11	73	—	10—50 (indoor)	36—66	—
Low-pressure sodium	90	140	—	35	100	—	20—45 (outdoor)	55—80	—
	180	180	27	30	155	23	20—45 (outdoor)	35—125	13—18.5
High-pressure sodium	250	100	—	33	—	—	20—45 (outdoor)	60—86	—
	400	120	29	39	108	24	20—45 (outdoor)	72—103	14—20
High-pressure mercury	80	44	—	8.5	39	—	20—45 (outdoor)	22—31	—
	250	54	—	18.5	50	—	20—45 (outdoor)	28—40	—
	400	57	15	26	53.5	14	20—45 (outdoor)	30—43	8—11
High-pressure metal-halide	400	85	23	26	75	21.5	30—40 (flood light)	45—52	13—15
	1000	90	—	43	86	—	30—40 (flood light)	51—60	—
	2000	100	—	68	97	—	30—40 (flood light)	58—68	—

From Smith, C. B., **Ed.**, *Efficient Electricity Use*, 2nd ed., Pergamon Press, New York, 1978. With permission.

TABLE 10

Color Reflectance Values

Color	Reflective value (%)	Color	Reflective value (%)
White	82	Brown tint	69
Light gray	65	Light brown	51
Medium light gray	50	Medium light brown	41
Medium gray	35	Medium brown	28
Medium dark gray	26	Medium dark brown	18
Dark gray	14	Dark brown	15
Blue tint	68	Orange tint	71
Light blue	63	Light orange	68
Medium light blue	49	Medium light orange	59
Medium blue	39	Medium orange	52
Medium dark blue	22	Medium dark orange	44
Dark blue	8	Dark orange	35
Aqua tint	61	Red tint	70
Light aqua	48	Light red	60
Medium light aqua	38	Medium light red	46
Medium aqua	20	Medium red	40
Medium dark aqua	13	Medium dark red	28
Dark aqua	7	Dark red	21
Green tint	73	Laminated plastic and wood:	
Light green	64	White bleached grain	77
Medium light green	55	Maple	37
Medium green	40	Teak	17
Medium dark green	25	Walnut	10
Dark green	12	Oak	9
Yellow tint	78		
Light yellow	75		
Medium light yellow	64	Miscellaneous: green chalkboard	15—20
Medium yellow	75		
Medium dark yellow	66		
Dark yellow	54	Desk tops	46—47

From Smith, C. B., **Ed.**, *Efficient Electricity Use,* 2nd ed., Pergamon Press, New York, 1978. With permission.

There is a great variation in the efficacy (a measure of light output per electricity input) of various lamps. Table 9 shows representative values and contains other lamp data. Incandescent lamps have the lowest efficacy. Wherever possible, fluorescent lamps should be substituted for incandescent lamps. This not only saves energy, but offers substantial economic savings as well, since fluorescent lamps last 10 to 50 times longer.

Still greater improvements are possible with high intensity discharge lamps such as mercury vapor, metal halide, and high pressure sodium lamps. While these are generally not suited to residential use (high light output and high capital cost) they are increasingly being used in commercial buildings.

Task-oriented lighting is another important lighting concept. In this, lighting is provided for work areas in proportion to the needs of the task. Hallways, storage areas, and other nonwork areas receive less illumination.

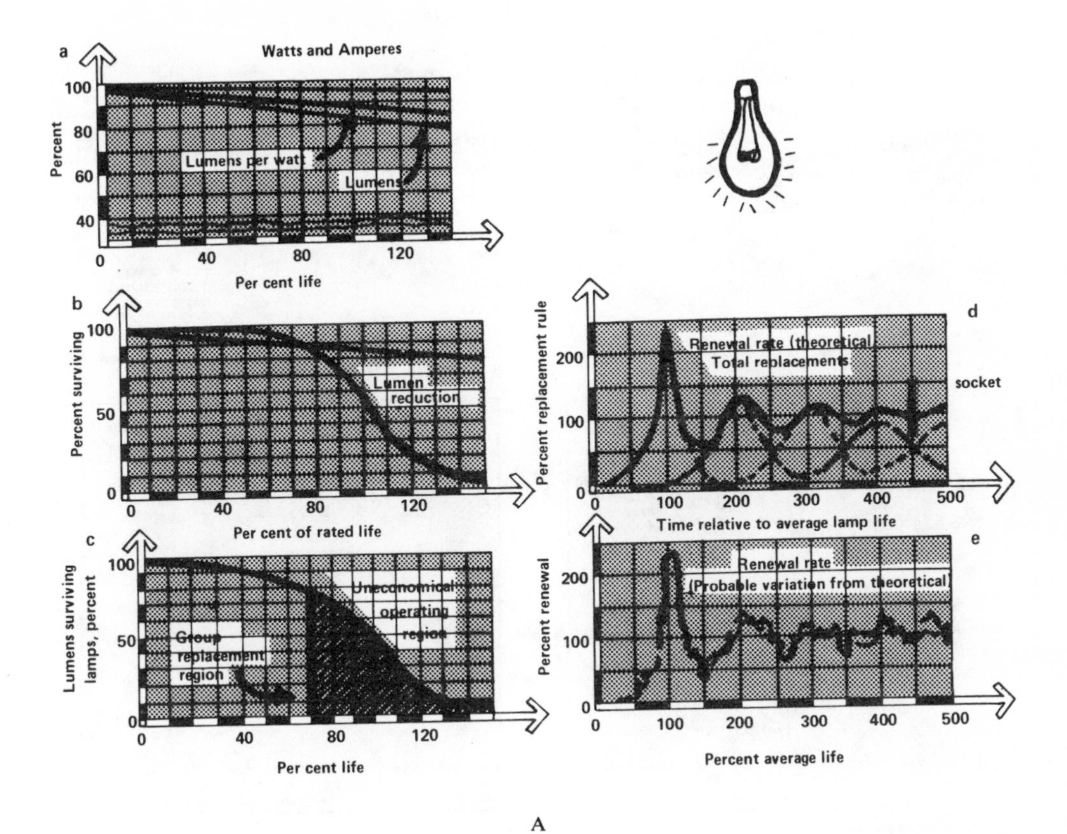

A

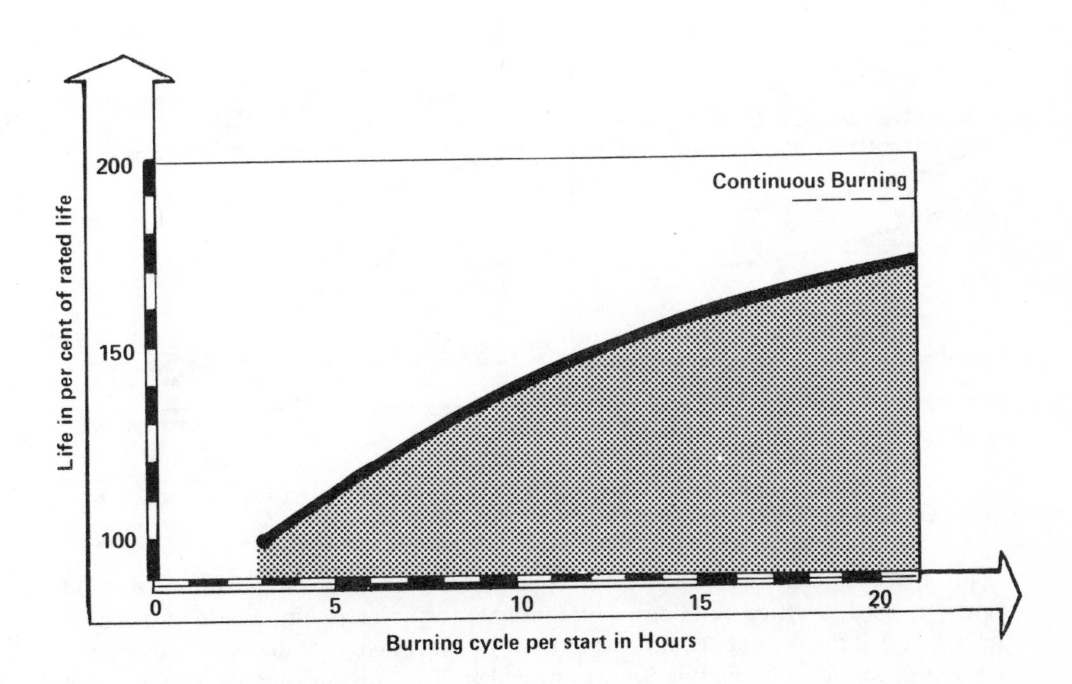

B

FIGURE 6. (A) Lamp life characteristics. (B) Fluorescent lamp life vs. burning cycle. (From Smith, Craig B., Ed., *Efficient Electricity Use*, 2nd ed., Pergamon Press, New York, 1978. With permission.)

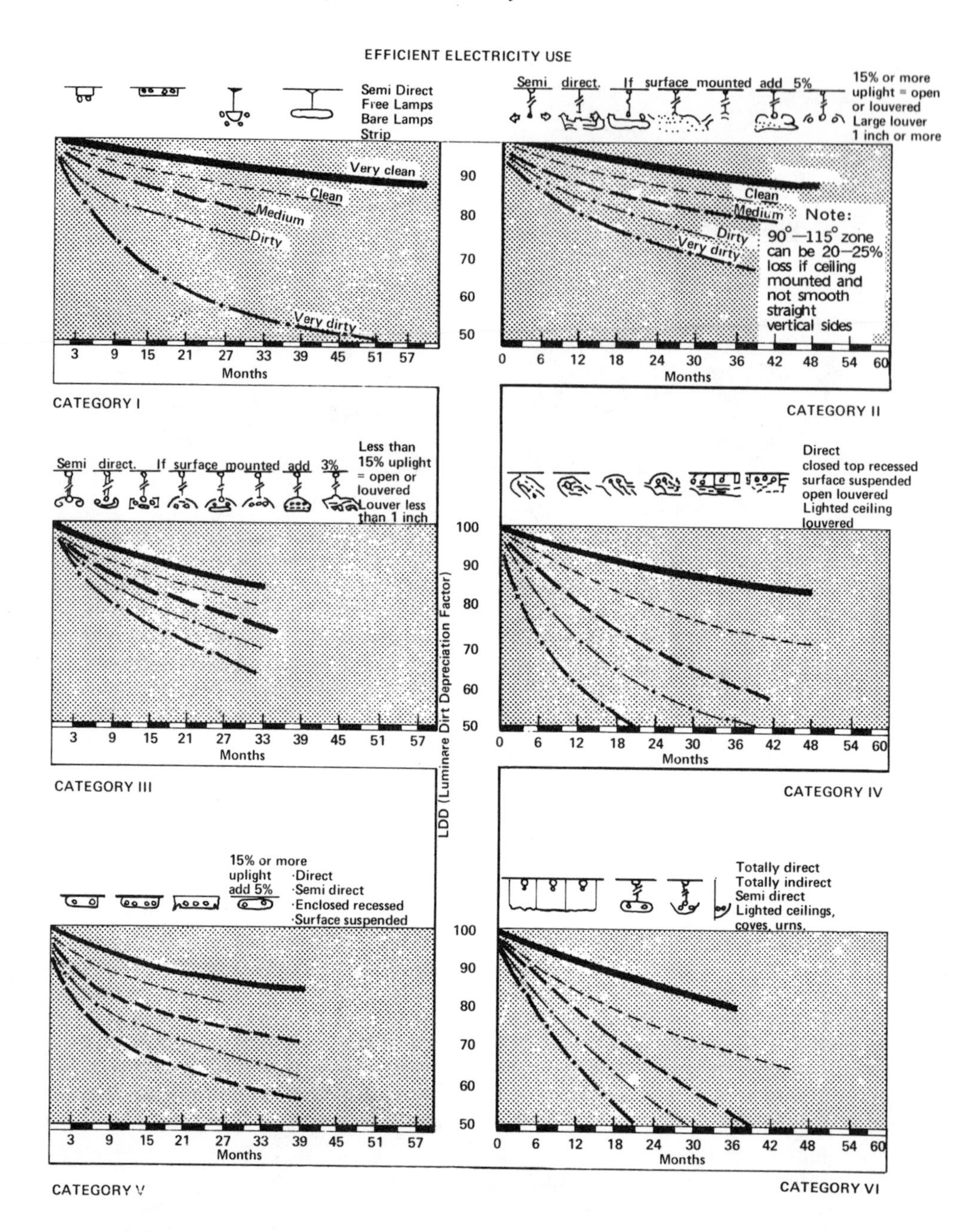

FIGURE 7. Luminaire maintenance — the effect of cleaning on light output. (From Smith, Craig B., Ed., *Efficient Electricity Use,* 2nd ed., Pergamon Press, New York, 1978. With permission.)

This approach can be contrasted to the so-called "uniform illumination" method sometimes used in office buildings. The rationale for uniform illumination was based on the fact that the designer could never know the exact layout of desks and equipment in advance, so uniform illumination was provided. This also accomodates revisions in the floor plan. Originally, electricity was cheap and any added cost was inconsequential.

TABLE 11

Heat Loss of a Typical Refrigerator/Freezer

Compartment	Heat transfer through walls	Door openings	Total
Freezer	42 W	4 W	46 W, 33%
Refrigerated compartment (GRC)	58 W	14W	72 W, 51%
	100 W, 71%	18 W, 13%	118 W, 84%
Other Items			
Anti-sweat heaters (40% of 20 W)			8 W, 5.5%
Air circulation fan			7 W, 5.0%
Defrost heaters (1.6% of 500 W)			8 W, 5.5%
			141 W, 100%

Note: GRC = general refrigerated compartment. 27°C (80°F) Ambient, 3°C (37°F) GRC, −18°C (0°F)freezer. 50 door openings/day for GRC. 25 door openings/day for freezer. GRC volume = 0.31 m^3 (10.8 ft^3), freezer volume 0.12 m^3 (4.2 ft^3). Door openings assume 60% relative humidity and complete air change for each opening.

From Smith, C. B., **Ed.**, *Efficient Electricity Use,* 2nd ed., Pergamon Press, New York, 1978. With permission permission.

Now, however, costs have doubled or tripled. Also, new developments in lamps and lighting systems have occurred, making it feasible to relocate lamps if the furniture has to be moved.

Daylighting was an important element of building design for centuries before the discovery of electricity. In certain types of buildings and operations today daylighting can be utilized to at least reduce (if not replace) electric lighting. Techniques include windows, an atrium, skylights, etc. There are obvious limitations such as those imposed by the need for privacy, 24-hr operation, and building core locations with no access to natural light.

The final step is to review building and room color schemes and decor. Table 10 lists typical color reflectance values. The use of light colors can substantially enhance illumination without modifying existing lamps.

An effective lamp maintenance program also has important benefits. (See Figure 6 and 7). Light output gradually decreases over lamp lifetime. This should be considered in the initial design and when deciding on lamp replacement. Dirt can substantially reduce light output; simply cleaning lamps and luminaires more frequently can gain up to 5 to 10% greater illumination, permitting some lamps to be removed.

5. Refrigeration

The refrigerator, at 100 to 200 kWh/month, is among the top six residential users of electricity. In the last 50 years the design of refrigerator/freezers has changed considerably, with sizes increasing from 0.14 to 0.28 m^3 (5 to 10 ft^3) to 0.34 to 0.68 m^3 (12 to 24 ft^3) today.[2] At the same time, the energy input has also increased, from roughly 210 to 490 W/m^3 (6 to 14 W/ft^3).

The theoretical heat loss from a modern refrigerator/freezer is tabulated in Table 11 for a unit with a freezer compartment of 0.12 m^2 (4.2 ft^3) and a refrigerator compartment of 0.31 m^3 (10.8 ft^3). The largest losses are due to heat losses through the walls.

Additional energy is used in modern refrigerators to provide for automatic defrosting (so-called "frost-free" operation) and to prevent condensation on the exterior of the refrigerator. These functions are conveniences and are not essential to the operation of the refrigerator. Some types have a "power saver" switch by which the anti-sweat heaters can be turned off during dry weather.

Since much of the energy used by a refrigerator depends on its design, care should be used in selection. Efficiency should be an important consideration. Units are available with added insulation and efficient compressors which decrease annual energy use by 25 to 45%. Information on the energy efficiency of specific models is available from the Association of Home Appliances Manufacturers.[5]

Purchase of a new, more efficient unit is not a viable option for many individuals who have a serviceable unit and do not wish to replace it. In this case, the energy management challenge is to obtain the most effective operation of the existing equipment.

More efficient operation of refrigeration equipment can be achieved by:

- Better insulation
- Disconnecting or reducing operation of automatic defrost and anti-sweat heaters
- Providing a cool location for the refrigerator coils (or reduce room temperature)
- Reducing the number of door openings
- Increasing temperature settings
- Precooling foods before refrigerating

Studies have been made of refrigerator operation based on these approaches.[6] Typical results will be described briefly.

A 0.4 m³ refrigerator (14 ft³) refrigerator/freezer was studied in a home. The refrigerator was located in the kitchen where the ambient temperature ranged from 19 to 22°C (66 to 72°F). The refrigerator temperature averaged 4.4°C (40°F) and the freezer −13.9°C (7°F).

Measurements indicated that in normal operation the compressor would cycle on for 7 to 8 min and then turn off for roughly an equal amount of time. This meant the compressor was on, on the average, 50% of the time. Under these conditions normal electricity use was about 0.17 kWh per hour or 4 kWh per day.

Better insulation and disconnecting anti-sweat heaters and defrost heaters could potentially reduce this by 50%. However, it is generally not practical to do this in the home.

Providing a cool location (such as a basement) can potentially save 5 to 10% by reducing the heat sink temperature to which the refrigerator must transfer heat. Calculations indicate a potential savings of 3.6% /°C (2%/°F).[7]

Holding the number of door openings to a minimum also saves energy — estimated to be 0.3% per opening. As an extreme case, the door of the experimental refrigerator described above was opened every 5 min for a 10-sec interval, resulting in a 47% increase in energy use (to 0.25 kWh/hr).

An effective means for saving energy is to increase temperature settings. This will save about 7 to 9%/°C (4 to 5%/°F). The correct temperature depends on the foods to be stored and the length of storage; for most household purposes 4°C for the refrigerator and −14°C for the freezer is adequate.

Precooling foods will also save refrigeration energy. Hot foods should be cooled to room temperature before placing in the refrigerator, to take advantage of as much "free" cooling as possible.

The same general concepts apply to commercial refrigeration systems. Many of the newer commercial refrigerators now come equipped with heat recovery systems which recover compressor heat for space conditioning or water heating.

It is common practice in some types of units to have the compressor and heat exchange equipment located remote from the refrigerator compartment. In such systems, a cool location should be selected, rather than locating the compressor next to other equipment which gives off heat.

Walk-in freezers and refrigerators lose energy through door openings; refrigerated display cases have direct transfer of heat. Covers, air curtains, or other thermal barriers can help mitigate these problems. The most efficient light sources should be used in large refrigerators and freezers; every watt eliminated saves three watts. Elimination of one watt of electricity to produce light also eliminates two additional watts required to extract the heat.

6. Cooking

Cooking accounts for only about 1% of total U.S. energy use but constitutes about 7% of residential use.[2,6] Improvements in energy use efficiency for cooking can be divided into three categories:

- More efficient use of existing appliances
- Use of most efficient existing appliances
- More efficient new appliances

The most efficient use of existing appliances can lead to substantial reductions in energy use. While slanted towards electric ranges and appliances, the following observations also apply to cooking devices using other sources of heat.

First, select the right size equipment for the job. Do not heat excessive masses or large surface areas which will needlessly radiate heat. Second, optimize heat transfer by ensuring that pots and pans provide good thermal coupling to the heat source. Flat-bottomed pans should be used on electric ranges. Third, be sure that pans are covered to prevent heat loss and to shorten cooking times. Fourth, when using the oven, plan meals so several dishes are cooked at once. Use small appliances (electric fry pans, "slow" cookers, toaster ovens, etc.) whenever they can be substituted efficiently for the larger appliances such as the oven.

Different appliances perform similar cooking tasks with widely varying efficiencies. For example, in a study of the electricity used and cooking time required for ten common food items, variations of ten to one in energy use and five to one in cooking times were found.[6] As an example, four baked potatoes required 2.3 kWh and 60 min in the oven (5.2 kW) of an electric range, 0.5 kWh and 75 min in a toaster oven (1.0 kW), and 0.3 kWh and 16 min in a microwave oven (1.3 kW). Small appliances are generally more efficient when used as intended. Measurements in a home indicated that a pop-up toaster cooks two slices of bread using only 0.025 kWh.[6] The toaster would be more efficient than using the broiler in the electric range oven, unless a large number of slices of bread (more than 17 in this case) were to be toasted at once.

Cooking several dishes at once can save energy. A complete meal, consisting of a canned ham (2.3 kg), frozen peas (0.23 kg), four yams, and a pineapple upside-down cake (23 cm × 23 cm) were cooked separately using a toaster oven and an electric range, together in an electric oven, and separately in a microwave oven. Cooked separately using the toaster oven and range required 5.2 kWh; together in the oven took 2.5 kWh, while the microwave required 1.2 kWh.

If new appliances are being purchased, select the most efficient ones available. Heat losses from a typical oven approach 1 kW, with insulation accounting for about 50%;

losses around the oven door edge and through the window are next in importance. These losses are reduced in certain models. Self-cleaning ovens are normally manufactured with more insulation. Careful design of heating elements can also contribute to better heat transfer.

Microwave cooking is highly efficient for many types of foods, since the microwave heat is deposited directly in the food. Energy input is minimized because there is no need to heat the cooking utensil. Although many common foods can be prepared effectively using a microwave oven, different methods must be used and certain foods are not suitable for microwave cooking.

Commercial cooking operations range from small restaurants and cafes where methods similar to those described above for residences are practiced, to large institutional kitchens in hotels, hospitals, and finally, to food processing plants.

Many of the same techniques apply. Microwave heating is finding increasing use in hotel and restaurant cooking. Careful scheduling of equipment use, and provision of several small units rather than a single large one, will save energy. For example, in restaurants, grills, soup kettles, bread warmers, etc., often operate continuously. Generally it is unnecessary to have full capacity during off-peak hours; one small grill might handle mid-morning and mid-afternoon needs, permitting the second and third units to be shut down. The same strategy can be applied to coffee warming stations, hot plates, etc.

In food processing plants where food is cooked and canned, heat recovery is an important technique. Normally, heat is rejected via cooling water at some step in the process. This heat can be recovered and used to preheat products entering the process, decreasing the amount of heating which eventually must be done.

7. Residential Appliances

A complete discussion of energy management opportunities associated with all the appliances found in homes is beyond the scope of this chapter. However, several of the major ones will be discussed and general suggestions applicable to the others will be given.

a. *Clothes Drying*

Electric clothes dryers operate most efficiently when fully loaded. To remove 97% of the moisture in a typical load of clothes (2 to 7 kg) requires 0.85 to 1.2 kWh/kg.[2] Operating with one third to one half load costs roughly 10 to 15% in energy efficiency.

Measurements were made on a 6 kW dryer used in a residence.[6] The dryer used 0.3 kWh to heat up before cycling off the first time. Electricity used to dry typical loads was found to range from 0.67 kWh/kg of clothes ("permanent press" items dried at a low setting) to 1.13 kWh/kg of heavy towels (dried at a "hot" setting). The average energy input for three different types of loads was 0.88 kWh/kg.

Theoretically, it should take an energy input of 0.73 kWh/kg of water removed. Averaging results from three typical loads for the dryer discussed above showed that it took 1.52 kWh/kg of water removed. Therefore about 50% of the energy input is dissipated as waste heat and does not go into water vaporization.

Locating clothes dryers in heated spaces could save 10 to 20% of the energy used by reducing energy needed for heating up. Another approach is to save up loads and do several loads sequentially, so the dryer does not cool down between loads.

The heavier the clothes the greater the amount of water they hold. Mechanical water removal (pressing, spinning, wringing) generally requires less energy than electric heat. Therefore, be certain the washing machine goes through a complete spin cycle (0.1 kWh) before putting clothes in the dryer.

Solar drying, which requires a clothesline (rope) and two poles or trees, has been practiced for millenia and is very sparing of electricity.* The chief limitation is, of course, inclement weather.

b. Clothes Washing

Electric clothes washers are designed for typical loads of 3 to 7 kg. Surprisingly, most of the energy used in clothes washing is for hot water; the washer itself only requires 1 to 2% of the total energy input. Typical machines use 1 to 2 kWh/kg of clothes. This is the total energy input, which is the sum of the machine energy plus the water heating energy. The machine energy is typically 0.0025 to 0.05 kWh/kg.[2]

For example, in a study conducted in a residence, a 68-ℓ washer used 0.16 kWh per load. Hot water usage was 68 ℓ for the wash cycle plus 34 ℓ during the rinse cycle. Total water heating energy was 23 MJ, equivalent to 6.5 kWh per load for hot water. These data indicate that the major opportunity for energy management in clothes washing is the use of cold water for washing. Under normal household conditions it is not necessary to use hot water. Clothes are just as clean (in terms of bacteria count) after a 20°C wash as after a 50°C wash.[2] If there is concern for sanitation (e.g., a sick person in the house), authorities recommend use of a chlorine bleach. If special cleaning is required, such as removing oil or grease stains, hot water (50°C) and detergent will emulsify oil and fat. There is no benefit in a hot rinse.

A secondary savings can come from using full loads. Surveys indicate that machines are frequently operated with partial loads, even though a full load of hot water is used.

Conversion to cold water washing (5 loads per week, electric water heater) would save approximately 1700 kWh per year, or 50 to 100 dollars per year.[6]

c. Dishwashers

The two major energy uses in electric dishwashers are the hot water and the dry cycle. The volume of hot water used ranges from 45 to 61 ℓ and can be varied on some machines depending on the load. The temperature requirement for dishwashing sometimes dictates the hot water temperature for the entire residence. Water at 55 to 60°C is needed for dishwashing, although most other residential functions can be satisfied at 38 to 43°C. Some dishwashers accept lower temperature water and use a booster heater to provide the proper temperature. This is a more efficient approach than maintaining the entire hot water system at the higher temperature with its attendant losses.

Since commercially available machines vary widely, the characteristics of the particular machine are important. In a study of one unit in a residence, the cycle used:[6]

	Electricity (kWh)	Hot water (liters)
Prewash	0.035 (6%)	21 (29%)
Wash	0.125 (20%)	51 (71%)
Dry	0.450 (74%)	—
Totals	0.610 (100%)	72 (100%)

By eliminating the prewash (e.g. by rinsing dishes in cold water prior to placing in the dishwasher) 6% of the electricity and 29% of the hot water can be saved. By eliminating the dry cycle (e.g., opening the door following the wash cycle and letting the dishes air dry), 74% of the electricity can be saved.

* In some countries, (Brazil for example) clothes are festooned on nearby shrubbery, eliminating the need even for the rope.

d. Television

The transition from black and white to color television roughly doubled or tripled electrical power requirements (from typically 100 to 240 W). More recently, however, the substitution of solid state electronics for vacuum tube technology reduced energy use by 40 to 50%.

Depending on the type and size, a television can account for 100 to 500 kWh per year in a typical residence. Energy use is directly proportional to hours of use except in two special cases.

Some sets have an "instant-on" feature by which the filaments of vacuum tubes are kept hot continuously. Although the power required is small (typically a few watts per tube), annual energy use can be significant due to the fact the load is on 24 hr per day. Energy can be saved by unplugging the set when not in use (or installing a switch).

Cable television is growing in popularity due to special programming, improved reception, etc. Some home receiving units have electronic amplifying and demodulating units which require electricity. These units require 5 to 10 W and unless unplugged or switched off will use 40 to 80 kWh per year.

e. General Suggestions for Residential Appliances and Electrical Equipment

Many electrical appliances (blenders, floor polishers, hand tools, mixers, etc.) perform unique functions which are difficult to duplicate. This is their chief value. In addition, their relatively infrequent use (a few hours per year) leads to annual energy use in the range of 1 to 10 kWh per year. These appliances are generally insignificant when measured against annual electricity use of 5,000 to 15,000 kWh per year.

Attention should be focussed on those appliances which use more than a few percent of annual electricity use. General techniques for energy management include:

- Reduce use of equipment (where feasible).
- Reduce losses (e.g., better insulation).
- Substitute for the same function (e.g., heat recovery).
- Perform maintenance to improve efficiency (e.g., clean filters).
- Reduce connected load (e.g., use smaller motors).
- Schedule use for off-peak hours (evening).

The last point requires further comment. Some utilities now offer "time-of-day" rates whch include a premium charge for usage which occurs "on-peak" (when the greatest demand for electricity occurs). Even if there is no direct economic benefit to the user, there is an indirect benefit. By scheduling energy intensive activities for off-peak hours (clothes washing and drying in the evening for example) the user helps the utility reduce its peaking power requirement, thereby reducing generating costs and helping ensure lower utility rates for all customers.

8. Computers and Office Equipment

A wide variety of equipment can be found in commercial buildings, depending on the size and function. Process equipment within buildings (e.g., clothes washers in laundries, printing presses, refrigerated display cases, etc.) will not be discussed due to the great diversity of these items. Some general remarks concerning various types of process equipment will be found in Volume III, Chapter 4.

Excluding process equipment, major energy using equipment in commercial buildings generally includes HVAC systems, lighting, and "other" equipment. Since energy management options for HVAC and lighting have already been described, the discussion here will be directed at "other" equipment which includes:

- Office equipment, e.g., typewriters
- Vending machines and water coolers
- Copying machines
- Elevators and escalators
- Computer equipment
- Miscellaneous pumps and fans

To assess the relative importance of these loads, consider a 15-story office building using approximately 65,000 kWh/week excluding HVAC.[8] Half (32,000 kWh/week) of this total was attributable to lighting. Of the balance, roughly 16,000 kWh/week (25%) was used by a computer center, 12,000 kWh/week (18%) was for miscellaneous power (typewriters, duplicating machines, water coolers, etc.), 4000 kWh/week (6%) was for elevators, and 1000 kWh/week (1%) was for miscellaneous exhaust fans, pumps, emergency lights, etc.

Computing equipment varies from small minicomputers using a few hundred watts to large central computer systems using hundreds of kilowatts. The computer center described above, which handled accounting and data processing for a large city, had 90 kW of computing equipment installed. This included about 35 separate equipment items, ranging from display stations rated at 200 W up to the central processing unit (512 k memory) rated at 16 kW. Included were card readers (0.5 kW), printers (1.5 kW), disk storage units (3.4 kW) and card punches (1.6 kW), etc.

In this example, special air conditioning equipment was installed in the computing facility to maintain proper temperature and humidity for the computer equipment. Total heat removal capacity of approximately 60 kW (200,000 Btu/hr) was required. According to the computer equipment supplier, acceptable environmental conditions ranged from 16 to 32°C with relative humidity in the range of 20 to 80%. Due to operational constraints, an effort was made to avoid either the high or lower limit on humidity. Equipment to both humidify and dehumidify was thus made part of the separate HVAC system which served the computer facility.

Several energy management opportunities were discovered in this facility. First, an energy audit revealed that overlapping control ranges on the humidify/dehumidify system permitted both units to be on simultaneously. In fact this mode of operation was observed during the audit. A minor modification to the control system would eliminate this duplication of energy waste and pay back the investment in 6 months.

The facility operates three shifts, 7 days per week. There are ten key punches which are left on throughout the day shift; many are left on throughout the second and third shift so personnel can avoid the delay caused by equipment "warm-up." By allocating one or two machines to "quick turnaround" jobs, many of the other machines could be turned off except when needed for production work.

Undoubtedly other energy management opportunities could be found in similar computer facilities. These examples suggest some of the considerations involved in evaluating energy use in computing equipment.

General office equipment includes electric typewriters (150 W), coffeemakers (1.2 to 6 kW), teletypes (0.5 kW), vending machines and calculators (10 to 100 W), tape recorders (100 W), and so forth.

The energy management opportunities with these types of equipment are more restricted, since they are generally designed to use energy only when operating. One obvious strategy is to insure that all equipment is turned off when not in use. Another is to size equipment with the right capacity to do the job, avoiding overcapacity which increases both energy and demand charges.

The four elevators in the building had a total capacity of 14.2 kW. During off hours one or more could be shut down. Efforts were made to encourage employees to walk on short trips — "one floor up, two floors down" — to avoid excessive use of the elevators. Overall energy use was small and no major savings opportunities were

In general the most likely opportunity with elevators, escalators, and similar equipment is to shut them down during off hours or other times when they are not needed.

There is a host of miscellaneous equipment in buildings which contributes only a small percentage of total energy use, is used infrequently, and is an unlikely candidate for improved efficiency. Their interaction with the building should be examined, however. For example, in one building stairwells and toilets were exhausted separately to the atmosphere, resulting in large and continuous heat losses. By adding heat recovery to these ventilation systems, a substantial savings resulted.

ACKNOWLEDGMENTS

I am grateful for the assistance of M. K. J. Anderson, K. M. Iyengar, and N. J. Smith, whose ideas and comments are reflected in this chapter.

REFERENCES

1. Electric Power Research Institute, *EPRI Journal,* 2(10), 40, 1977.
2. **Smith, Craig, B., Ed.,** *Efficient Electricity Use,* 2nd ed., Pergamon Press, New York, 1978.
3. **Ross and Baruzzini, Inc.,** Lighting Systems Study, Public Building Service, General Services Administration, Washington, D.C., March 1974.
4. **Kaufman, J. E., Ed.,** *IES Lighting Handbook,* 5th ed., Illuminating Engineering Society, New York, 1972.
5. Association of Home Appliance Manufacturers, *Directory of Certified Refrigerator/Freezers,* Chicago, 1976.
6. **Smith, Nancy J.,** Energy management and appliance efficiency in residences, in *Energy Use Management — Proceedings of the International Conference,* Vol. 1, Pergamon Press, New York, 1977.
7. **Rettburg, R. J. and Gratt, L. B.,** *Appliance Efficiency Program,* Final Rep. No. SAI76-551-LJ, Science Applications, La Jolla, California, March 1976.
8. Applied Nucleonics Company, Efficient Energy Use in the San Diego City Administration Building: A Study in Energy Management, Rep. No. 1168-1, Santa Monica, California, February, 1977.

Chapter 3

INDUSTRIAL ENERGY MANAGEMENT

Wesley M. Rohrer, Jr.

TABLE OF CONTENTS

I. INTRODUCTION

A. The Potential for Fuel Conservation

The potential for fuel conservation in the industrial sector of the economy is still considered to be large. Many companies, particularly the large corporations with abundant technical resources, have energy management programs that date back to the time of the first OPEC oil embargo. The early programs of the U.S. Department of Commerce enlisted the cooperation of the big corporations in conserving scarce fuels and helped the industrial and commercial trade organizations develop programs to monitor fuel consumption within their membership and to organize information on conservation to be used within the group. However, Ross and Williams[1] in 1976 estimated from their second law analyses of the consumption for energy in 1973 that the whole economy could do better by 42% and that the potential in industry amounted to 10.43 quads (1 quad = 10^{15} Btu = 2.93×10^{11} kWh = 1.055×10^{18} J) of a total of 29.65 or 40% of the total use. As shown in Table 1, the big potential is in improved housekeeping (37%) and in cogeneration of steam and electricity (25%). Waste-heat utilization contributed only 12% of the total potential savings. The most significant aspect of their study is the large savings which may be realizable with minimum capital requirement. Note also that substituting direct fired heating for electric heating is overall fuel savings but it quite probably would also result in the substitution of oil or gas consumption for the coal burned by most utilities. Until coal-burning equipment acceptable to the EPA is available for the small user, it is most likely that the substitution will go in the opposite direction — that is, from fossil fuels to electricity. This is thermodynamically unsound but environmentally more acceptable. Table 2, also from Ross and Williams,[1] shows the relatively high second law efficiencies for industrial heating processes as compared to space conditioning and water heating. Since some industries use up to one third of their total energy purchases for space conditioning, it can be seen that a large potential exists that is not considered in Table 1.

At this point it can be asked why, despite the fact that the attention of governmental and industrial leaders has been directed to energy conservation for almost a decade, such a small part of the total potential for fuel conservation has been realized. A great deal of the blame can be attributed to attitude alone. Both government and industrial leaders have been ambivalent about the role of conservation in achieving national energy independence. There has been the fear that conservation is tied inexorably to a forced change in American lifestyle and a degradation of the standard of living; there has been skepticism about the estimated potential for saving energy in the industrial sector; and there has been an even more widespread belief that the energy crisis has been contrived and will eventually go away when fuel prices are permitted to rise. Additionally, one finds that the internal technical resources of many medium- and small-sized companies are insufficient to take on additional burdens such as initiating and implementing conservation programs. However, hundreds of available case studies show that such programs are not only feasible for these companies but are profitable, not only because they result in energy cost savings but they often permit a company to live within its fuel allocations without having to install alternate fuel facilities or pay expensive penalties for exceeding allocations. The secret to a successful project is in carefully planning and organizing for it. The technical solutions are available to everyone. The energy problem must be viewed as an overall systems problem, and a systems analysis must be employed with everything considered that the systems approach entails. A random or scatter-gun approach no matter how enthusiastically and energetically entered into cannot produce satisfactory results. Only a carefully derived synthesis of good management practice and good engineering solutions will produce

TABLE 1

Hypothetical Potential for Energy Conservation in the Industrial Sector

Industrial sector	Potential savings (10^{18} J)
Improve housekeeping measures (better management practices with no charges in capital equipment)	4.06
Use fossil-fuel instead of electric heat in direct-heat applications	0.18
Adopt steam/electric cogeneration for half of process steam	2.73
Use heat recuperators or regenerators in half of direct-heat applications	0.78
Generate electricity from bottoming cycles in half of direct-heat application	0.52
Recycle aluminum in urban refuse	0.11
Recycle iron and steel in urban refuse	0.12
Use organic wastes in urban refuse for fuel	0.74
Savings from reduced throughput at petroleum refineries	0.92
Reduced field and transport losses associated with reduced use of natural gas	0.84
Total savings	11.00
Total demand in 1973	31.28×10^{18} J

Hypothetical energy demand with savings — 20.28×10^{18} J

From Ross, M. H. and Williams, R. H., The potential for fuel conservation, *Technol. Rev.*, 79(4), 50, 1977. With permission.

optimum results. As essentially presented in the EPIC Guide,[2] the following elements are necessary for success:

- Gaining management commitment
- Setting up the organization
- Carrying out the energy audit
- Educating employees
- First-level economic studies
- Engineering studies
- Second-level economic analyses
- Investment analysis
- Installation of energy conservation measures
- Developing and maintaining the energy information system.
- Installation of submeters
- Exploiting program successes

Each of these essentials will be discussed in greater detail in the remainder of this chapter.

TABLE 2

Second Law Efficiencies of Thermal Systems

Energy consuming activities (current technology)	Second law efficiency (%)
Residential and commercial	
Space heating	
Fossil-fuel-fired furnace	5
Electric resistive	2.5
Air conditioning	4.5
Water heating	
Gas	3
Electric	1.5
Refrigeration	4
Transportation	
Automobile	9
Industrial	
Electric power generation	33
Process steam production	33
Steel production	23
Aluminum production	13

From Ross, M. H. and Williams, R. H., The potential for fuel conservation, *Technol. Rev.*, 79 (4), 51, 1977. With permission.

B. Gaining Commitments to the Energy Conservation Program

The greatest impediment to a successful program in energy conservation is the lack of commitment at all levels of the organizational structure. Generally the difficulties in gaining the necessary commitments are related to disbelief in the seriousness of the national energy problem, or to a belief that the only solution to that problem is in a more vigorous program of exploration for conventional fuels or in future technologic breakthroughs which will result in boundless energy supplies. Also related to these kind of concerns are fears that energy conservation means public deprivation and sacrifice, a decline in economic activity, and a degradation of the standard of living. Our own position, that the energy situation is now critical and will continue to worsen, and that energy conservation (or put in a different idiom — the most efficient utilization of available fuels consistent with environmental requirements) will benefit local, national, and world economic conditions with no adverse effects on the quality of life in our society, is one that must be argued persuasively in order to attain full cooperation in energy conservation programs. The highest priority, of course, rests with top management, for without top-level commitment the necessary resources cannot be made available and the program will fail by starvation. Furthermore, a great part of the necessary cooperation from all lower echelons can be secured in some measure by management dictum, although the potential for success is enhanced by the addition of every single willing and enthusiastic soul. Horror stories are legion concerning mighty programs being made impotent by the stubborn resistance of one supposedly minor character who believes that the energy problem has been contrived for someone else's personal gain. To counter these attitudes and arguments one must use as inducement the beneficial effects of energy conservation on the economic self-interests of everyone concerned. Using the methods that shall be developed below, management and labor must be convinced that energy conservation measures are indeed economically feasible, avoid costs, and thereby put money in the bank — money that is just as good as that derived from profits on product sold. Furthermore, this additional income is earned regularly and perpetually, and as fuel costs rise increases in value. Additionally, the savings in fuel may help to forestall future fuel allocation cuts or to avoid exceeding present fuel allocations. During the hard winter of 1976—77 one plant in Northwest Pennsylvania, by taking Draconian steps in energy conservation, was actually able to maintain full production on maintenance fuel levels and thus emerge from the emergency with their business unscathed, while neighboring plants lost several weeks' production. Thus, good energy management can be sold on the basis that: (1) overall profits go up, (2) the possibility of maintaining full production is increased, (3) jobs are saved, and (4) a step is taken toward the fulfillment of national energy goals. Investment in cost containment programs has not hitherto been a popular management tool. However, a growing list of case studies involving successful energy conservation programs is educating us to the acceptance of that possibility.

C. Organizing for Energy Conservation Programs

The most important organizational step which will effect the success of an energy conservation (e.c.) program is the appointment of one person who has full responsibility for its operation. Preferably he should report directly to the top management position and be given substantial discretion in directing technical and financial resources within the bounds set by the level of management commitment. His background need not be technical because he can control sufficient in-house or purchased technical resources to carry out his directive. It is, however, difficult to stress enough the importance of making the position of plant energy coordinator a full-time job. Any diversion of his interests and attention to other aspects of the business are bound to badly affect the e.c. program. The reason is that the greatest opportunity for conservation is in improved housekeeping. The improvement and maintenance of good housekeeping procedures is an exceedingly demanding job and requires a constant attention and a dedication to detail that is rarely found in corporate business life. The coordinator should be energetic, enthusiastic, dedicated, and political.

The second step is the appointment of the plant e.c. committee. This should consist of one group of persons who are able to and have some motivation for cutting fuel costs and a second group who have the technical knowledge or access to data needed for the program department managers or their assistants. Union stewards and/or other labor representatives often make up the first group, while the second should include the maintenance department head, a manager of finance or data storage, some engineers, and a public relations person. The coordinator should keep himself up to date on the energy situation daily, he should convene the committee weekly, and he should present a definitive report to top management at least monthly and at other times when required by circumstance. It is suggested also that several subcommittees be broken out of the main committee to consider such important aspects as: capital investments, employee education, operator-training programs, external public relations, etc. The committee will define strategy, provide criticism, publish newsletters and press releases, carry out employee programs, argue for the acceptance of feasible measures before management, represent the program in the larger community, and be as supportive as possible to the energy coordinator. This group has the most to risk and the most to gain. They must defend their own individual interests against the group but at the same time must cooperate in making the program successful and thus be eligible for rewards from top management for their good work and corporate success.

As the e.c. program progresses to the energy audit and beyond, it will be necessary to keep all employees informed as to its purposes, its goals, and how its operation will impact on plant operations and employee routine, comfort, and job performance. The education should proceed through written and oral channels as best benefits the organizational structure. Newsletters, posters, and employee meetings have been used successfully.

In addition to general education about energy conservation, it may prove worthwhile to offer specialized courses for boiler and mechanical equipment operators and other workmen whose jobs can affect energy utilization in the plant. The syllabuses should be based on thermodynamic principles applied to the systems involved and given on an academic level consistent with the workmen's backgrounds. Long-range attempts to upgrade job qualifications through such training can have very beneficial effects on performance. The courses can be given by community colleges, private enterprises, or by in-house technical staff, if available.

The material presented here on organization is based on the presumption that a considerable management organization already exists and that sufficient technical and financial resources exist for support of the energy conservation program as outlined. Obviously, very small businesses cannot operate on this scale and some are so small that energy conservation cannot be made a realistic goal. However, we have found companies with energy costs below $50,000 annually with capabilities for carrying out effective energy conservation efforts.

D. Setting Energy Conservation Goals

It is entirely appropriate and perhaps even necessary to select an energy conservation goal for the first year of the program very early in the program. The purpose is to gain the advantage of the competitive spirit of those employees that can be aroused by a target goal. Unfortunately, the true potential for conservation and the investment costs required to achieve it are not known until the plant energy audit is completed and a detailed study made of the data. Furthermore, a wide variety of energy-use patterns exists even with a single industry. A number of industry studies have shown that energy consumption per unit of production does not fit a normal population curve. Individual plants are often clustered around low, average, and high values for consumption with two or three peaks in the distribution curve. However, it may be of interest to know where your plant stands vis-à-vis national averages in your sector of industry. Table 3 lists the primary energy consumption data for the nine most energy-intensive industries for the census year 1970.[3] The notes following the table explain some of the rationale used and lists several data sources. Table 4 gives unit energy costs for the same industries. It should be noted particularly that electrical energy is accounted for at the source rather than at the point of end use and consequently, each kWh is counted as 10,443 Btu or 11.018 MJ since the average utility plant has a thermal efficiency of 0.33. The energy use in office buildings varies from 34 MJ/m^2 to 340 MJ/m^2 per year. Analyses of available data prior to 1976 have been published by the FEA.[4]

It is suggested that an intuitive selection be made, perhaps in the range of 5 to 25% for the first year. More realistic goals will be forthcoming as the work progresses. For every additional time period the potential savings and the required investment will be known with increasing certainty. Thus, in subsequent years, goals will be set and will be expected to be achieved as a matter of course, just as are other management goals such as those for production and sales.

E. Unit Process Efficiency

It has been customary to determine industrial process efficiency on the basis of the simple ratio of useful energy to total energy input or

$$\eta = \frac{\text{Energy used}}{\text{Input energy}} \quad (1)$$

and in the case of an industrial furnace, oven, or boiler, this becomes

$$\eta = \frac{\text{Energy absorbed in product}}{\text{Fuel input} \times \text{heating value}} \quad (2)$$

This quantity, often called the first-law efficiency, ranges in value from 0.05 or less in the case of a high temperature heat treatment furnace to 0.85 for a high efficiency steam boiler. However, this measure is misleading because a heat pump can absorb energy from the ambient surroundings and deliver it to a process with less expended energy than a fuel-fired device operating at 100% efficiency based on the calculation from Equation 1. A heat pump has a theoretical coefficient of performance.

$$\text{COP} = \frac{T_H}{T_H - T_L} \quad (3)$$

where the COP is defined as the ratio of useful heat delivered to the energy required to deliver it, T_H is the temperature of heat delivery, and T_L is the temperature at which heat is absorbed. T_L is ordinarily taken to be T_o or the ambient temperature. Thus the minimum energy that could supply the process, A_{min}, is the work required for a reversible heat pump or

$$A_{min} = \frac{E_p (T_H - T_L)}{T_H} = E_p\left(1 - \frac{T_L}{T_H}\right) \quad (4)$$

where E_p is the energy required to carry out the process. A so-called second-law efficiency[1,4] is defined as

$$\epsilon = \frac{A_{min}}{A_{act}} \quad (5)$$

where A_{min} is the minimum possible change in available energy needed to complete the process and A_{act} the observed change in available energy. Although the first-law process efficiency allows us to make comparisons among presently available devices, it does not indicate the extent to which those devices approach the thermodynamically optimum one. A note of caution must be raised in this regard, however. Thermodynamically optimum devices operate in absence of irreversibilities; that is without mechanical or fluid friction and with heat transfers across infinitesmal temperature differences. This implies slow speed machines, small fluid velocities, and extremely large heat exchangers. Thus the cost of increasing second-law efficiency may well be unacceptable in economic terms. The concept of second-law efficiency is of extreme theoretic interest, but will not necessarily lead to the best solution in the real economic world. In practice the first-law efficiency often conveys sufficient information to make an economic analysis of the process, which gives the answer to the question of how great a

TABLE 3

Primary Energy Consumption Data for Nine U.S. Industries

	Notes	Primary energy consumption (GJ/t product)	Breakdown of primary energy use by type of resource (%)					Total energy used in 1970 in making product (10^{15} J)
			Coal	Oil	Gas	Purchased electricity	Derivative fuels	
Low-density polyethylene resin	a,b	108.73	0	23.6	67.3	18.2	(9.1)	212
High-density polyethylene resin	a,b	103.09	0	28.1	73.1	8.4	(9.6)	79
Polystyrene resin	a,b	136.56	1.1	100.4	27.1	6.9	(35.5)	208
Polyvinyl-chloride resin	a,b	96.44	9.1	19.4	55.6	23.4	(7.5)	138
Petroleum refinery products	a,e,d	(0.51)	—	—	—	—	—	1841
Portland cement — wet process	a,b	9.35	30.4	13.7	39.9	16.0	—	399
Portland cement — dry process	a,b	8.43	42.6	8.0	32.4	17.0	—	236
Primary copper	a,b	130.07	10.1	13.5	38.4	38.0	—	179
Primary aluminum	a,c	201.50	0.5	15.1	9.0	72.2	3.2	728
Raw steel	a,b	22.35	81.1	6.6	13.5	8.4	(9.6)	2667
Glass containers	a,b	21.12	35.8	7.3	48.8	14.5	(6.4)	216
Newsprint	a,b	25.53	6.6	12.8	13.5	67.1	—	77
Writing paper	a,b	28.46	19.3	27.8	23.5	29.4	—	60
Corrugated containers	a,b	25.28	26.2	26.2	42.1	6.9	(1.4)	228
Folding boxboard	a,b	25.47	17.1	25.8	40.6	16.5	—	22
Virgin styrene butadiene rubber	a,b	155.41	0.1	47.8	53.9	9.7	(11.5)	210

Note: The total energy figures quoted in this table correspond to the primary energy consumption shown in the first column of numbers. These consumption figures were multiplied by the total tonnage of product manufactured in 1970. The energy totals therefore correspond to the specific definition given in this report. The numbers do not necessarily correspond to any one particular standard industrial classification code number. Total energy use represented by these products was 7604×10^{15} J in 1970.

a. The figures shown are based on average industrial practice in the U.S. during 1970—1971.

b. For all process steps other than alumina smelting and petroleum refining, electric energy was derived from the following mix of primary energy sources: 48.5% coal, 16.1% oil, 26.8% natural gas, 2.6% nuclear fuels, 6.0% hydroelectric. Taking generation and transmission losses into account, this is equivalent to 10286 Btu/kWh.

c. For alumina smelting, electric energy was derived from the following mix of primary energy sources: 39.0% coal, 12.9% oil, 21.5% natural gas, 2.1% nuclear fuels, 24.5% hydroelectric.

d. For petroleum refining, the primary energy consumption in generating electricity was taken as equivalent to 12000 Btu/kWh, which is derived from the breakdown given in Note b above with the exclusion of the hydroelectric contribution.

e. Petroleum refining industry data expressed as million Btu per barrel of crude oil processed. Data taken from typical refinery calculations. Breakdown by resource type not determined for overall industry. Typically, all energy is derived from oil, gas, and purchased electricity. Electricity use in the refinery is typically 11%. Total energy use is based on 0.44 million Btu/bbl and the 1970 total of crude runs to stills, 3.967×10^9 bbls (*API Annual Statistical Review*).

From The Data Base, The Potential for Energy Conservation in Nine Selected Industries, NTIS PB-243-611/A, Federal Energy Administration, Washington, D.C., 1975. With permission.

TABLE 4

Cost of Primary Energy as a Percentage of Product Price (U.S. 1970)

	Cost of primary energy consumed ($ per ton product)	Product selling price ($ per ton)	Primary energy cost as % of selling price
Low-density polyethylene resin	40.74	265	15.4
High-density polyethylene resin	40.71	280	14.5
Polystyrene resin	72.68	240	30.3
Polyvinyl-chloride resin	34.86	290	12.0
Petroleum refinery products	—	—	—[a]
Portland cement-wet process	3.37	17.5	19.3
Portland cement — dry process	2.90	17.5	16.6
Primary copper	43.94	1148	3.8
Primary aluminum	56.40	574	9.8
Raw steel	9.36	87	10.8
Glass containers	7.20	158	4.6
Newsprint	7.94	150	5.3
Writing paper	11.02	245	4.5
Corrugated containers	10.20	207	4.9
Folding boxboard	9.99	200	5.0
Virgin styrene butadiene rubber	68.79	460	15.0

[a] Highly dependent on product slate specifications and market prices. Refinery energy use typically represents about 5%.

change in second law efficiency we can afford. In the world of thermal-process design the law of diminishing returns begins to control very early in the process of trying to improve the efficiency of a system.

F. The Question of Alternate Fuels

In an economy with abundant supplies of all types of fuels, the choice of fuel would depend first on the requirements of process and product quality and secondly on cost. In a fuel-scarce economy the choice may be reduced to what is available regardless of cost, perhaps with sharply reduced technical specifications. Experience has shown that it is usually more economic to pay higher energy costs than to shut down or even reduce production. Thus many managers are considering the use of all electric production facilities, even though the present price ratio of electric to fossil fuels ranges from 3:1 to 15:1 on an energy basis, depending on local market conditions and annual consumption levels.

The reasons for the large price differentials are related mainly to the attainable thermal efficiency of fixed-station power plants, which average about 35% in this country. However the price difference may be partly offset by a potentially more efficient use of electrically derived heat over that from combustion processes. For example, the substitution of radiant electric heat for a gas-fired oven could involve the following statistics: First-law efficiency of gas oven is 10%; first law efficiency of electric oven is 85%. The ratio of cost of electric energy to gas-derived energy is 4:1; the assumed electric generating station efficiency is 35%.

$$\text{Percent fuel savings using electricity} = 100\left(\frac{\dfrac{1}{0.1} - \dfrac{1}{0.85 \times 0.35}}{\dfrac{1}{0.1}}\right) = 66\%$$

On the other hand the cost of the electrical energy would be:

$$\left(\frac{1}{0.85}\Big/\frac{1}{0.10}\right) \times 4 = 0.47$$

times that of the gas-derived energy. The use of electrical energy for heating purposes is sometimes in poor taste from a thermodynamic outlook and usually costs more per unit of heat delivered, but it does have some demonstratable advantages. It alleviates the most pressing national energy problem because it substitutes (in most areas of the nation) coal as a fuel for scarcer liquid and gaseous fuels. In some cases coal can be used as a local industrial fuel, but in the majority of plants in this country coal is unacceptable because of process requirements, environmental restrictions, or the availability of small-scale coal-fired equipment. This is not to say that the present outlook for coal as an industrial fuel will continue indefinitely. Should suitable coal-fired equipment with the capability of removing sulfur and/or its oxides from the liquid and gaseous emissions become available to industrial users, or should the environmental constraints on sulfur-oxide emissions be relaxed, then coal can be expected to take over much more of the industrial fuel market. In the meantime each enterprise might well review its position vis-à-vis alternate fuel sources and make provisions for more flexibility in fuel use. Such flexibility should pay off in enhanced ability to stay in production and to use the cheapest available fuel, as market factors cause relative changes in per-unit fuel costs. Recently, small-scale industrial boilers have been introduced which can burn gaseous, liquid, or solid fuels. This is clearly a step in the right direction. Furthermore, more attention needs to be given to the possibility of using as fuels locally generated waste which is presently being discarded. We are referring here to such materials as waste paper and paperboard, used lubricants, and contaminated liquid solvents which can often be used as boiler fuels when trifuel burners have been installed.

II. THE ENERGY AUDIT AND THE ENERGY INFORMATION

A. Carrying Out the Energy Audit

1. Describing the Audit

The energy audit consists of a complete study of the consumption and costs of all purchased energy, the way it is distributed throughout the plant, how it is used in individual systems and processes, the potential for energy saving and where it exists, and the capital costs involved in recovering that potential. Additionally, it provides the basis for creating a very inexpensive energy information system (EIS) which can and should be incorporated into the present management information system (MIS). It represents in effect the total body of knowledge needed by management to institute an effective energy conservation program. The particular uses to which the audit can be applied are found in the list below.

- To inform management of the growing cost of energy and to provide motivation for carrying out an energy conservation program as a cost-containment measure
- To inform the engineering staff of the plants energy utilization characteristics so that intelligent conservation measures may be planned
- To provide management with the information needed to make wise investment decisions concerning energy conservation measures

- To provide part of the basis for planning and installing alternate fuels
- To provide base-line energy consumption data to which future energy consumption can be compared
- To provide the basis for a perpetual energy information system (EIS) which can be integrated with existing management information systems
- To uncover poor housekeeping practices which can be quickly remedied to give almost immediate energy and cost savings.

In other words, the energy audit provides the rationale for an energy management program, as well as all the information for design, implementation, and evaluation of energy conservation measures and in addition provides the base upon which to build the EIS which is needed to complete the MIS.

The first step to be taken is always to construct a retrospective record of energy purchased monthly from all sources for a consecutive 2-year period. Monthly consumption and monthly costs are recorded. This includes separate records for electricity, natural gas, each grade of oil, propane, steam at each pressure purchased, and each waste fuel. It does not include steam generated on site or waste heat recovered, but does include by-products used as waste fuel. Purchased water is usually included because it is closely associated with energy utilization and is a controllable expense. Along with the energy consumption data, one must gather month-by-month measures of weather severity and level of plant activity. The most useful weather data are the monthly degree-days for the heating season. The level of plant activity can be measured in a number of ways and one cannot predetermine the most useful measure. In the brewing industry barrels of beer produced suffices. In complicated fabrication industries in which highly diversified products are manufactured and then assembled, direct labor hours are often a better measure. The chronology of records is often a big problem in constructing the consumption audit. Billing dates for electric and natural gas bills lag behind the consumption and the records must be compensated. For oil, propane, and coal, one usually has records of delivery dates and quantity delivered. Depending upon the ratio of the rate of use to the storage volume, the delivery dates may or may not be a significant measure of monthly use. Happy is the auditor who finds records of monthly inventory of these fuels or better still meter records.

The next step is to construct process flow charts for the total operation. These should be arranged so that indications of energy added to the product can be entered at each process stage. These allow the quick determination of the most energy-intensive operations on which one should presumably concentrate first, as well as permit studies of possible waste-heat interchange and process modifications for purposes of energy conservation.

The third step is to derive an in-plant distribution tree for each source of energy purchased. This should be done by plant, building, department, production line, and units process. This step is probably the most difficult one in the audit but is most important for the successful operation of your program. If submetering is done at any of the above-mentioned levels, the time and effort required is correspondingly reduced. Furthermore, as one suffers through the attempt to derive a realistic in-plant distribution, his pain should motivate him toward mounting a program to add submetering for all large energy flows. The reason that such a program is so important has to do with establishing responsibility for energy utilization within the plant. If top management has no way of knowing how much energy departments A, B, and C consume, then they can hardly be in a position to reward the managers of those department for good energy management or punish them for poor energy management. If manager A is aware that no way exists for him to become a hero by way of improved energy management, he will expend as little effort as possible on that activity for any improvements that he brings about himself may be credited more or less equally to all of the departments.

The next activity to be taken is to produce energy balances on each integrated system and unit process which consumes appreciable amounts of energy. In order to prevent undue emphasis on arbitrating the division of large and small energy users, it is best to select the 5 or 10 or 20 largest energy users as your first focus for analysis. The information that must be provided for each system includes the mass flow rates, identification of materials and the temperature for each entering and exiting stream. This should include the flows of raw materials as well as finished product. In addition, any surface heat losses must be estimated or measured. Time-of-use profiles must also be derived when possible. An alternative to typical daily, weekly, or monthly time profiles is an estimate of the annual load factor for each system or process unit studied.

A data bank which stores the information from the initial audit must be established and fed data on at least a monthly basis. This energy information system should be maintained in perpetuity. The data bank may be computerized or kept by hand. If it is complex enough to be computerized, the data acquisition, handling, and retrieval system can be used for carrying out the numerical analysis required by the audit and the associated energy conservation program. It is relatively easy and much less expensive to develop the energy information system while constructing the energy audit than it is to establish it later. A well designed EIS can be maintained at a fraction of the cost of making annual energy audits. A detailed log of energy conservation activities must be maintained. The log must include changes in scheduling, production techniques, operating set points, additions or replacement of equipment, and retrofit of energy conservation features to existing equipment.

During the energy audit one must make an energy-saving survey. This consists of a walking plant tour designed to uncover poor housekeeping practices which result in energy waste. The defects which one looks for are fuel, steam, compressed air or water leaks; uninsulated, poorly insulated, or damaged insulation on building, pipes, or equipment; excess use of any utility, particularly due to lighting or power in use when not needed; heat loss due to broken or open windows and open doors; control temperatures set too high or too low, and fuel burners in need of maintenance or poorly adjusted. A sample form is shown in Figure 1.* A recapitulation of the content of the energy survey appears below.

- Retrospective survey of energy consumption and cost for each source of purchased energy and monthly measures of weather severity and level of plant activity
- Process flow charts
- In-plant energy distribution
- Heat balances and time-of-use profiles for all integrated systems and unit processes
- An energy information system
- Energy conservation activity log
- Energy-saving survey

The contents missing from the above list, which are often considered to be part of the audit, are the engineering and economic analyses leading to specification of energy conservation measures requiring capital expenditures. This author has chosen to regard these analyses as integral parts of the energy conservation program of which the audit

* Figures are set at the end of the text. See p. 140.

is also an integral part but not to specifically include it in the content of the audit itself. The reason is that energy conservation is much more a matter of good management practices than it is of capital spending. Thus it deemphasizes the importance of capital spending by making the feasibility and engineering studies separate and discretionary activities. This point of view is supported by the estimates of Ross and Williams[1] who reported that their second-law analysis showed a potential savings of 35% in the industrial sector. However, when broken down to the specific measures that will recover that potential, it is seen that 37% of the total exists in better housekeeping and only 7% in waste-heat recovery. It might be added, however, that the potential for waste-heat recovery is probably overconcentrated in a few energy-intensive industries. In primary metals, glass, paper, etc., there exists a much larger potential for fuel savings from capital intensive projects. The topics of engineering and economic studies are treated later in this chapter.

Governmental attitudes toward the energy audit are revealed in the rules set out in the Code of Federal Regulations.[5]

2. Data Collection

The data collection is always a task of considerable magnitude. A summary outline of the data requirements is given below.

First a retrospective, month-by-month record of the consumption and cost of every form of energy purchased or sold out of plant for at least a 2-year period. This includes energy in the form of electricity, natural gas, fuel oils, gasoline, propane, coal, purchased steam, and waste or by-product fuels. Separate records should be used for each form of energy including each grade of fuel oil and each steam pressure. The data for purchased energy are usually derived from utility bills and other financial records. Gas and electric utilities often keep computerized records which are available for audit purposes. However, bulk purchases of oil, propane, and coal cause problems unless monthly inventories are taken of fuel storage. If this has not been the practice it should be instituted at the earliest possible time. Audit records for electricity should include monthly peak demands, power factor multipliers, and any other available data that affect the cost of energy and at the same time are susceptible to some control.

Second, monthly measures of production level and of weather severity are required for the same period of time as the purchased energy records. Measures of production level are needed in order to assess the effects of production on energy use. It has been found, however, that the best measure to use for this purpose differs widely from industry to industry and perhaps from plant to plant. Measures of production (or of business activity) range from tonnage produced to direct labor hours used per month. In general, it is found that basic industries such as steel, cement, oil, and coal can use monthly units of production while more complex assembly-type producers can best use direct labor hours. Once the proper measure is found, then records should be kept of the total energy use per unit of activity. These will then constitute a measure of efficiency of fuel utilization for the plant.

The best measures of weather severity are the degree-days for the heating season and the degree-days for the cooling season. These are available monthly from the National Oceanographic and Atmospheric Administration.

Third, a complete inventory must be kept of energy-consuming equipment with as much information as possible about energy and material flows, and entrance and exit pressures, and temperatures. This information should preferably be derived from actual inspections and from experienced operators and engineers.

Fourth, time-of-use profiles for all systems and unit processes and the corresponding operating conditions should also be kept.

A set of audit forms is appended to this section which were reproduced from the Energy Information Workbook.[6] These are shown as Figures 2 through 6. Additional information concerning energy audits and audit forms can be found in References 2, 5, and 6.

3. Extracting Information from the Gross Energy Consumption Data

This author's approach to wringing out the energy consumption data is to plot it chronologically in as many ways as seems useful for each case. This is illustrated with some plots of both raw and derived data. In Figure 7 there is a block diagram of the energy distribution within a commercial laundry and in Table 5 the annual energy and cost data.

This rather atypical facility uses no direct-fired laundry equipment. Instead, natural gas is used as a boiler fuel to supply steam-heated washers, dryers, and ironers, for firing a hot-water furnace for space heating and to supply four direct gas-fired air make-up heaters. Figure 8 shows the monthly energy consumption for a 2-year period with the gas and electric used converted to thermal units. The seasonal variation is quite apparent, but the base load is some 55% of the peak consumption. The base load is shown by the dashed lines on the figure as 2.796 TJ per month. The average annual gas consumption is known from utility bills to be 40.778 TJ. The difference between the total and the base load is the heating load, where annual heating energy $= 40.778 - 2.796 \times 12 = 7.226$ TJ.

The plant operates on one 10-hr split shift a day for 5 days a week. Thus the total production time is 2600 hr per year and

$$\text{Base load consumption} = \frac{2796 \times 12}{2600} = 13 \text{ GJ/hr}$$

Since the rated input of each of the 250 hp fire-tube boilers is 11.1 GJ/hr, a single boiler will hardly be able to maintain this base load, which averages 16% overload year round. Thus a study is required to determine the proper load distribution for the boilers in order to minimize fuel consumption for each system energy requirement. Figure 9 is a plot of the total energy consumption and cost on a monthly basis.

Costs seem to follow consumption closely except for the last 6 months of 1976 where costs seem to rise faster than consumption compared to the previous 18 months. The gas cost rose slowly and the electric rate was volatile, as shown in Figure 10.

The variation in electric rate points to either large and variable peak demands or poor power factor control. Figure 11 was then plotted to show that during 1975 and the first quarter of 1976, power factor multipliers ranged from 1.1 to 1.5 and then improved to a consistent number close to 1.0. Demand averaged about 400 kW until the fall of 1976 when it rose to a peak at 540 kW, 35% over the previous base.

In May of 1975 the fuses on the power-factor-correcting-capacitor banks had blown causing the spike in the curve. That event triggered a concern for the penalties being incurred by the poor power factor, and additional capacitors were installed in the spring of 1976. The peak monthly demands at the end of that year were eventually attributed to an electrical hookup permitted to be used by the contractor installing a new continuous washer. Large variation in the power factor multiplier usually signals a defect in a system which should be corrected immediately to end an unnecessary expense. In fact, any plant load with power factor less than 0.95 should be compensated for. In most parts of the country the cost of capacitors is less than the expense

TABLE 5

Annual Energy and Cost Data for Laundry[a]

Total annual gas consumption	40.8 TJ
Total annual electric consumption	5.0 TJ
Gross annual energy consumption	45.8 TJ
Production	4,445,200 kg
Degree-days (C), heating	3786
Annual energy cost	$113,032
Average unit energy cost — all sources	$2.47/GJ
Per unit energy consumption	10,311 kJ/kg

[a] This table summarizes the information known about gross consumption and production for the year 1976.

of paying increased demand charges. On the other hand, increases in demand are usually caused by increased connected capacity and may not be susceptible to control

Figure 12 illustrates the close relation between demand and production for a foundry, but it also shows a dramatic increase in demand during 1976 when production was well below previous peaks. The increase in demand was discovered to be the result of the mandated installation of a cupola emissions control system which raised energy requirements and increased product costs.

Demand charges can be as much as one half of the total electrical utility bill. Ways to reduce them may be suggested by a study of daily demand variations. The utility will usually provide continuous records of demand over periods as long as a week at no cost to the consumer.

It is instructive to view the energy use per unit of production activity seen in Figure 13. The laundry data are plotted as energy consumed per kilogram of laundry processed. This index goes up in the winter due to the heating load but otherwise is quite constant, which indicates a constant production rate. One sees more often a variation of the sort shown for a commercial heat treatment plant in Figure 14. Energy per unit of product treated is shown to be inverse to the quantity processed. This points out the advantages in energy efficiency of keeping your plant at full capacity. In the case of this job shop, and in most other businesses, production is scheduled to fit product demand. Businesses that have seasonal demand, such as breweries, can study the implications of leveling production and stockpiling to meet peak demands. Some considerations that control the economics are start-up and shut-down losses, electrical demand costs, incremental costs for refrigerated storage, overtime costs for labor, and a host of others.

Unit energy costs for each energy source plotted against the month of the year for several years are useful motivational tools. The unit costs of gas and electricity for a metal fabrication plant are plotted in Figures 15 and 16 for a 2-year period. The 40% annual increase in electricity rates and the 14.5% annual increase in natural gas costs sharpened the company president's interest in energy and resulted in a program that reduced energy use by 23% in a year's time. An energy audit revealed the extent to which insulation, reduced ventilation, heat recovery, and better machine scheduling could put money in the bank.

4. The Internal Audit

Up to this point, the discussion has been concentrated mainly on the gross consumption audit and the information about the plant that can be derived from it. Although the benefits derived from it can be large, the gross consumption audit is really just the beginning of the whole program. The biggest and the most difficult portion is still ahead. At this point an understanding of the grand design must be developed, both of the plant and its processes which lead from the receipt of raw materials to the shipping of the final products. The more thoroughly the plant operation is studied, the greater the amount of knowledge that can be derived about smaller and still smaller parts of the overall scheme. The reason for carrying out the energy audit is to produce an understanding that is as complete as possible of the ways that energy is involved in the activity of the plant, particularly as it is essential to the plant's primary purposes.

The laundry referred to in the previous section will be used as an example of how one proceeds with distributing the gross energy consumption over unit processes within the plant. Table 6 is an equipment list with nameplate ratings, excluding the process boilers and hot-water furnace.

From Figure 6 the annual heating load had been previously estimated to be 7.227 TJ. Subtracting this from the total natural gas consumption we get (40.776 − 7.227) TJ or 33.549 TJ as the estimated process load. Note that this is also the boiler input since no other direct-fired gas equipment is used.

Because of the energy losses from liquid and solid surfaces, very little if any space heating is required during production hours. Therefore almost all heating during that time is due to the heating of make-up air. However, the ventilation is reduced to a low value during the remaining 14 hr of the day, so that during off-production hours the largest energy loads are the unit space heaters and the office heating system.

An air-flow survey for the plant gave the following information:

Ventilation	905.6m³/min
Continuous dryer exhaust	277.3
Batch dryer exhaust (average)	244.8
Boiler	148.6
Furnace	29.7
Total outdoor air	1606 m³/min

The average outdoor temperature during the heating season is −3°C. Therefore, the approximate output of the make-up heaters in a heating season, Q_v is

$$Q_v = 1606 \text{ m}^3\text{/min} \times 60 \times 10 \ \frac{\text{min}}{\text{day}} \times 1.2 \ \frac{\text{kg air}}{\text{m}^3}$$

$$\times\ 0.519 \ \frac{\text{kJ}}{\text{kg}\ {}^\circ\text{C}} \times 3786 \text{ DD}/\left(10^6 \ \frac{\text{kJ}}{\text{GJ}}\right) = 2273 \text{ GJ}$$

The annual input to the heaters, Q_h, assuming a 60% efficiency, is

$$Q_h = 2.273/0.60 = 3.788 \text{ TJ}$$

Since the total heating load was 7.227 TJ the annual furnace input is 7.227 − 3.788 = 3.439 TJ. That is, half the heating load is make-up air heating and the other half is due to heat leaks. The total process heating load;

$$q_p = (40.776 - 7.227) = 33.549 \text{ TJ/year}$$

For 80% boiler efficiency, the heat supplied to the steam, q_{st}, is $q_{st} = 0.80 \times 33.349 = 26.839$ TJ.

The only batch processes are the flat ironers and the batch washers and dryers. The ironers are never valved off and when not in use they heat the plant. The batch washers

TABLE 6

Laundry Equipment List

Equipment	Number	Energy source	Input rating	Output rating
Air make-up heaters	4	Natural gas	332 MJ/hr	—
Batch dryers	2	Steam	10 BHP	—
Automatic washers	3	Steam	1.15×10^6 kJ/hr	—
Conditioners	3	Steam	1.75×10^6 kJ/hr	—
Continuous dryers	2	Steam	3.0×10^6 kJ/hr	—
Flat ironers	4	Steam	25 BHP	—
Unit heaters	18	Steam	60 MJ/hr	—
Domestic hot water heater (850 gal)	1	Steam	527,500 kJ/hr	—
50-lb batch-washer motors	1	Electric	—	1 hp
100-lb batch-washer motors	2	Electric	—	2.5 hp
Batch-dryer motors	2	Electric	—	2.5 hp
Automatic-washer motors	3	Electric	—	2.5 hp
Conditioner fans	3	Electric	—	1 hp
Continuous dryer fans	2	Electric	—	1.5 hp
Ironer motors	4	Electric	—	¼ hp
Return pump	1	Electric	—	1.5 hp
Make-up air fans	4	Electric	—	3 hp
Ventilating fans	8	Electric	—	¾ hp
Air conditioner	1	Electric	—	10 ton
Air conditioner	1	Electric	—	15 ton

average five batches of 488 kg of linen each day using approximately 8.345 ℓ of hot water for each kilogram or 20,363 ℓ per day. Each of the 70 employees uses an average of 37.85 ℓ per day for another 2650 ℓ or a total of 23,013 ℓ per day. The heat used to heat this water, q_{hw}, is

$$q_{hw} = 23{,}013\ \ell \times 1\ \text{kg}/\ell \times 4.186\ \text{kJ/kg}$$
$$-{}^\circ\text{C} \times (71.1 - 12.8)^\circ\text{C}/(10^6\ \text{kJ/GJ}) = 5.62\ \text{GJ}$$

or $0.0056 \times 52 \times 5 = 1.458$ TJ per year. That leaves $(26.839 - 1.458) = 25.381$ TJ per year. This is distributed among all other steam-heated units in direct proportion to the nameplate ratings. This allows the preparation of a new schematic diagram of the heating processes shown below in Figure 17. The same procedure is used to distribute the electrical load but will not be carried out in this example. The crude estimating done here to distribute the total energy load over the unit processes is necessary so that approximate heat balances for the large-unit energy users can be derived. The heat balances will then aid in the design of economic waste heat recovery systems or other measures which will increase the efficiency of energy utilization. The crude estimating is unnecessary when all large process units are submetered.

B. In-Plant Metering

Submetering reduces the work and time required for an energy audit; indeed it does much more than that. Because meters are tools for assessing production control and for measuring equipment efficiency, they can contribute directly to energy conservation and cost containment. Furthermore, submetering offers the most effective way of

TABLE 7

Cost and Benefits of Gas Meters

Gas metering flows up to (m³/hr)	Costs as of November 1977 ($)	Estimated minimum gas bill reduction ($ per year)
57	310	300
113	580	600
141	920	750
283	1500	1500
424	1440	2250
566	1440	3000
1132	1750	6000
2264	2500	12000

evaluating the worth of an energy conservation measure. Too many managers accept a vendor's estimate of fuel savings after buying a recuperator. The may scan the fuel bills for a month or two after the purchase to get an indication of savings — usually in vain — and then relax and accept the promised benefit without ever having any real indication that it exists. It may well be that, in fact, it does not yet exist. The equipment may not be adjusted correctly or it may be operated incorrectly, and there is no way of knowing without directly metering the fuel input. This author estimates that at least 2.5% waste is recoverable by in-plant metering. The list in Table 7 gives approximate costs for gas meters and the minimum savings for a load averaging one half of full-scale operating for 6000 hr per year for natural gas priced at $2.00/mcf.

Oil meters are just as effective as gas meters used in the same way and are even less expensive on an energy-flow basis. Electric meters are particularly helpful in monitoring the continued use of machines or lighting during shutdown periods and for evaluating the efficacy of lubricants and the machineability of feed stock. The use of in-plant metering can have its dark side too. The depressing part is the requirement for making periodic readings. It does not stop even there. Someone must analyze the readings so that something can be done about them. If full use is to be made of the information contained in meter readings, it must be incorporated into the energy information portion of the management information system. At the very least each subreading must be examined chronologically to detect malfunctions or losses of efficiency. Better still, a derived quantity such as average energy per unit of production should be examined.

In-plant metering need not be as accurate as that required for billing. It is often possible to buy out-of-tolerance billing meters from your utility for a fraction of the cost of a new meter. If these are installed by plant personnel is slack periods, the overall cost can be very modest indeed.

III. THE TECHNOLOGY OF ENERGY CONSERVATION

A. Overview

Energy conservation can be accomplished in two general ways categorized as "belt-tightening" and "leak-plugging". Belt-tightening consists of those measures that represent some sacrifice or restraint on normal operations such as lowered living space temperatures, less ventilating air flow, lower lighting levels, the elimination of aesthetic architectural features which are energy intensive, the purchase of smaller and less comfortable company autos, and the substitution of telephone conferences for meetings requiring air travel. This category can be considered almost an emergency category, since it requires complete employee cooperation and can have negative effects on morale and organization harmony if resisted. On the other hand, the range of conditions for human comfort are quite wide[7] and physiological and psychological adjustments can be made given enough time and proper communications. Furthermore, a great deal of waste is presently encountered in situations where excess lighting, ventilation, and temperature levels produce human discomfort that can easily be alleviated. However, the real impact of energy conservation efforts over the long range will come about by leak-plugging measures. These are measures that reduce the amount of heat loss to the surroundings, thus increasing the efficiency of utilization of industrial heat. Obvious examples include better building and piping insulation, repair of gas, oil, and steam leaks, use of waste heat and waste fuels, and improved control of combustion processes. These and other measures will be discussed in greater detail later in this section.

1. Improving Housekeeping

Excellent housekeeping is a measure of good management yet it is relatively rare in industry. That is so because of the dedication to purpose and the attention to detail that is required to accomplish it. However, it is well worth striving for because, in terms of energy management, it pays off so well. Since the major part of industrial energy waste results from poor housekeeping, the biggest savings can be had by improving it. An equally attractive feature of housekeeping improvement is that it involves little or no capital expenditure. The combination of big savings at small cost should prove irresistible, but one must face up to the demands on management attention that are entailed.

The procedure requires that one person or a small party make a detailed inspection of the plant looking for energy leaks. When found, these are duly noted in an inspection form and orders for corrective action are issued as soon as possible following the inspection. Figure 1 is an example of a suitable inspection form.[2]

The initial inspection is just the beginning of a perpetual routine which is used to evaluate the corrective actions generated by previous inspections and to find new energy leaks as soon as possible after they occur. The best person to carry out this part of the program is the plant energy coordinator. It would not be unreasonable for him to dedicate the beginning of every working day to a thorough plant inspection. The danger is in making it so routine that it becomes a purely perfunctory checkoff.

The EPIC Guide[2] provides a detailed list of energy leaks to look for. A sampling of those is found below.

- Leaks in air, gas, oil, steam, or water piping
- Deteriorated insulation
- Broken or poor-fitting windows and doors
- Electric motors running when not in use
- Lights on in unused areas
- Overheated/overcooled, overventilated and overlighted spaces
- Gas torches lighted when not in use
- Unused transformers, motor generators or rectifiers excited
- Furnaces and boilers idling unnecessarily long
- Steam coils heated when not in use
- Doors and windows of conditioned spaces left open

Many of the corrective actions that are suggested face potential opposition from the work force. The opposition can be overcome with patient explanation, estimates of potential annual savings, and offers to change back if the corrective action disrupts effective plant operation.

A 1.59-mm hole in a gas line at 413,700 Pa will waste 2.3 TJ of energy worth approximately $4000 in 1977's fuel market.

One 1.59-mm hole in a compressed air line operating at 689,000 Pa will waste 62,826 m^3 per year which requires 6060 kWh costing $151.50 at an average electric rate of $.025/kWh.

As much as 0.01 TJ per year costing over $24.50 can be wasted by 1 m^2 of missing insulation from a 20-cm steam main operating at 138,000 Pa.

Five 100-W fixtures more than required can waste more than 0.046 TJ per year and cost $109.50 in excess electrical billing when the rate is $0.025/kWh.

2. Combustion Control

The stoichiometric equation for the combustion of methane, the principal constituent of natural gas, with air is:

$$CH_4 + 2\,O_2 + 7.52\,N_2 \rightarrow CO_2 + 2\,H_2O + 7.52\,N_2 \qquad (6)$$

The stoichiometric equation is the one representing the exact amount of air necessary to oxidize the carbon and hydrogen in the fuel to carbon dioxide and water vapor. However, it is necessary to provide more than the stoichiometric amount of air since the mixing of fuel and air is imperfect in the real combustion chamber. Thus the combustion equation for hydrocarbon fuels becomes

$$\begin{aligned} &C_x H_y + \phi\,(x + y/4)\,O_2 + 3.76\,\phi\,(x + y/4)\,N_2 \\ &\rightarrow xCO_2 + y/2\,H_2O + (\phi - 1)\,(x + y/4)\,O_2 \\ &+ 3.76\,\phi\,(x + y/4)\,N_2 \end{aligned} \qquad (7)$$

Note that for a given fuel nothing in the equation changes except the parameter ϕ, the equivalence ratio, as the fuel-air ratio changes.

As ϕ is increased beyond the optimal value for good combustion, the stack losses increase and the heat available for the process decreases. As the equivalence ratio increases for a given flue temperature and a given fuel, more fuel must be consumed to supply a given amount of heat to the process.

The control problem for the furnace or boiler is to provide the minimum amount of air for good combustion over a wide range of firing conditions and a wide range of ambient temperatures. The most common combustion controller uses the ratio of the pressure drops across orifices, nozzles, or Venturis in the air and fuel lines. Since these meters measure volume flow, a change in temperature of combustion air with respect to fuel, or vice-versa, will affect the equivalence ratio of the burner. Furthermore, since the pressure drops across the flow meters are exponentially related to the volume flow rates, and control dampers must have very complicated actuator motions. All the problems of ratio controllers are eliminated if the air is controlled from an oxygen meter. These are now coming into more general use as reasonably priced, high-temperature oxygen sensors become available. It is possible to control to any set value of percent oxygen in the products, i.e.:

$$\%\,O_2 = \frac{\phi - 1}{\dfrac{x + y/2}{x + y/4} + 4.76\,\phi - 1} \qquad (8)$$

Figure 18 is a nomograph from the Bailey Meter Company[16] that gives estimates of the annual dollar savings resulting from the reduction of excess air to 15% for gas-, oil-, or coal-fired boilers with stack temperatures from 300 to 700°F. The fuel savings are predicted on the basis that as excess air is reduced, the resulting reduction in mass flow of combustion gases results in reduced gas velocity and thus a longer gas residence time in the boiler. The increased residence time increases the heat transfer from the gases to the water. The combined effect of lower exhaust gas flows and increased heat exchange effectiveness is estimated to be 1.5 times greater than that due to the reduced mass flow alone.

As an example assume the following data pertaining to an oil-fired boiler:

Burner capacity	63 GJ/hr (60×10^6 Btu/hr)
Annual operating hours	6200
Fuel cost	\$0.11/ℓ (\$0.42/gal)
Heating value fuel	42.36 GJ/ℓ (152,000 Btu/gal)
Percent O_2 in exhaust gases	6.2%
Stack temperature	327°C (620°F)

Entering the graph at the top abscissa with 6.2% O_2, we drop to the oil fuel line and then horizontally to the 327°C (620°F) flue gas temperature line. Continuing to the left ordinate we can see that 6.2% O_2 corresponds to 37.5% excess air. Dropping vertically from the intersection of the flue gas temperature line and the excess air line we note a 3.4% total fuel savings. Fuel costs are

$$10^9 \text{ J/GJ}(10^6 \text{ Btu/million Btu}) \times \frac{\$0.11/\ell}{42{,}365{,}000 \text{ J}/\ell}\left(\frac{\$0.42/\text{gal}}{150{,}000 \text{ Btu/gal}}\right)$$
$$= \$2.60/\text{GJ } (\$2.80/\text{million Btu})$$

Continuing the vertical line to intersect the \$2.60/million Btu and then moving to left ordinate shows a savings of \$72,000 per year for 8000-hr operation, 100×10^6 Btu/hr input and \$2.50/million Btu fuel cost. Adjusting that result for the assumed operating data:

$$\text{Annual savings} = \frac{6200}{8000} \times \frac{60 \times 10^6}{100 \times 10^6} \times \frac{2.80}{2.50} \times \$72{,}000$$
$$= \$37{,}498 \text{ per year}$$

This savings could be obtained by installing a modern oxygen controller, an investment with approximately 1-year payoff, or from heightened operator attention with frequent flue gas testing and manual adjustments. References 8 through 11 are valuable sources of information concerning fuel conservation in boilers and furnaces.

3. Waste-Heat Management

Waste heat as generally understood in industry is the energy rejected from any process to the atmospheric air or to a body of water. It may be transmitted by radiation, conduction, or convection, but often it is contained in gaseous or liquid streams emitted to the environment. Almost 50% of all fuel energy used in the U.S. is transferred as waste heat to the environment causing thermal pollution as well as chemical pollution of one sort or another. It has been estimated that one half of that total may be economically recoverable for useful heating functions.

What must be known about waste heat streams in order to decide whether they can become useful? The list appears below along with a parallel list of characteristics of the heat load which should be matched by the waste-heat supply.

Waste Heat Supply

Quantity
Quality
Temporal availability of supply
Characteristics of fluid

Heat Load

Quantity required
Quality required
Temporal availability of load
Special fluid requirements

Let us examine the particular case of a plant producing ice-cream cones. All energy quantities are given in terms of a 15.5°C reference temperature. Sources of waste heat include:

- Products of combustion from 120 natural gas jets used to heat the molds in the carrousel-type baking machines. The stack gases are collected under an insulated hood and released to the atmosphere through a short stack. Each of six machines emits 236.2 m^3/hr of stack gas at 160°C. Total source rate is 161,400 kJ/hr or 3874 MJ/day for a three-shift day.
- Cooling water from the jackets, intercoolers, and aftercoolers of two air compressors used to supply air to the pneumatic actuators of the cone machines; 11.36 ℓ/min of water at 48.9°C is available. This represents a source rate of 96 MJ/hr. The compressors run an average of 21 hr per production day. Thus this source rate is 2015 MJ/day.
- The water chillers used to refrigerate the cone batter make available — at 130°F — 264 MJ/hr of water heat. This source is available to heat water to 48.9°C using desuperheaters following the water chiller compressors. The source rate is 6330 MJ/day.
- 226 M^3/min of ventilating air is discharged to the atmosphere at 21.1°C. This is a source of rate of less than 22.2 MJ/hr or 525 MJ/day.

Uses for wate heat include:

- 681 ℓ/hr of hot water at 82.2°C is needed for cleanup operations during 3 hr of every shift or during 9 of every 24 hr. Total daily heat load is 4518 MJ.
- Heating degree-days total in excess of 3333 annually, thus any heat available at temperatures above 21.1°C can be utilized with the aid of run-around systems during the 5 ½-month heating system. Estimated heating load per year is 4010 GJ.

Total daily waste heat available — 12.74 GJ/day
Total annual waste heat available — 3.19 TJ/year
annual worth of waste heat (at $1.90/GJ for gas) — $6040.00
Total daily heat load — this varies from a maximum of 59.45 GJ/day at the height of the heating season to the hot-water load of 4.52 GJ/day in the summer months.

Although the amount of waste heat from the water chillers is 40% greater than the load needed for hot-water heating, the quality is insufficient to allow its full use, since the hot water must be heated to 82°C and the compressor discharge is at a temperature of 54°C.

However the chiller waste heat can be used to preheat the hot water. Assuming 13°C supply water and a 10° heat exchange temperature approach, the load that can be supplied by the chiller is

$$\frac{49 - 13}{82 - 13} \times 4.52 = 2.36 \text{ GJ/day}$$

Since the cone machines have an exhaust gas discharge of 3.87 GJ/day at 160°C, the remainder of the hot-water heating load of 2.17 GJ/day is available. Thus a total saving of 1129 GJ/year in fuel is possible with a cost saving of $2381 annually based on $2.11/GJ gas. The investment costs will involve the construction of a common exhaust heater for the cone machines, a desuperheater for each of the three water chiller compressors, a gas-to-liquid heat exchanger following the cone-machine exhaust heater, and possibly an induced draft fan following the heat exchanger, since the drop in exhaust gas temperature will decrease the natural draft due to hot-gas buoyancy.

It is necessary to almost match four of four characteristics. Not exactly, of course, but the closer the match the more economic the waste-heat recovery. The term quality used in the list really means thermodynamic availability of the waste heat. Unless the energy of the waste stream is sufficiently hot, it will be impossible to even transfer it to the heat load, since spontaneous heat transfer occurs only from higher to lower temperature.

The quantity and quality of energy available from a waste heat source or for a heat load are studied with the aid of a heat balance. Figure 19 shows the heat balance for a steam boiler. The rates of enthalpy entering or leaving the system in fluid streams must balance with the radiation loss rate from the boilers external surfaces. Writing the first-law equation for a steady flow-steady state process

$$\dot{q}_L = \dot{m}_f h_f + \dot{m}_a h_a + \dot{m}_c h_c - \dot{m}_s h_s - \dot{m}_g h_g \tag{9}$$

and referring to the heat-balance diagram, one sees that the enthalpy flux $\dot{m}_g h_g$, leaving the boiler in the exhaust gas stream, is a possible source of waste heat. A fraction of that energy can be transferred in a heat exchanger to the combustion air, thus increasing the enthalpy flux $\dot{m}_g h_a$ and reducing the amount of fuel required. The fraction of fuel that can be saved is given in the equation

$$\frac{\dot{m}_f - \dot{m}_f'}{\dot{m}_f} = 1 - \left[\frac{K_1 - (1 + \phi)\,\overline{C}_p\,T_g}{K_1 - (1 + \phi')\,\overline{C}_p'\,T_g'}\right] \tag{10}$$

where the primed values are those obtaining with waste-heat recovery. K_1 represents the specific enthalpy of the fuel-air mixture, $h_f + \phi\, h_a$, which is presumed to be the same with or without waste-heat recovery. ϕ is the molar ratio of air to fuel, and $\overline{C}_p$ is the specific heat averaged over the exhaust gas components. Figure 20, which is derived from Equation 10, gives possible fuel savings from using high temperature flue gas to heat the combustion air in industrial furnaces.

It should be pointed out that the use of recovered waste heat to preheat combustion air, boiler feed water, and product to be heat treated, confers special benefits not necessarily accruing when the heat is recovered to be used in another system. The preheating operation results in less fuel being consumed in the furnace, and the corresponding smaller air consumption means even smaller waste heat being ejected from the stacks.

Table 8 shows heat balances for a boiler with no flue gas-heat recovery, with a feedwater preheater (economizer) installed and with an air preheater, respectively.

It is seen that air preheater alone saves 6% of the fuel and the economizer saves 9.2%. Since the economizer is cheaper to install than the air preheater, the choice is easy to make for an industrial boiler. For a utility boiler, both units are invariably used in series in the exit gas stream.

Figure 21 is an example of a heat balance for a high temperature industrial furnace. A comparison with the boiler diagrams reveals an extremely large difference in efficiencies. The tube furnace extracts only $39.47 \times 10^9/43.88 \times 10^{10}$ or 9% efficiency while the simple boiler shows an efficiency of 84%.

Table 9 is an economic study using 1975 prices for fuel, labor, and equipment. At that time it was estimated that a radiation reruperator fitted to a fiberglass furnace would cost $162,475 and effect a savings of $129,909/year making for a payoff period of approximately 1¼ years. This assumes of course that the original burners, combustion control system, etc., could be used without modification. It is unlikely that this would be the case for such a high-temperature system. However, one could estimate with a good deal of confidence that the cost of additional equipment would not exceed the cost of the recuperator itself. This would mean a maximum payoff period of 18 months, which still offers a very interesting investment opportunity.

At this point, we can take some time to relate waste-heat recovery to the combustion process itself. We can first state categorically that the use of preheated combustion air improves combustion conditions and efficiency at all loads, and in a newly designed installation permits a reduction in size of the boiler or furnace. It is true that the increased mixture temperature that accompanies air preheat results in some narrowing of the mixture stability limits, but in all practical furnaces this is of small importance.

In many cases low-temperature burners may be used for preheated air, particularly if the air preheat temperature is not excessive and if the fuel is gaseous. However the higher volume flow of preheated air in the air piping may cause large enough increases in pressure drop to require a larger combustion air fan. Of course larger diameter air piping can prevent the increased pressure drop, since air preheating results in reduced quantities of fuel and combustion air. For high preheat temperatures alloy piping, high-temperature insulation and water-cooled burners may be required. Since many automatic combustion-control systems sense volume flows of air and/or fuel, the correct control settings will change when preheated air is used. Furthermore, if air preheat temperature varies with furnace load, then the control system must compensate for this variation with an auxiliary temperature-sensing control. On the other hand, if control is based on the oxygen content of the flue gases, the control complications arising from gas volume variation with temperature is obviated. This is the preferred control system for all furnaces, and only cost prevents its wide use in small installations. Burner operation and maintenance for gas burners is not affected by preheat, but oil burners may experience accelerated fuel coking and resulting plugging from the additional heat being introduced into the liquid fuel from the preheated air. Careful burner design, which may call for water cooling or for shielding the fuel tip from furnace radiation, will always solve the problems. Coal-fired furnaces may use preheated air up to temperatures that endanger the grates or burners. Again, any problems can be solved by special designs involving water cooling if higher air temperatures can be obtained and/or desired.

TABLE 8

Heat Balances for a Steam Generator

Case	Input streams				Output streams			
	Name	Temperature (°C)	Flow rate	Energy	Name	Temperature (°C)	Flow rate	Energy (GJ/hr)
Without economizer or air preheater	Natural gas	26.7	3611 m³/hr	134.78 GJ/hr	Steam	185.6	45,359 kg/hr	126.23
	Air	26.7	37,574 m³/hr	560.32 MJ/hr	Flue gas	372.2	41,185 m³/hr	19.49
	Make-up water	10.0	7076 kg/hr	297.25 MJ/hr	Surface losses	—	—	1.21
	Condensate return	82.2	41,277 kg/hr	14.21 GJ/hr	Blow down	185.6	2994 kg/hr	2.36
With air preheater	Natural gas	26.7	3395 m³/hr	126.72 GJ/hr	Steam	185.6	45,359 kg/hr	126.23
	Air	232.2	32,114 m³/hr	478.90 GJ/hr	Flue gas	260.0	35,509 m³/hr	11.42
	Make-up water	10	7076 kg/hr	297.25 MJ/hr	Surface losses	—	—	1.21
	Condensate return	82.2	41,277 kg/hr	14.21 GJ/hr	Blow down	185.6	2994 kg/hr	2.36
With economizer	Natural gas	26.7	3278 m³/hr	122.37 GJ/hr	Steam	185.6	45,359 kg/hr	126.23
	Air	26.7	31,013 m³/hr	462.48 MJ/hr	Flue gas	176.7	34,281 m³/hr	7.08
	Make-up water	101	7076 kg/hr	297.25 GJ/hr	Surface loss	—	—	1.21
	Condensate return	82.2	41,277 kg/hr	12.21 GJ/hr	Blow down	185.6	2994 kg/hr	2.36

TABLE 9

Cost — Fuel Savings Analysis of a Fiberglass Furnace Recuperator (Cost as per 1975)

Operation	Continuous
Fuel input	19.42 GJ/hr
Fuel	No. 3 fuel oil
Furnace temperature	1482°C
Flue gas temperature entering recuperator	1204°C
Air preheat (at burner)	552°C
Fuel savings = 37.4%	
Q = 7.26 GJ/hr or 173.6 ℓ/hr of oil	
Fuel cost savings estimation	
per GJ = $2.24	
per hour = $16.24	
per year (8000 hr) = $129,909	
Cost of recuperator	$ 83,475
Cost of installation, related to recuperator	$ 79,000
Total cost of recuperator installation	$162,475
Approximate payback time 1¼ years.	

The economics of waste-heat recovery today range from poor to excellent depending upon the technical aspects of the application as detailed above. But the general statement can be made that, at least for most small industrial boilers and furnaces, standard designs and/or off-the-shelf heat exchangers prove to be the most economic. For large systems one can often afford to pay for special designs, construction, and installations. Furthermore, the applications are often technically constrained by material properties and space limitations, and as shall be seen later, always by economic considerations.

Figures 22 and 23 show, respectively, the temperature profiles along the length of single pass parallel and counterflow heat exchangers. Note that the parallel flow exchanger, since it involves such a high initial temperature difference to drive heat transfer, has the potential for being the cheapest unit, but it limits the highest temperature of the heated fluid leaving to the lowest temperature of the hot fluid leaving. On the other hand, the counterflow exchanger permits the temperature of the heated fluid leaving to equal or exceed the temperature of the hot fluid leaving possibly at the expense of some additional transfer area, since the mean temperature difference across the exchanger may be smaller than in the case of parallel flow exchanger.

The heat transfer rate, q, in the exchanger for a recuperator or air preheater, is

$$\begin{aligned} q &= \dot{m}_g \, \bar{C}_{p,avg,g} \, (T_g - T'_g) \\ &= \dot{m}_a \, C_{p,avg,a} \, (T'_a - T_a) \end{aligned} \tag{11}$$

where $C_{p,avg,g}$ = specific heat of flue gases averaged over all components and over the temperature range $(T_g - T'_g)$; $C_{p,avg,a}$ = specific heat of combustion air averaged over the temperature range $(T'_a - T_a)$; T_g = temperature of flue gas entering the heat exchanger; T'_g = temperature of flue gas leaving the heat exchanger; T_a = temperature of combustion air entering the heat exchanger; T'_a = temperature of combustion air leaving the heat exchanger.

$$\epsilon = \frac{\bar{C}_{p,avg,g} (T_g - T'_g)}{C_{min} (T_g - T_a)} \qquad \epsilon = \frac{C_{p,avg,a} (T'_a - T_a)}{C_{min} (T_g - T_a)} \tag{12}$$

where C_{min} is the smaller of the quantities $(\dot{m}_a \overline{C_{p,avg,a}})$ or $(\dot{m}_g \overline{C_{p,avg,g}})$.

The heat exchanged can now be expressed as

$$q_x = \epsilon \, C_{min} \, (T_g - T_a)$$

The relationships between the effectiveness of the heat exchanger and the heat transfer area are seen in Figure 24 taken from Kays and London[12] where effectiveness in percentage is plotted against the NTU parameter. The NTU or Number of Transfer Units is defined by NTU = AU/C_{min} where A = area of heat exchange surfaces and U = overall average thermal conductance including both hot and cold boundary layers. Overall thermal conductance is controlled by the fluid temperatures and the fluid velocities (and thus is related to fluid stream pressure drops). See Volume I, Chapter 2 for more details. The cost of the exchanger is almost directly proportional to the heat exchange area A, while the cost of operation is a complex function of the conductance.

Note from the graphs that in all cases the effectiveness curves knee over and quickly became nearly asymptotic as the number of heat transfer units exceed 1.0 or 2.0. It is quite obvious that the quest for high effectiveness runs into the law of diminishing returns rather quickly. This explains the previous statement to the effect that special heat exchanger designs are rarely economic in smaller sizes. For larger projects marginal analysis should be used. This analysis determines an investment which will maximize net savings.

4. Heating, Ventilating, and Air Conditioning

Heating, ventilating, and air conditioning (HVAC), while not usually important in the energy intensive industries, may be responsible for the major share of energy consumption in the light manufacturing field, particularly in high technology companies and those engaged primarily in assembly.

Because of air pollution from industrial processes many HVAC systems require 100% outside ventilating air. Furthermore, ventilating air requirements[13] are often much in excess of those in residential and commercial practice. An approximate method for calculating the total heat required for ventilating air in Btu-per-heating season is given by

$$E_v\,(\mathrm{kJ}) = 60 \times 24 \left(\frac{\mathrm{min}}{\mathrm{day}}\right) \times (1.2 \times 0.519) \left(\frac{\mathrm{kJ}}{\mathrm{m^3 - K^\circ}}\right) \times \mathrm{SCMM} \times$$

$$\mathrm{DD} = 896.8 \times \mathrm{SCMM} \times \mathrm{DD} \qquad (13)$$

where SCMM = standard cubic meter per minute of total air entering plant including unwanted infiltration; DD = heating degree (C) days.

This underestimates the energy requirement, because degree-days are based on 18.33°C reference temperature and indoor temperatures are ordinarily held 1.6 to 3.9° higher. For a location with 3333 degree-days each year the heating energy given by Equation 13 is about 17% low.

Savings can be effected by reducing the ventilating air rate to the actual rate necessary for health and safety and by ducting outside air into direct-fired heating equipment such as furnaces, boilers, ovens, and dryers. Air infiltration should be prevented through a program of building maintenance to replace broken windows, doors, roofs, and siding and by campaigns to prevent unnecessary opening of windows and doors.

Additional roof insulation is often economic, particularly because thermal stratification makes roof temperatures much higher than average wall temperatures. Properly installed vertical air circulators can prevent the vertical stratification and save heat loss through the roof. Windows can be double glazed, storm windows can be installed, or windows can be covered with insulation. Although the benefits of natural lighting are eliminated by this measure, it can be very effective in reducing infiltration and heat transfer losses.

Waste heat from ventilating air itself, from boiler and furnace exhaust stacks, and from air-conditioning refrigeration compressors can be recovered and used to preheat make-up air. Consideration should also be given to providing spot ventilation in hazardous locations instead of increasing general ventilation air requirements.

As an example of the savings possible in ventilation air control, a plant requiring 424.5 cmm outside air flow is selected. A gas-fired boiler with an energy input of 0.0165 TJ is used for heating and is supplied with room air.

$$\text{Combustion air} = \frac{16.5 \times 10^6}{37{,}281} \times \frac{12}{60}$$

$$= 88.52 \text{ CMM}$$

for a fuel with 37,281 kJ/m^3 heating value and an air fuel ratio of 12 m^3 air per m^3 fuel. The annual degree-days was 3175.

A study showed that the actual air supplied through infiltration and air handlers was 809 m^3/min. An outside air duct was installed to supply combustion air for the boiler and the actual ventilating air supply was reduced to the required 424 m^3/min. The fuel saving that resulted was

$$896.8\ (809 - 424)\ 3175 = 1096.2 \text{ GJ}$$

worth \$2083 in fuel at \$1.90/GJ for natural gas.

5. Changes in Unit Processes

The particular process used for the production of any item affects not only the cost of production but also the quality of the product. Since the quality of the product is critical in customer acceptance and therefore in sales, the unit process itself cannot be considered a prime target for the energy conservation program. That does not say that one should ignore the serendipitous discovery of a better and cheaper way of producing something. Indeed, one should take instant advantage of such a situation, but that clearly is the kind of decision that management could make without considering energy conservation at all. Some new basic industrial processes are nearing readiness for plant use. The aluminum-chloride process for basic aluminum production is an example. It is true that if this process comes into use, that fuel costs for producing aluminum will decrease substantially. However, the decision to changeover or to retain the old electrolytic process will not depend in any way on the existence of formal energy conservation programs in the industry.

6. Optimizing Process Scheduling

Industrial thermal processing equipment tends to be quite massive compared to the product treated. Therefore, the heat required to bring the equipment to steady-state production conditions may be large enough to make start-ups fuel intensive. This calls for scheduling this equipment so that it is in use for as long periods as can be practically scheduled. It also may call for idling the equipment (feeding fuel to keep the temperature close to production temperature) when it is temporarily out of use. The fuel rate for idling may be between 10 and 40% of the full production rate for direct-fired equipment. Furthermore, the stack losses tend to increase as a percentage of fuel energy released. It is clear that over-frequent start-ups and long idling times are wasteful of energy and add to production costs. The hazards of eliminating some of that waste through rescheduling must not be taken lightly. For instance a holdup in an intermediate heating process can slow up all subsequent operations and make for inefficiency down the line. The breakdown of a unit that has a very large production backlog is much more serious than in one having a smaller backlog. Scheduling processes in a complex product line is a very difficult excercise and perhaps better suited to a computerized PERT program than to an energy conservation engineer. That does not mean that the possibilities for saving through better process scheduling should be ignored. It is only a warning to move slowly and take pains to find the difficulties that can arise thereby.

A manufacturer of precision instruments changed the specifications for the finishes of over half of his products, thereby eliminating the baking required for the enamel that had been used. He also rescheduled the baking process for the remaining products so that the oven was lighted only twice a week instead of every production day. A study is now proceeding to determine if electric infrared baking will not be more economic than using the gas-fired oven.

B. Cogeneration of Process Steam and Electricity

In-plant (or on-site) electrical energy cogeneration is nothing new. It has been used in industries with large process steam loads for many years, both in the U.S. and Europe. It consists of producing steam at a higher pressure than required for process use; expanding the high-pressure steam through a back-pressure turbine to generate electrical energy; and then using the exhaust steam as process steam. Alternatively, the power system may be a diesel engine which drives an electrical generator. The diesel engine exhaust is then stripped of its heat content as it flows through a waste-heat boiler where steam is generated for plant processes. A third possibility is operation of a gas turbine-generator to supply electric power and hot exhaust gases which produce process steam in a waste-heat boiler. As shall be seen later, the ratio of electric power to steam-heat rate varies markedly from one of these type systems to the next. In medium to large industrial plants the cogeneration of electric power and process steam is economically feasible provided certain plant energy characteristics are present. In small plants or in larger plants with small process steam loads, cogeneration is not economic because of the large capital expenditure involved. Under few circumstances is the in-plant generation of electric power economic without a large process steam requirement. A small industrial electric plant cannot compete with an electric utility unless the generation required in-plant exceeds the capacity of the utility. In remote areas where no electric utility exists, or where its reliability is inferior to that of the on-site plant, the exception can be made.

Cogeneration if applied correctly is not only cost effective, it is fuel conserving. That is, the fuel for the on-site plant is less than that used jointly by the utility to supply the plant's electric energy and that used on-site to supply process steam. Figures 25 and 26 illustrate the reasons for and the magnitude of the savings possible. However, several conditions must be met in order that an effective application be possible. First, the ratio of process steam heat rate to electric power must fall close to these given in the table below:

Heat engine type	E_{steam}/E_{elect}
Steam turbine	2.3
Gas turbine	4.0
Diesel engine	1.5

The table is based upon overall electric plant efficiencies to 30, 20, and 40% respectively, for steam turbine, gas turbine, and diesel engine. Secondly, it is required that the availability of the steam load coincide closely with the availability of the electric load. If these temporal availabilities are out of phase, heat-storage systems will be necessary, and the economy of the system destroyed. Thirdly, it is necessary to have local electric utility support. Unless backup service is available from your utility, the cost of building in redundancy is too great. This may be the crucial factor in the majority of cases. Public utilities understandably do not like to lose base-load, although they are delighted to lose opportunities for supplying peak loads. If they decide to refuse interconnection in order to discourage the idea of on-site cogeneration, they have effectively killed that option for the plant owner; in most states the utility management may have that legal prerogative. It will not be exercised, however, if the utility management can be persuaded that it is to their present or future advantage to encourage on-site cogeneration. Recent studies have shown[1,14] that the potential for cogeneration is as great as 50% of the industrial steam load and that by 1985 as much as 135,000 MW of electrical power may be produced by cogeneration, which is 40% of 1977's utility capacity. If that prediction is correct, then obviously U.S. industry and the national electric power grid are soon to be joined in a tremendous cooperative effort to share the U.S. electric power load. The benefit to industry will be the reduction of fuel costs that cogeneration brings, while the private utilities will benefit from the reduced obligation to build high-cost fixed-station plants. Potential weaknesses in the U.S. capital market will promote utility cooperation, while escalating electric costs relative to fuel costs will provide incentives for industrial retrofit of on-site cogeneration. It is suspected that in the case of a very large on-site cogeneration plants — larger than 50 MW capacity — private utilities may be interested in building and operating the plants much as the air-reduction companies now build and operate oxygen plants in U.S. steel operations. This is scarcely a possibility for the average-sized cogeneration facility however.

In order to bring about an understanding of the advantages of cogeneration, one typical industrial plant will be selected and the economics of retrofitting a cogeneration electric plant will be examined. Figure 25 is a schematic of a steam plant as it existed before cogeneration was installed.

In order to save fuel, the decision was made to install steam turbine-generators in the plant in order to produce as much on-site electrical energy as was consistent with the process steam load. The decision then had to be made as to whether the present boiler, which had operated at 150 psia would be suitable, or if a new high-pressure boiler would be required. The insurance carrier conducted tests which indicated that the present boiler could be operated safely at 350 psia. Two turbine-generators were then installed as shown in Figure 26. Bypass steam lines were also provided so that process steam would be available in the event of turbine shutdown. The electrical capacity of the new plant was capable of providing 56% of the annual electrical energy normally purchased and could handle only 36% of the short-term peak-power requirement. The utility interconnection was not modified so that purchased power would be available in case both turbine-generators went out simultaneously. Although this latter event seemed improbable, the plant management made the decision that full backup should be available, although substantial increases in average rates for purchased power resulted. The higher charges occurred because the utility collected a minimum monthly demand charge based on the distribution-line capacity. Table 10 gives the data on annual energy consumption and costs for the plant operating with on-site

TABLE 10

Annual Energy Use and Energy Costs Before and After Installation of Generator

	Before installation	After Installatin
	Annual Energy Use	
Purchased electricity	39.945×10^6 kWh	17.576×10^6 kWh
Maximum demand	7397 kW	7397 kW
Electricity generated	—	22.37×10^6 kWh
Steam produced	347.8×10^6 kg	348.3×10^6 kg
Boiler input	985.34 TJ	1154.82 TJ
Tons coal burned for co-generation	33,620 t	39,403 t
	Energy Costs	
Electricity		
Demand charges	$165,012	$120,720
Energy charges	454,368	199,920
Fuel adjustment charges	291,600	128,304
Total electric	$910,980	$448,944
Cost of coal	896,563	1,050,777
Total	$1,807,543	$1,499,721
Total fuel consumption[a]	1.396×10^{15} J	1.336×10^{15} J
Net annual savings = $307,822		

[a] Includes fuel burned by utility power plant at a heat rate of 10,443 Btu/kWh (11018 MJ/kWh).

cogeneration of process steam and electricity compared to the same data prior to the new installation.

The investment costs are calculated below:

Cost of superheater and installation	$176,000
Cost of turbines, generators, gear reducers, and controls	141,540
Cost of piping values and installation	76,400
Cost of engineering and supervision	31,515
	$425,455

Assuming a 15-year plant life, 15% interest costs, $50,000 salvage value, and using the net present worth method (the multipliers are the appropriate present-worth factors, see Volume I, Chapter 3).

Engineering (15% − 1 year)

$$P = -\$31{,}515 \times 1.15 = -\$36{,}242$$

Purchase and installation equipment

$$P = -\$393{,}940$$

Net increase in annual operating and maintenance costs ($23,150 per year)

$$P = -23{,}150 \times 5.847 = -\$135{,}358$$

Salvage values after 15 years

$$P = 50{,}000 \times 0.1229 = \$6145$$

Annual savings

$$P = \$307{,}822 \times 5.847 = \$1{,}799{,}835$$

Net present worth = $1,240,440 which represents a return of $82,696 per year over 15 years on an investment of $425,455, or 19% per year.

It should be recalled that the worth of an investment in cogeneration depends on the favorable ratio of electrical energy rates to fuel costs. Had natural gas or oil been used as a boiler fuel, the project would not have been such an attractive investment. However a relevant factor in the economics of the project developed above was deliberately overlooked. If one refers to the fuel-price projections derived from FEA's PIES program it is noted that electrical prices are estimated to increase at a faster annual rate than coal by a factor of almost two. Gas prices are projected as escalating even faster than electricity. Thus the economics of the coal-fired cogeneration plant are even better than indicated above.

C. Computer-Controlled Energy Management

Minicomputers are available or will soon become available for accurately controlling boilers, furnaces, and unit industrial processes. The largest application thus far has been in the control of mechanical and electrical systems in commercial buildings. These have been used to control lighting, electric demand, ventilating fans, thermostat setbacks, air-conditioning systems, and the like. These same computer systems can also be used in industrial buildings with or without modification for process control. Several manufacturers of the computer systems will not only engineer and install the system but will maintain and operate it from a remote location. Such operations are ordinarily regional. For large systems in large plants one may be able to have the same service provided in-plant.

After installing a demand limiter on an electric-arc foundry cupola, the manager found that he was able to reduce the power level from 7100 kW to 4900 kW with negligible effect on the production time and no effect on product quality. The savings in demand charges alone were $4400 per month with an additional savings in energy costs.

D. Commercial Options in Waste-Heat Recovery Equipment

The equipment that is useful in recovering waste heat can be categorized as: heat exchangers, heat storage systems, combination heat storage-heat exchanger systems, and heat pumps.

Heat exchangers certainly constitute the largest sales volume in this group. They consist of two enclosed flow paths and a separating surface which prevents mixing, supports any pressure difference between the fluids of the two fluids, and provides the means through which heat is transferred from the hotter to the cooler fluid. These are ordinarily operated at steady state-steady flow condition. The fluids may be gases, liquids, condensing vapors, or evaporating liquids, and occasionally fluidized solids. A great deal of detail concerning commercial heat-recovery devices is found in the NBS Handbook No. 121.[15] Only brief descriptions are given here.

Radiation recuperators are high-temperature combustion-air preheaters used for transferring heat from furnace exhaust gases to combustion air. As seen in Figure 27, they consist of two concentric cylinders, the inner one acting as a stack for the furnace and the concentric space between the inner and outer cylinders the path for the combustion air which ordinarily moves upward and therefore parallel to the flow of the exhaust gases. With special construction materials it can handle 1355°C furnace gases and save as much as 30% of the fuel otherwise required. The main problem in their use is damage due to overheating for reduced air flow or temperature excusions in the exhaust gas flow.

Convective air preheaters are tubular or corrugated metal devices which are used to preheat combustion air in the moderate temperature range (121°C to 649°C) for ovens, furnaces, boilers, and gas turbines or preheat ventilating air from sources as low in temperature as 21°C. Figures 28 and 29 illustrate typical construction. These are often available in modular design so that almost any capacity and any degree of effectiveness can be obtained by multiple arrangements. The biggest problem is keeping them clean.

Economizer is the name traditionally used to describe the gas-to-liquid heat exchanger used to preheat the feedwater in boilers from waste heat in the exhaust gas stream. These often take the form of loops, spirals, or parallel arrays of finned tubing through which the feedwater flows and over which the exhaust gases pass. They are available in modular form to be introduced into the exhaust stack or into the breeching. They can also be used in reverse to heat air or other gases with waste heat from liquid streams.

A shell and tube heat exchanger is illustrated in Figure 30. These are ordinarily used as liquid-to-liquid heat exchangers where the high-pressure liquid flows in the tubes and the low-pressure liquid in the shell over the tubes in one or more baffled passes. Almost any combination of materials can be used to solve special problems of pressure, temperature, and chemical corrosion.

Heat pipe arrays are often used for air-to-air heat exchangers because of their compact size. Heat-transfer rates per unit area are quite high. A disadvantage is that a given heat pipe (that is, a given internal working substance) has a limited temperature range for efficient operation. The heat pipe as shown in the diagram of its internal structure, Figure 31, transfers heat from the hot end by evaporative heating and at the cold end by condensing the vapor. Figure 32 is a sketch of an air preheater using an array of heat pipes.

Waste-heat boilers are water-tube boilers, usually prefabricated in all but the largest sizes, used to produce saturated steam from high-temperature waste heat in gas streams. The boiler tubes are often finned to keep the dimensions of the boiler smaller. They are often used to strip waste heat from diesel-engine exhausts, gas-turbine exhausts, and pollution-control incinerators or afterburners. Figure 33 is a diagram of the internals of a typical waste-heat boiler.

Heat storage systems, or regenerators, once very popular for high-temperature applications, have been largely replaced by radiation recuperators because of the relative simplicity of the latter. Regenerators consist of twin flues filled with open ceramic checkerwork. The high-temperature exhaust of a furnace flowing through one leg of a swing valve to one of the flues heated the checkerwork while the combustion air for the furnace flowed through the second flue in order to preheat it. When the temperatures of the two masses of checkerwork were at proper levels, the swing valve was thrown and the procedure was continued, but with reversed flow in both flues. Regenerators are still used in some glass- and metal-melt furnaces where they are able to operate in the temperature range 1093 to 1649°C. It should be noted that the original application of the regenerators was to achieve the high-melt temperatures required with low heating value fuel.

A number of ceramic materials in a range of sizes and geometric forms are available for incorporation into heat-storage units. These can be used to store waste heat in order to remedy time discrepancies between source and load. A good example is the storage of solar energy in a rock pile so that it becomes available for use at night and on cloudy days. Heat storage, other than for regenerators in high temperature applications, has not been used a great deal for waste-heat recovery as of this writing but will probably become more popular as more experience with it accumulates.

Combination heat-storage unit-heat exchangers called heat wheels are available for waste-heat recovery in the temperature range 0 to 982°C. The heat wheel is a porous flat cylinder which rotates within a pair of parallel ducts, as can be observed in Figure 34. As the hot gases flow through the matrix of the wheel they heat one side of it, which then gives up that heat to the cold gases as it passes through the second duct. Heat-recovery efficiencies range to 80%. In low and moderate temperature ranges the wheel is composed of an aluminum or stainless steel frame filled with a wire matrix of the same material. In the high-temperature range the material is ceramic. In order to prevent cross-contamination of the fluid streams, a purge section, cleared by fresh air, can be provided. If the matrix of the wheel is covered with a hygroscopic material, latent heat as well as sensible heat can be recovered. Problems encountered with heat wheels include freeze damage in winter, seal wear, and bearing maintenance in high temperature applications.

The heat pump is a device operating on a refrigeration cycle which is used to transfer energy from a low-temperature source to a higher temperature load. It has been highly developed as a domestic heating plant using energy from the air or from a well but has not been used a great deal for industrial applications. The COP (or the ratio of heat delivered to work input) for an ideal Carnot refrigeration cycle equals $T_H/(T_H - T_L)$, where T_H is the load temperature and T_L is the source temperature. It is obvious that when the temperature difference $T_H - T_L$ becomes of the order of T_H, the heat could be derived almost as cheaply from electric resistance heating. However for efficient refrigeration machines and a small temperature potential to overcome, the actual COP is favorable to moderate-cost waste energy. The heat pump can be used to transfer waste heat from and to any combination of liquid, gas, and vapor. See Volume II, Chapter 9 for a detailed discussion of heat pumps.

IV. SAMPLE REPORT ON AN ENERGY STUDY IN A FOUNDRY

The following report is synthesized from several actual energy studies made in gray iron foundries. The characteristics of any one of those locations are not recognizable from the description of the fictitious Iron Brothers Foundry. However, the energy consumption and fuel costs are not unrepresentative for the location selected in the years 1976 and 1977. Likewise, the investment costs proposed for the waste-heat recovery systems are well within the wide range possible for those systems in that same time frame and region of the country. However, no actual quotations from vendors have been used in the following report.

Iron Brothers Foundry located at 145th Street and West Newland in Cleveland, Ohio is a foundry wholly owned by Smith and York Turbine Company. It employs 320 production workers on three shifts. In addition there are 60 management and office workers. Production in 1976 was 13,600 metric tons of castings, down 27% from 1975. Production in 1977 is estimated to be up 30% over 1976. In 1976 total energy consumption was 298,000 GJ at a cost of $837,520 or an average unit cost of $2.81 per GJ. The plant is housed in three adjacent buildings totaling 156,000 ft^2.

Annual sales are in the vicinity of $10 million. Annual degree-days (C) in 1976 were slightly over 3889 up 11% over 1975.

Figure 35 gives us the production flow chart in block diagram form. Iron and steel scrap and carbon are the raw materials. The gas-fired cupola produces iron during one shift only. This is held in an electric furnace until the molds are ready for pouring in the second and third shifts. Sand is used to form the molds, many of which are dried and thermally conditioned in gas-fired ovens. After cooling, the castings are ground and machined, heat treated, assembled if necessary, painted if necessary, and packed for shipping. An electric motor-driven air compressor supplies air for packing and compressing molds, shaking out castings, and for driving pneumatic hand tools and mold-preparation machines. Electricity is used to drive fans and blowers for combustion air, to supply heat for the holding furnace, for miscellaneous drive motors, and for lighting. Table 11 shows the estimated electrical consumption division among the major user systems.

Space and hot-water heating is accomplished with natural gas. An average of 23.06 GJ is used per ton of product with 12% derived from coke, 75% from natural gas, and 13% from electricity.

Figures 36 and 37 show, respectively, the monthly consumption and monthly cost of natural gas and electric energy. The consumption of gas, although obviously seasonal, shows a drop in 1976 over that used in 1975, which is consistent with the lower production in 1976. Figure 38 shows clearly the relationship between the gas consumption and the heating degree-days. The use of electric energy is not seasonally dependent and is not clearly production dependent. In fact electric consumption increased 6% in the first 10 months of 1976 over the same period of 1975, while production dropped some 27%. Figure 39 is a graph of monthly production and monthly electric billing demand. It shows that in December of 1975 electric demand rose over 500 kW above the average of 1975 and continued at that level over the greater part of the year. It is assumed that this increase in both electric billing demand and in the electric energy consumption previously mentioned was due to the installation of air-pollution control systems. Figure 40 graphs total monthly energy consumption and total monthly energy costs as well as unit energy cost by month. Total consumption on the average is dropping while the cost is rising, giving a definite upward trend to unit energy costs. These are seen to increase from a low of $2.18 per GJ in January 1975 to $3.79 per GJ in September of 1976. Figure 41 is an attempt to plot energy efficiency and energy-cost efficiency on a monthly basis. The graphs indicate a strong seasonal influence on gas consumption. There is as well a strong inverse relationship between total energy efficiency and production that can be seen by comparing the curves of Figure 41 with the monthly production curves of Figure 39. The introduction of the Kleen-Air System for cupola emissions control raised the energy requirements and increased the energy costs per unit of product. During the latter part of 1976 the average unit cost for gas and electric energy was $3.80 per GJ. This results in a cost for utilities of $77.08 per metric ton. At $154.00 per metric ton for metallurgical coke, the total energy cost per metric ton of iron is $90.78.

V. ENERGY CONSERVATION OPPORTUNITY

The following paragraphs comprise a small list of measures which may provide significant energy savings and at the same time promise a suitable return on the required investment. It must be emphasized, however, that more detailed engineering and economic studies should be undertaken to provide a firmer basis for making investment decisions.

TABLE 11

Distribution of Electrical Energy Over Major Systems

Type system	Total capacity hp or kW	% Total load
Collector and combustion air fans	1025 hp	35.5
Air Compressors	450 hp	33.2
Electric furnace	545 kW	20.4
Mold shakers	225 hp	5.0
Lighting	65 kW	4.0
Miscellaneous		1.9
Total		100.0

On the block diagram, Figure 35, we look for the energy intensive parts of the operation. Fifty-seven percent of the natural gas is consumed in the eight heat-treat furnaces, 21% in the boiler used exclusively for space heating, 11.7% in the cupola, and 8.6% in the mold-baking ovens. Less than 2.0% is used for other purposes throughout the plant. On the basis that the biggest energy users have the biggest potential for energy conservation, we look at the operation of the heat-treating furnaces first.

A. Recuperation of a Heat-Treat Furnace

A battery of eight front-loading bottom-fired furnaces operate on staggered 5 and ½-hr treatment cycles, four cycles per day. New ratio controllers have recently been installed to replace the original controls. As can be detected from the data in Table 12 the air-fuel ratio is maintained in an inverse ratio to the firing rate so that the volume of combustion products are more or less constant with varying temperature. This is necessary in order to retain the proper flow path and velocities of the hot gases in the furnace so that the castings are heated uniformly at each control temperature.

Figure 42 is a profile of the furnace temperatures over a typical 5 and ½-hr cycle. After the castings are loaded into the furnace on steel pallets by a forklift truck, the furnace goes on high fire for approximately 45 min to reach 893°C, which is then maintained constant for 1 and ¾ hr. At the end of the austinitizing process, the castings are quenched in a water pool and returned to the furnaces for drawing at 693°C for 2 and ½ hr under a low fire (25 to 30%) condition. During periods of production holdup the furnace is maintained at a stand-by temperature of 621°C. Each of the eight identical furnaces has five burners with a total rating of 5.3 GJ/hr. The average firing rate as determined from the firing chart in Figure 42 is 47%. The availability factor for the furnaces is 87.5%. Production hours run 8000 hr annually. Figure 43 is an annual heat-balance diagram of the eight furnaces. It shows an enthalpy flux of 60.53 TJ per year for the exhaust gases or 52.6% of the 133.7 TJ fuel-input energy. To heat the iron 5.75 TJ or 4.3% are required, −401 GJ per year or 0.3% for heating outside air to reference temperature (15.6°C) while the remainder equaling 57.49 TJ per year or 43% of the fuel-input energy is assumed to be surface heat loss and radiation losses from the interior through cracks and door openings.

Studies showed that recuperation of the exhaust gases using ceramic heat wheels could represent the most economic heat-recovery solution. It was decided that the furnace battery be divided into two groups of four, and that two heat wheels be employed each to provide the combustion air for four adjacent furnaces. With a heat recovery efficiency of 70% representing an annual saving of 42.30 TJ, a fuel reduction of 32% was predicted. Because of the problem of gas flow and temperature uniformity, no additional furnace efficiency was assumed because of the longer gas residence time in

TABLE 12

Flue Gas Analysis of Heat-Treat Battery

Furnace	1	2	3	4	5	6	7	8
% Firing rate[a]	25	50	N.T.[b]	N.T.	25	25	50	25
Tstack (°C)	693	915	N.T.	N.T.	693	693	915	693
% Excess air	145.5	53.3	N.T.	N.T.	168.0	154.0	35.0	130.0
% O_2 flue	13.0	7.8	N.T.	N.T.	13.7	13.3	5.9	12.4

[a] Percent of rated input.
[b] N.T. = not tested.

the furnace. Instead, it was assumed that the excess air quantities would have to be increased to compensate for the smaller fuel flows. However it was recognized that recuperation could result in shortened full-fire periods in the cycle. Table 13 is a list of specifications for the heat wheels. Since the combustion air fan had been overdesigned as evidenced by considerable throttling at the air damper occurring under all observed firing conditions, it was concluded that the fan was capable of overcoming the additional pressure drop caused by the heat wheels. However, the burners and the air piping would have to be replaced by high-temperature burners and stainless steel air piping, respectively.

At 70% recovery efficiency from the enthalpy flux of 60.4 TJ per year the energy savings are

$$\text{Savings} = 0.7 \times 60.4 \text{ TJ/year} = 42.3 \text{ TJ/year}$$

The life of the units is expected to be 8 years. From Table 13 of Reference 5 (also see Volume I, Chapter 7 for another projection of natural gas prices) the deflated dollar values of natural gas from 1979 to 1986 are found and the annual savings are calculated.

Year	Energy Price $ per GJ	$ Savings
1979	1.88	79,503
1980	2.07	87,538
1981	2.12	89,653
1982	2.19	92,613
1983	2.26	95,573
1984	2.31	97,688
1985	2.38	100,612
1986	2.46	104,031

The costs of the project are estimated to be — cost heat wheel, $48,775 each × 2 = $97,500; cost high-temperature burners, $1135 × 40 = $45,400; cost installation, including cost of manifold ducting and hot-air piping, $241,600; engineering, $22,445; with a total investment of $406,995.

Assuming an interest rate of 8.75% and no salvage value for the recuperator, present value of savings = $514,043. Present value of heat wheel, burners, installation is $353,609, present value (P.V.) of engineering cost, $24,409. Total P.V. of costs = $378,081, and net present value = $136,025.

This represents a return on investment of 36% for an 8-year period or 4.5% per year.

TABLE 13

Specifications for the Heat Wheel Recuperator

Exhaust gas flow	60 SCM/min
Average hot gas temperature	749°C
Maximum hot gas temperature	916°C
Pressure drop — hot side	3 cm H_2O
Cold air flow	54 SCM/min
Combustion air temperature	624°C
Pressure drop-cold side	1.25 cm H_2O
Energy recovery efficiency	70%

If, as seems likely, natural gas will become a prohibited furnace fuel, then the unit cost of fuel increases by 30% if oil is substituted, and the present value of the annual savings for 8 years becomes $757,574. That would represent a net present value of the investment of $379,556 or 100.4% return on investment for 8 years or 12.5% per year. The same would be true if controls were lifted on natural gas and it assumed parity with oil in the energy market. It should be noted that the present burners have dual-fuel capabilities and that in the recent past oil has been used for short periods in the cold months to stay within mandatory gas allocations.

B. Recuperation of Cupola

The second largest energy consumer is the cupola. The melting furnace is a hot-blast continuous production, water-cooled cupola with external air heaters, an afterburner to accelerate combustion of CO following the air sweep, a spray cooler and a dust-collecting system. The operating data are given in Table 14. Figure 44 is a block diagram of the same system.

$$\frac{\text{heat in steel}}{\text{preheat + energy coke}} = (9091 \text{ kg/hr} \times 0.7 \text{ kJ/kg}-°\text{C})$$

$$\times\ (1510 - 27)\ °\text{C}) / (9 \times 10^6 + 822 \text{ kg coke/hr} \times 30{,}500 \frac{\text{kJ}}{\text{kg}})$$

The system appears to be 27.6% efficient. That is not a low efficiency for a furnace operating at 1510°C. On the other hand one cannot help to notice that the heat content of the coke is many times that required to supply the energy needed to carry out the reaction. It is presumed, however, that the air blast and the raw gas feed increase the production rate and perhaps provide a more uniform product. If so the cost does not seem excessive, for the total expense for the additional 18.39 GJ/hr is about $39 or just $3.90 per ton of product. If this expense is indeed justified, then we need to give thought to ways that we may recover some of the waste heat that leaves in the various fluid streams. There is available some 31.7 GJ/hr in the cooling water at a low temperature and 19.93 GJ/hr in the exhaust gas streams. Some additional waste heat might become available if high-temperature insulation is added to the preheaters and after burners. The following two figures illustrate ways to recover some of that waste heat. Figure 45 illustrates the retrofit installation of a recuperator which will recover 50% of the exhaust gas energy for preheating combustion air. The recuperator is a finned-tube-and-shell heat exchanger with automatic internal soot blowing because of the dirty gas stream. Note that a cyclone dust separator is required upstream of the heat exchanger to remove the large particulate load. The rest of the equipment is essentially unchanged from the present system. The specifications for the recuperator are given in Table 15. The annual energy savings based on a 22% recovery of the waste heat, as derived from the specifications, will be

$$19.93 \text{ GJ/hr} \times 0.22 \times 1950 \text{ hr/year} = 8573 \text{ GJ/year}$$

The additional costs and investment required are

Price for recuperator	$60,843
Engineering	9,775
Piping and fittings	4,240
Installation	43,230
Cost of additional fan energy per year	367
Write-off period	5 years
Interest rate	8.75%
Operating hours	1950 hr
Electric rate	$.026/kWh
Expected salvage value	0
Natural gas cost	From Reference 5

	Year	Natural Gas Rate ($ GJ)	Net annual savings ($)	Present value of savings ($)
	1979	1.88	16,116	14,819
15,004	1980	2.07	17,745	
	1981	2.12	18,174	14,131
	1982	2.19	18,774	13,423
	1983	2.26	19,374	12,737
			Total PV savings	70,114

Present value of costs and savings

5 years savings	$70,114
Cost engineering (1976)	−10,630
Cost piping and fittings	−4,240
Cost recuperator	−60,843
Cost installation	−43,230
Present value	$−48,829

This is obviously not a good investment opportunity. The reasons are that (1) the cupola operates only 22% of the time during a year and (2) a high efficiency recovery system is impossible because of the very dirty exhaust gas.

C. Heat Recovery from Cupola and Air Compressor Cooling Water

The steel cupola is cooled with 3028 ℓ of water which is recirculated after cooling in a cooling tower. It enters the cooling tower at 57°C and is cooled to 29°C. The heat recovery could be

$$3028 \text{ } \ell\text{/min} \times 60 \text{ min/hr} \times 1 \text{ kg/}\ell \times 4.185 \text{ kJ/kg-°C}$$

$$\times (57.2 - 29.4)\text{°C} = 21.1 \text{ GJ/hr}$$

Since the cupola operates 1950 hr per year the monthly waste heat available is

$$\frac{1950 \times 21.1}{12} = 3429 \text{ GJ}$$

This is sufficient heat to take care of the total maximum space heating load (excluding

TABLE 14

Cupola Operating Data

Capacity	13,636 kg/hr
Production	9.091×10^3 kg/hr
Temperature of iron	1510°C
Coke consumption	826 kg/hr
Preheater fan capacity	127.5 SCMM
Hot-blast temperature	482°C
Preheater capacity	9 GJ/hr
Exhaust gas temperature	316°C
Afterburner capacity	2 each — 3.17 GJ/hr
Temperature — following after-burning	454°C one burner 607°C two burners
Composition of exhaust gas	
CO	15.65%
CO_2	11.50%
Air-to-carbon mass ratio	1.90
Exhaust gas to carbon mass ratio	9.22

TABLE 15

Recuperator Specifications

Clean air volume	127.5 SCMM
Clean air temperature	21°C
Clean air outlet temperature	482°C
Clean air inlet pressure	114 cm water column[a]
Dirty gas volume	292 SCMM
Dirty gas inlet temperature	766°C
Dirty gas inlet pressure	−10 cm water column

[a] Pressure in height of water column.

heating make-up air), but because it is only available during the melting shifts each day, it can supply only 27% of the heat required for space heating. The cooling towers would be required to operate during those times when less heat was required. Since the air compressor cooling water can supply about 16% of the base heating load over the year, a combination system utilizing the compressor and cupola cooling water is worth investigating. This is particularly appropriate since the compressor cooling water flow exceeds the normal evaporation loss from the cupola cooling system.

The monthly space heating load is shown in the graph of Figure 46 where the constant supply of waste heat from the compressors is shown as a dashed line.

Waste heat from the compressor intercoolers averages 44.3 kJ/min/hp as 49°C cooling water.

$$450 \text{ hp} \times 44.3 \text{ kJ/min} \times 60 \text{ min/hr} \times 8000 \text{ hr/year}$$

$$= 9570 \text{ GJ annually}$$

This waste heat is presently carried to the sewer with the 133.5 ℓ/min or 8010 ℓ/hr of cooling water. Since 3400 ℓ/hr of cupola cooling water is evaporated into the air, we propose the system of Figure 47 where the cooling water in the sump at 57°C is supplemented with the 49°C cooling water from the compressor.

Because the cupola cooling water is very dirty and because the make-up air preheater that constitutes the principal waste-heat load for the system would freeze during the coldest weather, the system consists of a run-around secondary system where a water-glycol mixture picks up heat in coils immersed in the sump and the heated liquid is circulated to the preheater coils and to unit space heaters. Any waste heat that is not used by the space heating system must be removed by the existing cooling tower. Because of the time discrepancies in availability of the cupola waste heat and the heating load, the present heating system must be retained and additional hot-water unit-space heaters purchased for use in the run-around loop. If the immersed coils of that loop have a heat exchange effectiveness of 75%, then the compressor waste heat can supply a maximum of $0.75 \times 9570/12$ or 598 GJ per month. From the space heating graph of Figure 46, it was determined that 76.7% of the compressors' contribution to the heating system could be utilized or 7177 GJ per year. The remaining heating load of

$$(18{,}462 - 7177)\ \text{GJ} = 11{,}285\ \text{GJ/year}$$

cannot be supplied totally by the cupola. Since the cupola operates only about 22% of the time, then it can supply approximately 22% of that load or $11{,}285 \times 0.22 = 2{,}512$ GJ per year. The total waste-heat utilization is then $(7177 + 2512)\ \text{GJ} = 9689$ GJ per year. If we assume that the cupola will last another 10 years, then the cost of the waste-heat system can be amortized over that period. Using fuel prices from Reference 5 and an interest rate of 8.75% we find the present value of the savings to be

Year	Cost natural gas ($/GJ)	Savings in $	Present value of savings $
1979	1.88	18,215	16,749
1980	2.07	20,056	16.959
1981	2.12	20,541	15.971
1982	2.19	21,219	15,171
1983	2.26	21,897	14,396
1984	2.31	22,381	13,530
1985	2.38	23,050	12,819
1986	2.46	23,835	12,184
1987	2.56	24,804	11,659
1988	2.64	25,579	11,056
		Total P.V. savings	140,494

	Present Cost	Present Value
Cost of heat exchangers	11,640	10,703
Cost of pipe and fittings	24,897	22,894
Cost of controls	1,400	1,287
Cost of pumps	3,880	3,568
Cost of unit heaters	15,520	14,271
Cost of labor	18,792	17,280
Cost of engineering	4,613	5,017
Net present value of costs		75,020

Net present value of investment $= 140{,}494 - 75{,}020$
$= 65{,}474$ or 87.3%

This is a return on investment of $87.3\%/10 = 8.73\%$ per year.

This is a break-even investment now, but should fuel prices inflate faster than the PIES estimate, or if natural gas were to be denied to industrial customers, it would be a very sound investment. It is even now a very good hedge against cuts in gas allocation or expansions in market.

VI. RECOMMENDATIONS

The economic analyses provided above in the previous section of this report indicate the following characteristics of the suggested investments listed in descending order of present value:

Project	Present value cost of investment	Net present value of investment	Annual rate of return	Life of project
Recuperation of heat-treat furnaces	$378,018	$136,025	4.5%	8
Waste heat for space heating	75,020	65,474	8.73	10
Recuperation of cupola	118,943	−48,829	—	5

The second investment listed, although it does not have the largest net present value is obviously the most attractive. Not only is the size of investment the smallest of the three, but the annual rate of return is almost double that of any other.

It is recommended, however, that the first two investments be seriously considered because together they offer a total annual savings in natural gas of 23.3% of the natural gas used in 1976. This will go a long way in permitting an expansion of production without exceeding natural gas allocations. Furthermore, if the plant is denied natural gas for furnace fuel, then the above returns on investment will increase approximately by a factor of 1.47. It is probable, also, that natural gas may be deregulated and in that case, the improvement factor can be expected to be almost the same.

ENERGY SAVING SURVEY

Surveyed by: ________________

Department: ________________

Date: ________________

Fuel Gas or Oil Leaks	Steam Leaks	Compressed Air Leaks	Condensate Leaks	Water Leaks	Damaged or Lacking Insulation	Excess Lighting	Excess Utility Usage	Equipment Running & Not Needed	Burners Out of Adjustment	Leaks of or Excess of HVAC	Location	Date Corrected

FIGURE 1. Energy saving survey. (From Gatts, R., Massey, R., and Robertson, J., NBS Handbook 115, U.S. Superintendent of Documents, Government Printing Office, Washington, D.C.,)

19__/19__

1	2	3	4	5	6	7	8	9	10	11	12	13	14	15	16
Month	Demand MCF	Demand Charge Rate	Demand Charge	MCF	MMBTU	Net Energy Charge	Pch'd Gas Surcharge	Tax Surcharge	Gross Energy Charge	Total Cost	Year-to-Date MCF	Year-to-Date Cost	Last Year Cost	% Diff	Cost/ MMBTU
Total															
19__/19__															
Total															

FIGURE 2. Historical quantity and cost data for natural gas. (From Colbert, M., Looney, Q., Rohrer, W., Rudoy, W., and Skoglund, T., Office of Energy Programs, U.S. Department of Commerce, Washington, D.C., to be published.)

19__/19__

1	2	3	4	5	6	7	8	9	10	11	12	13	14	15	16	17
Month	Actual Demand KW	Power Factor Multiplier	Billing Demand KW	Demand Charge Rate	Demand Charge	KWH	MMBTU	Net Energy Charge	Fuel Surcharge Rate	Tax Surcharge Rate	Gross Energy Charge	Total Cost of Electricity	Year-to -Date Cost	Last Year Cost	% Diff	Cost MMBTU
Total																
19__/19__																
Total																

FIGURE 3. Historical quantity and cost data for electricity. (From Colbert, M., Looney, Q., Rohrer, W., Rudoy, W., and Skoglund, T., Office of Energy Programs, U.S. Department of Commerce, Washington, D.C., to be published.)

19__/19__

1	2	3	4	5	6	7	8	9	10
	DEGREE-DAYS		SIZE OF FACILITY			LEVEL OF ACTIVITY			
Month	Heat	Cool	sq. ft.	cu. ft.	Production Units	Operating Cost	Operating Level		
Total									
19__/19__									
Total									

FIGURE 4. Major factors affecting energy use. (From Colbert, M., Looney, Q., Rohrer, W., Rudoy, W., and Skoglund, T., Office of Energy Programs, U.S. Department of Commerce, Washington, D.C., to be published.)

19__/19__

1	2	3	4	5	6	7	8	9	10	11	12	13	14	15
Month	Elec.	Natural Gas	#___Oil	#___Oil	Coal	Steam ___ PSIG	Steam ___ PSIG	Water	Waste	Other	Total	Year-to-Date	Last Year	% Diff.
Total														
Last Year's Total														
% Diff														

FIGURE 5. Summary of all energy, quantity. (From Colbert, M., Looney, Q., Rohrer, W., Rudoy, W., and Skoglund, T., Office of Energy Programs, U.S. Department of Commerce, Washington, D.C., to be published.)

19__/19__

Month	Elec.	Natural Gas	#___Oil	#___Oil	Coal	Steam ___ PSIG	Steam ___ PSIG	Water	Waste	Other	Total	Year to Date	Last Year Cost	% Diff.
Total														
Last Year														
% Diff.														

FIGURE 6. Summary of all energy, cost. (From Colbert, M., Looney, Q., Rohrer, W., Rudoy, W., and Skoglund, T., Office of Energy Programs, U.S. Department of Commerce, Washington, D.C., to be published.)

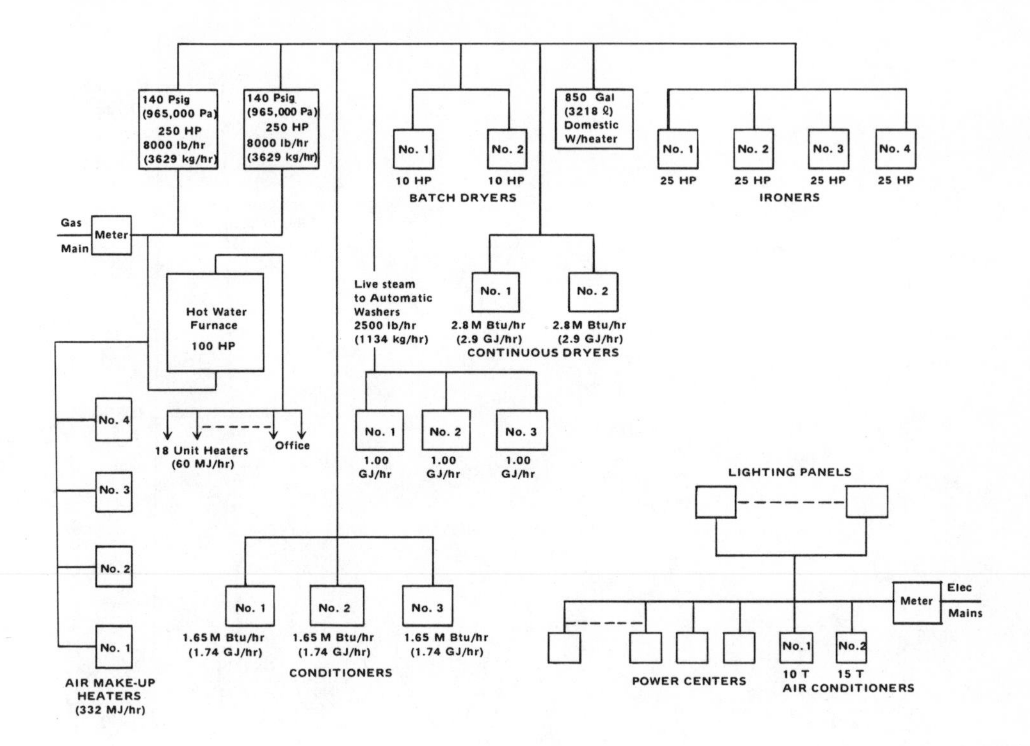

FIGURE 7. Laundry process diagram.

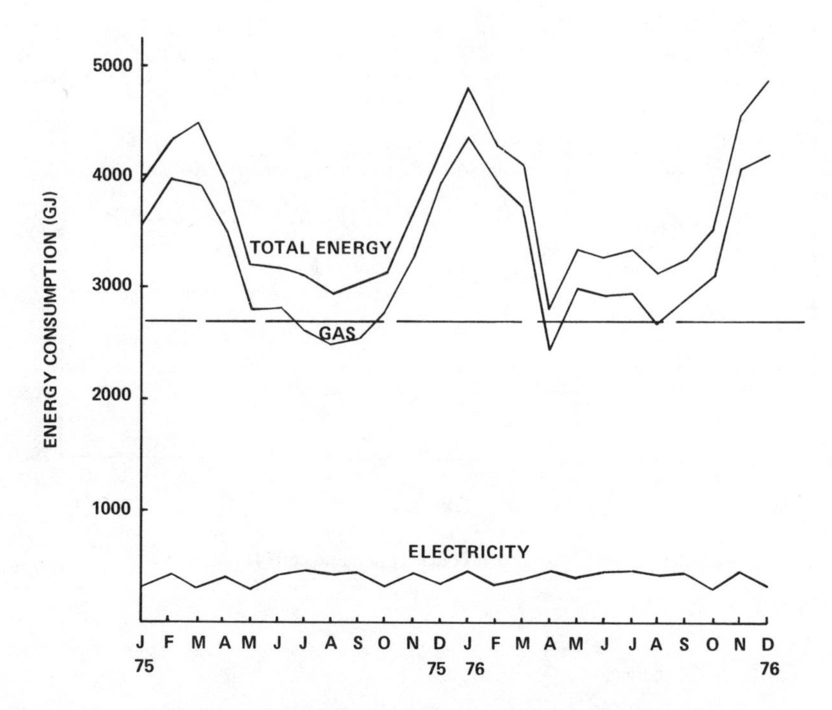

FIGURE 8. Monthly energy consumption for laundry.

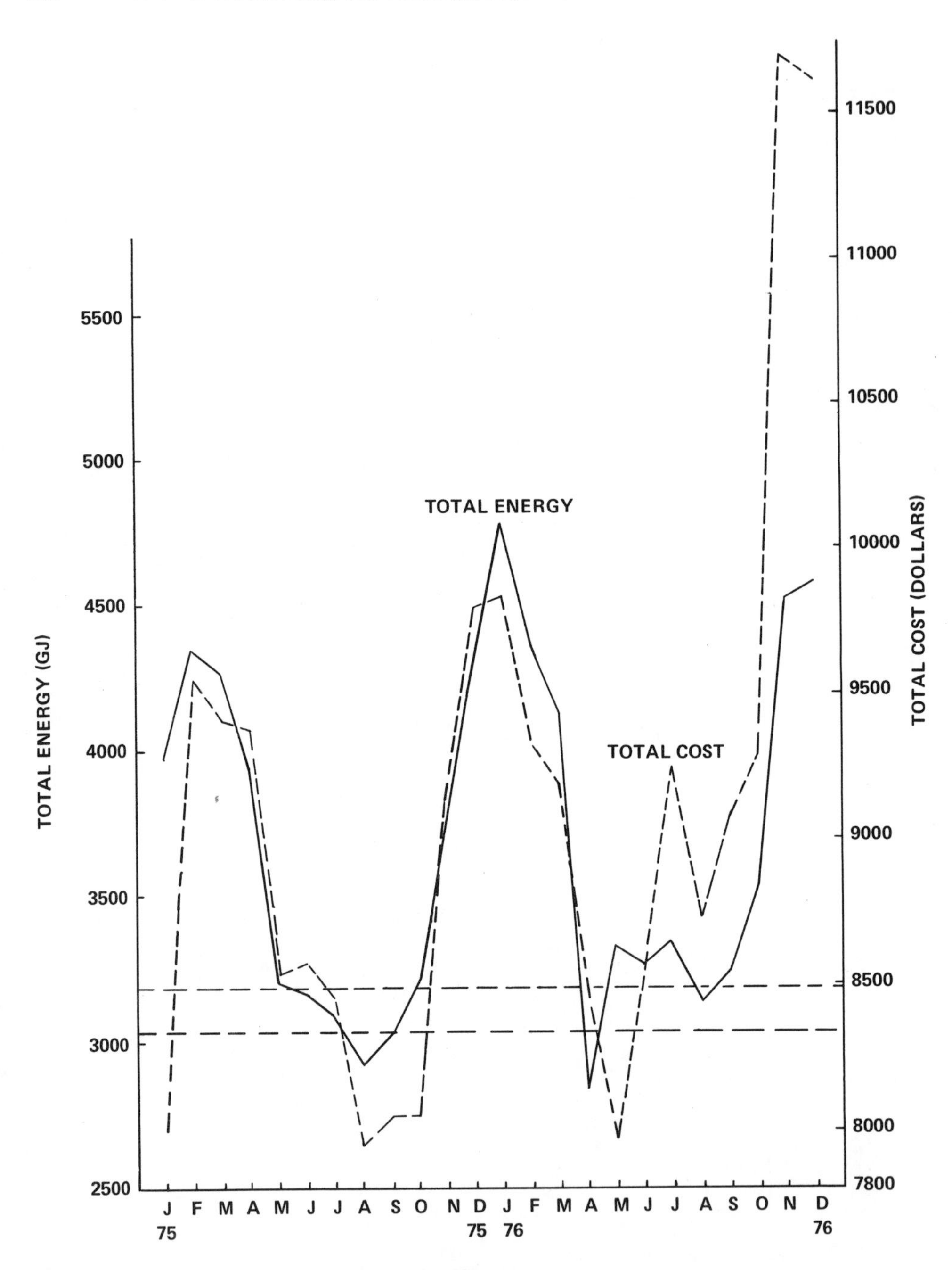

FIGURE 9. Monthly energy consumption and costs for laundry.

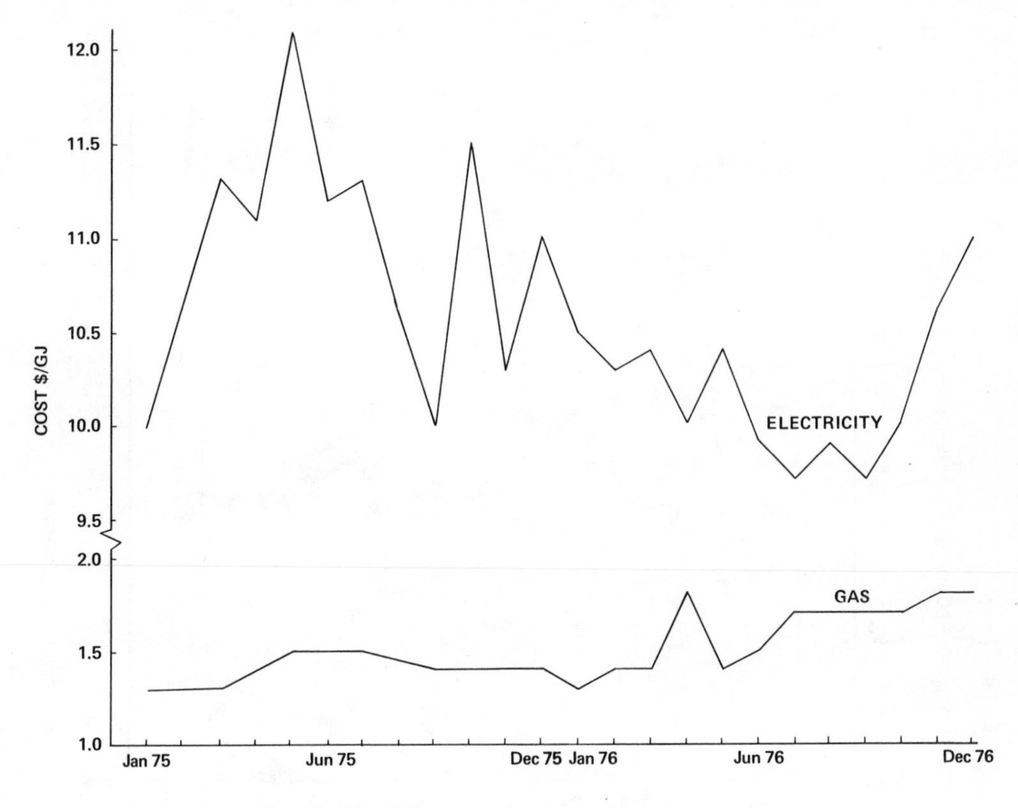

FIGURE 10. Monthly gas and electric costs for laundry.

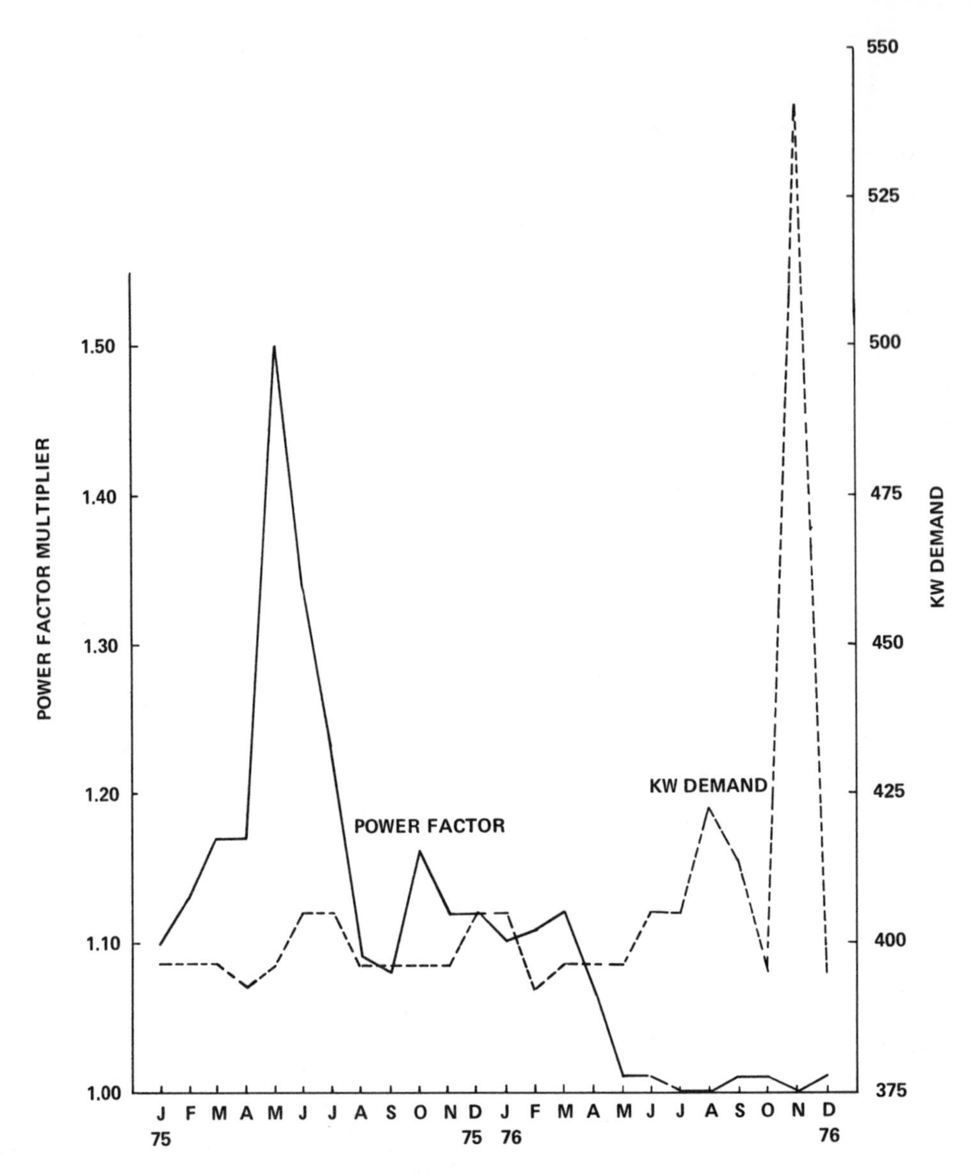

FIGURE 11. Monthly electrical demand and power factor for laundry.

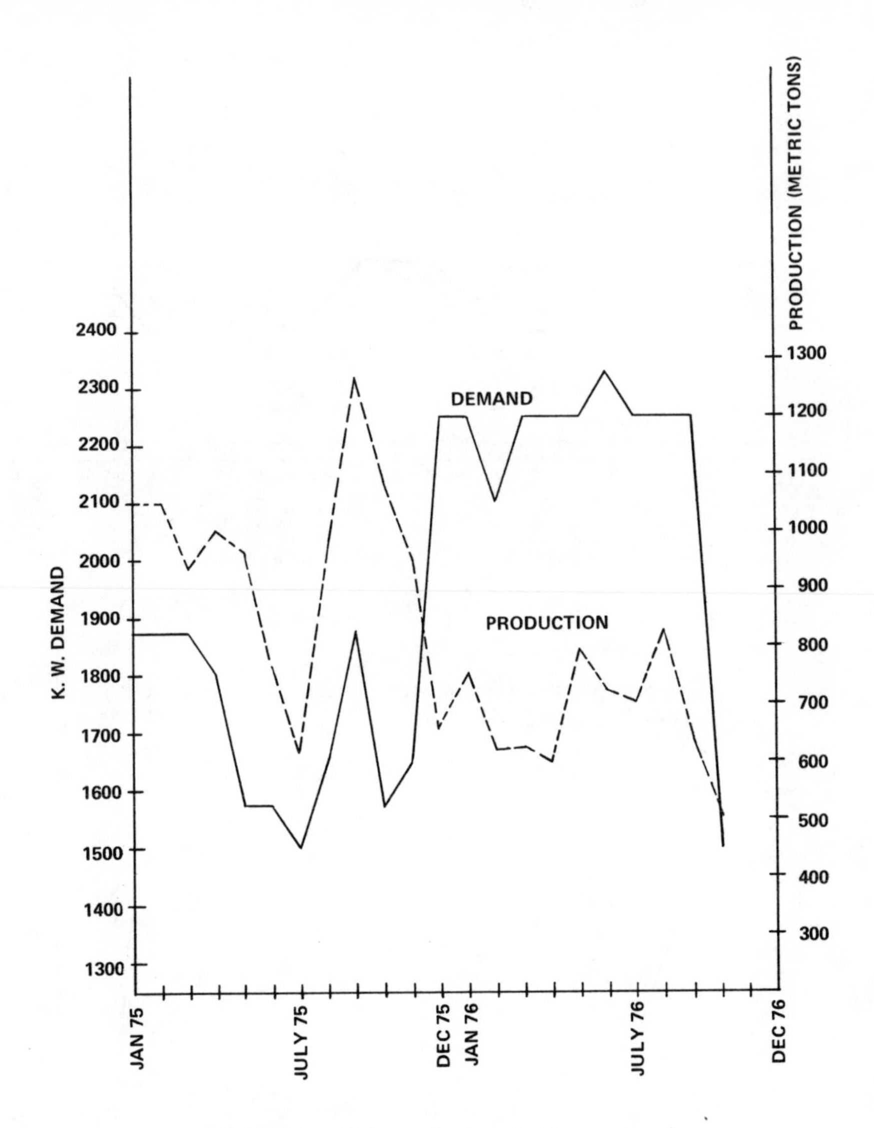

FIGURE 12. kW demand and production in a foundry.

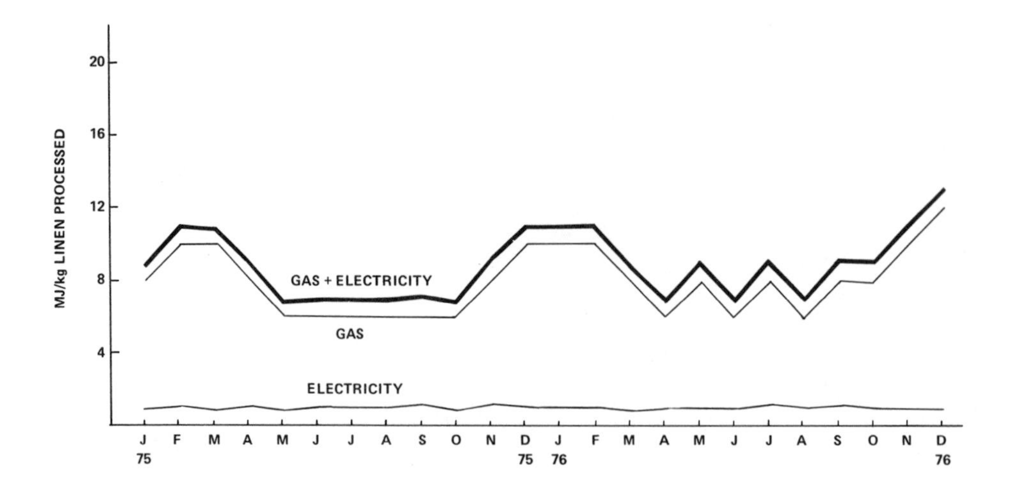

FIGURE 13. Per unit energy consumption for a laundry.

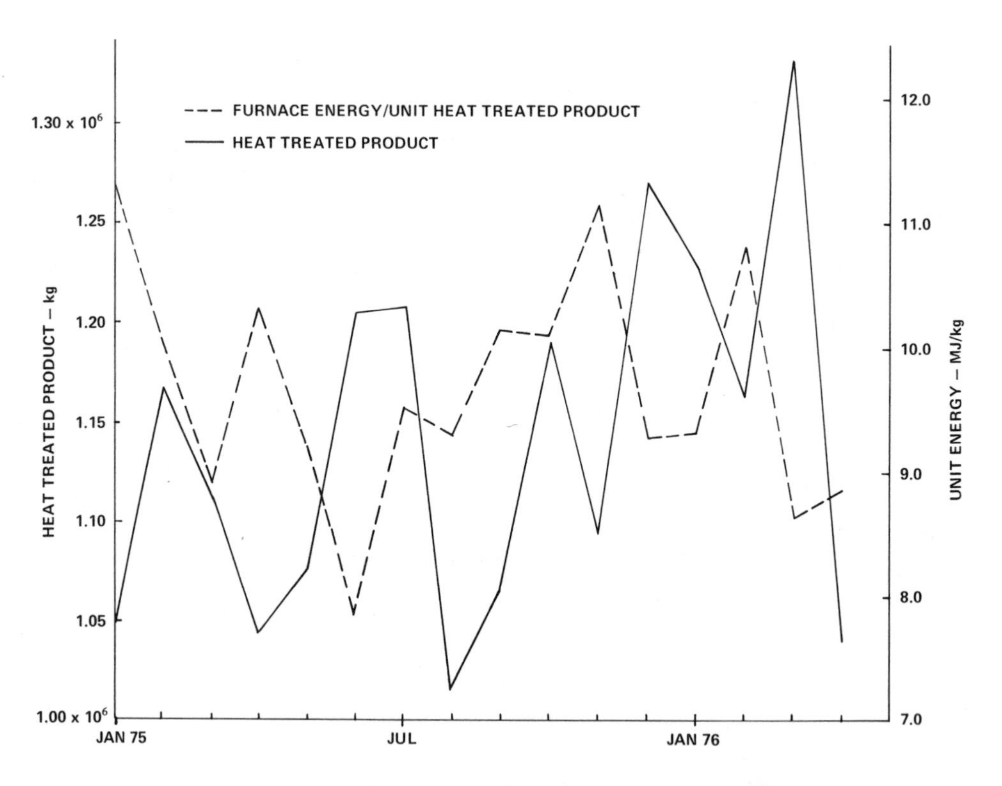

FIGURE 14. Per unit energy consumption and production for a heat treatment plant.

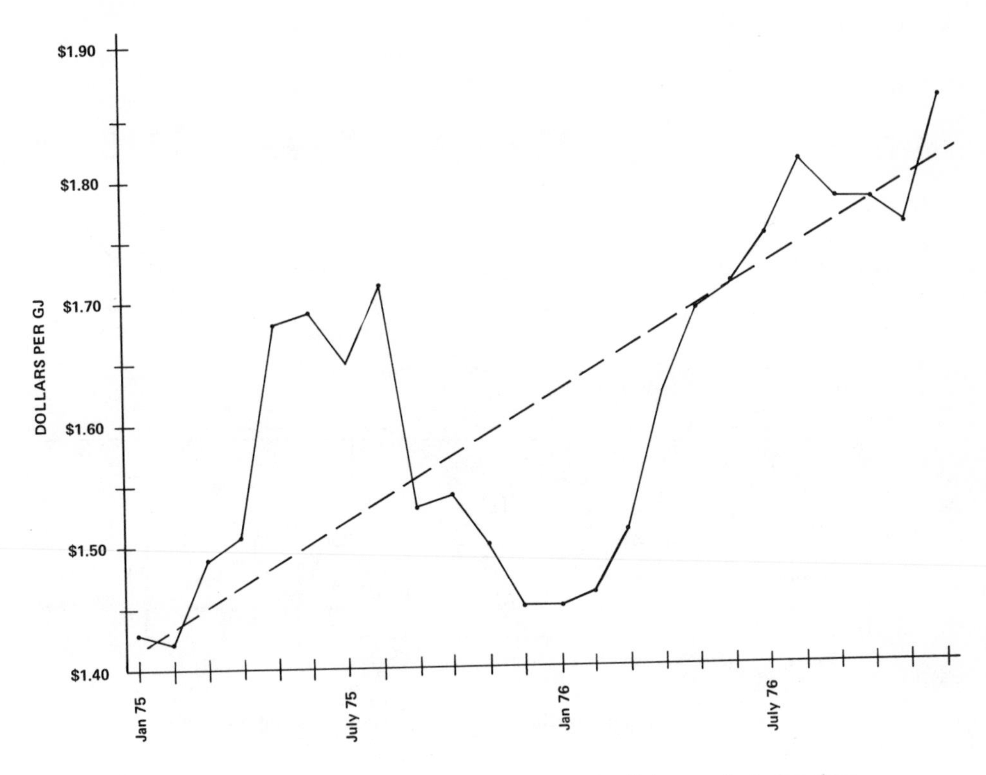

FIGURE 15. Average energy rates for natural gas in a light manufacturing plant.

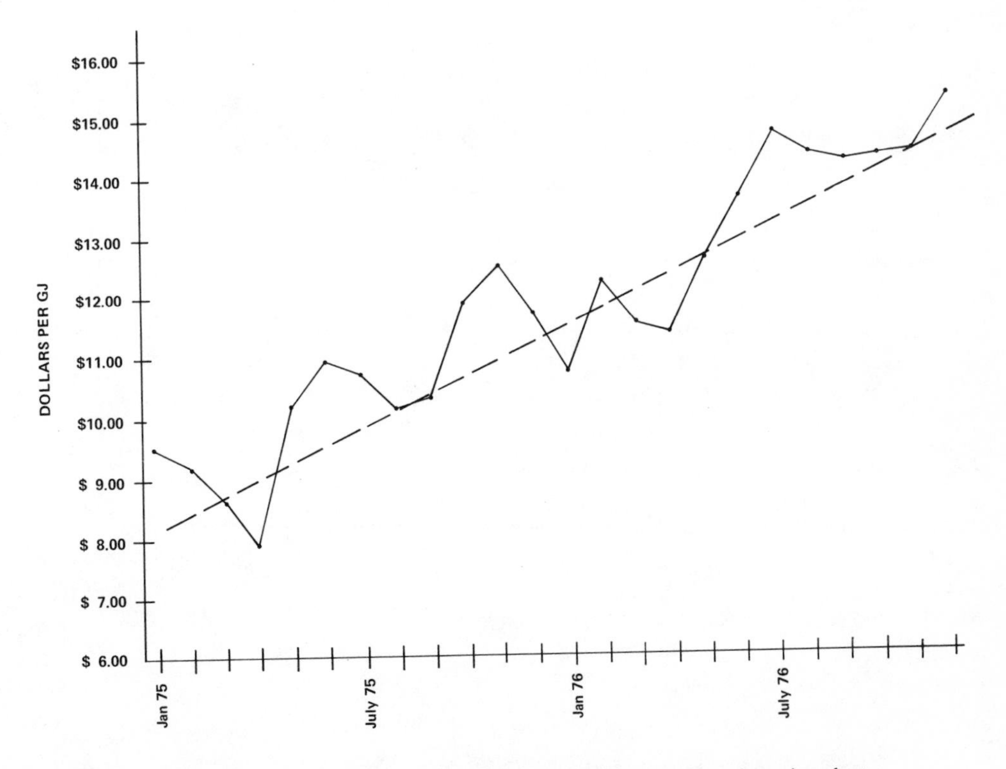

FIGURE 16. Average energy rates for electricity in a light manufacturing plant.

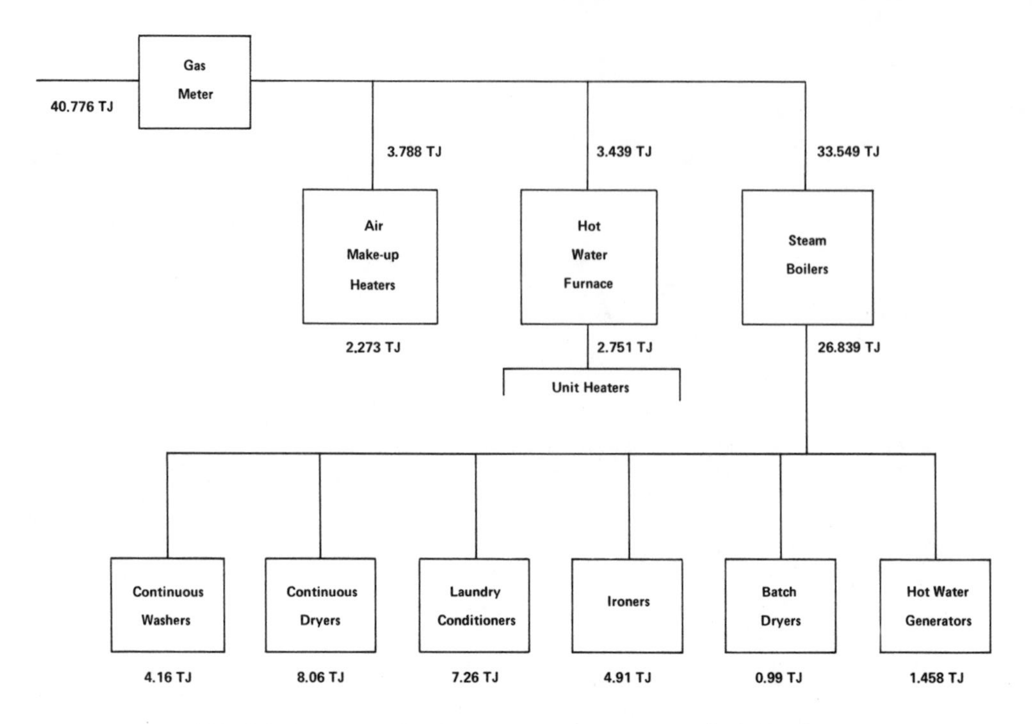

FIGURE 17. Energy distribution over thermal process units in laundry.

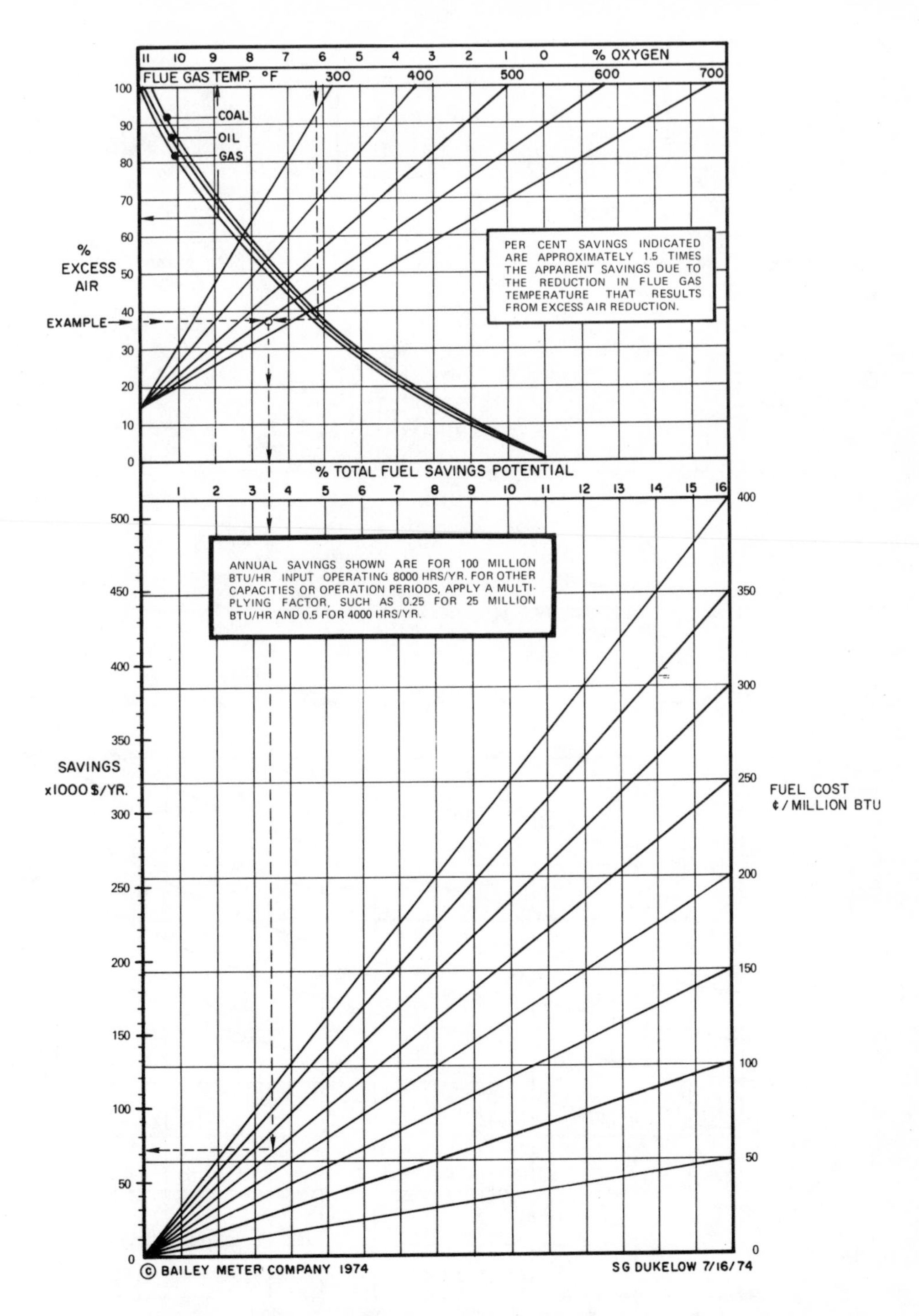

FIGURE 18. Nomograph for estimating savings from adjustment of burners. (From Dukelow, S. G., Bailey Meter Company, 29801 Euclid Avenue, Wickliffe, Ohio, 44092. With permission.)

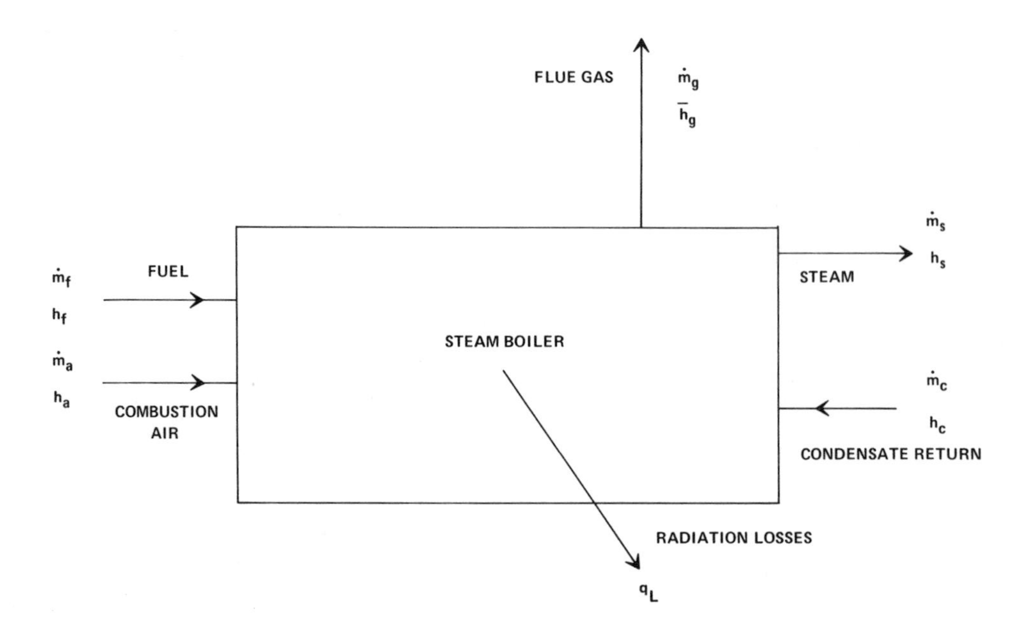

FIGURE 19. Heat balance on steam boiler.

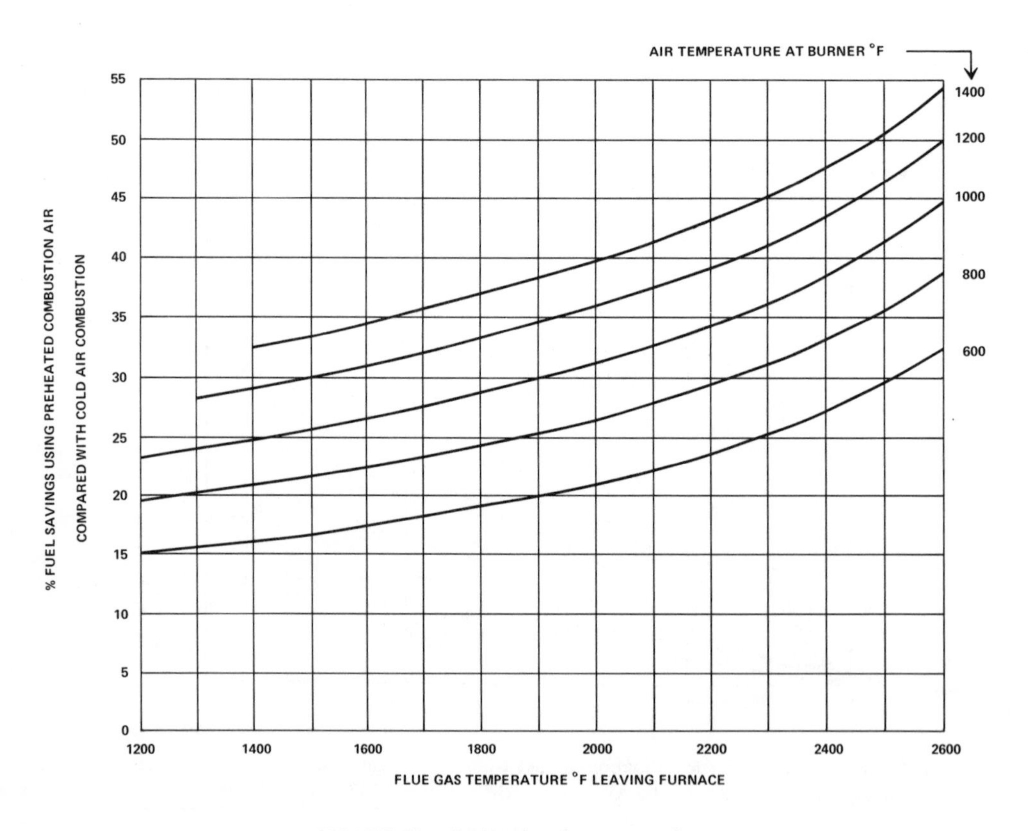

FIGURE 20. Fuel savings by recuperation.

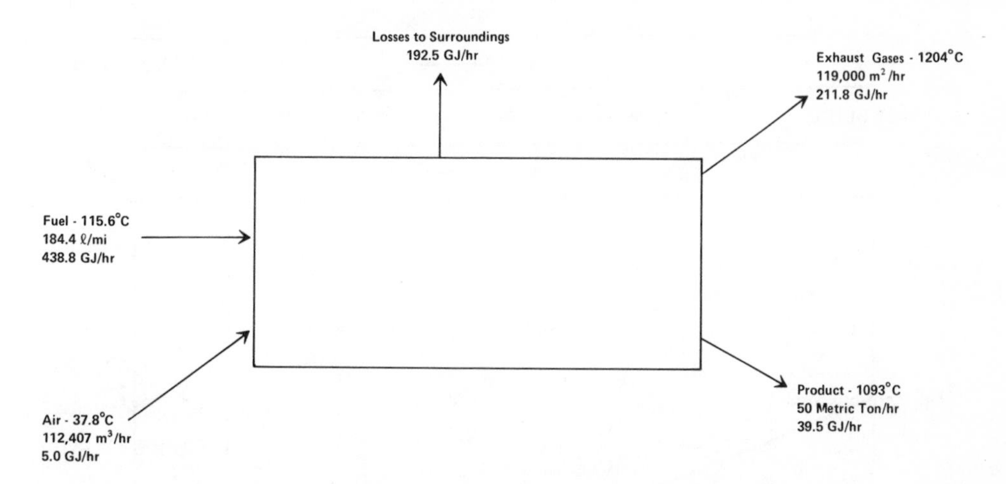

FIGURE 21. Heat balance for a simple continuous steel tube furnace.

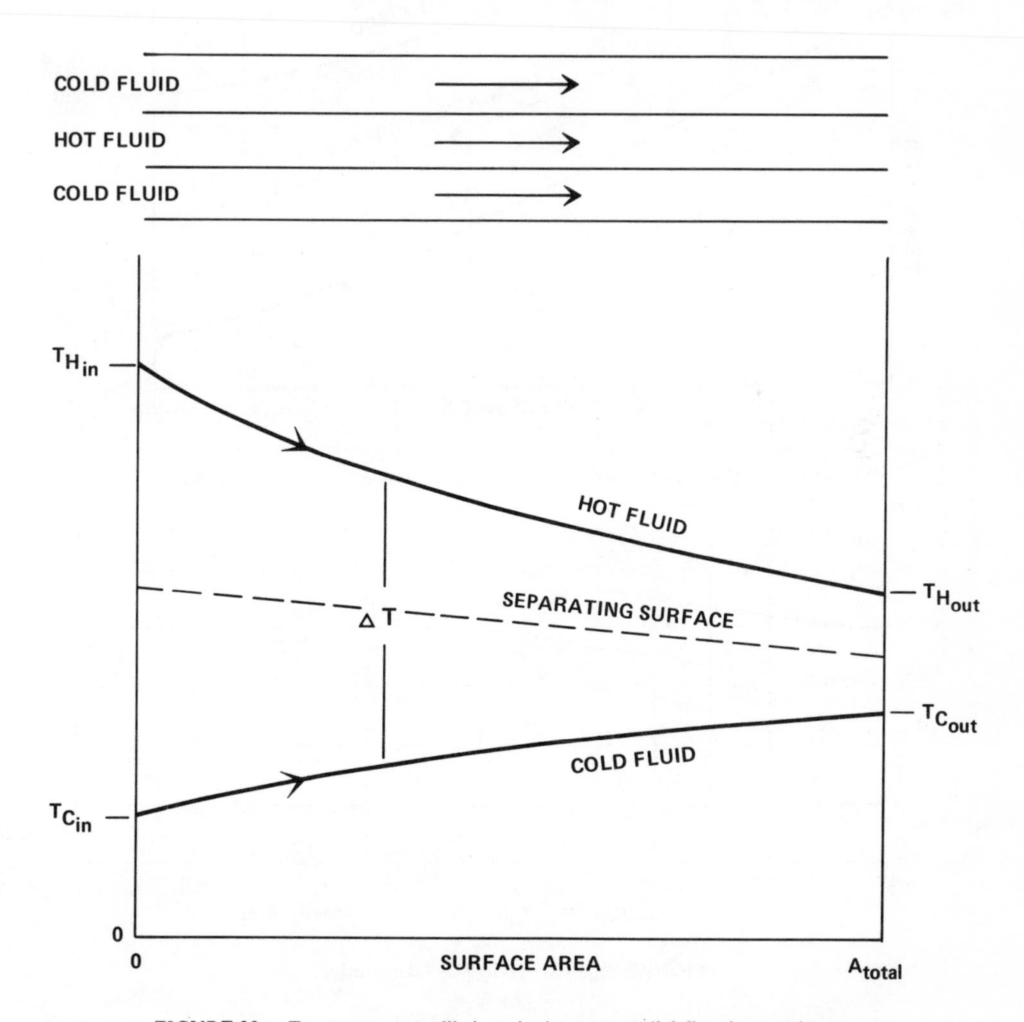

FIGURE 22. Temperature profile in a single-pass parallel-flow heat exchanger.

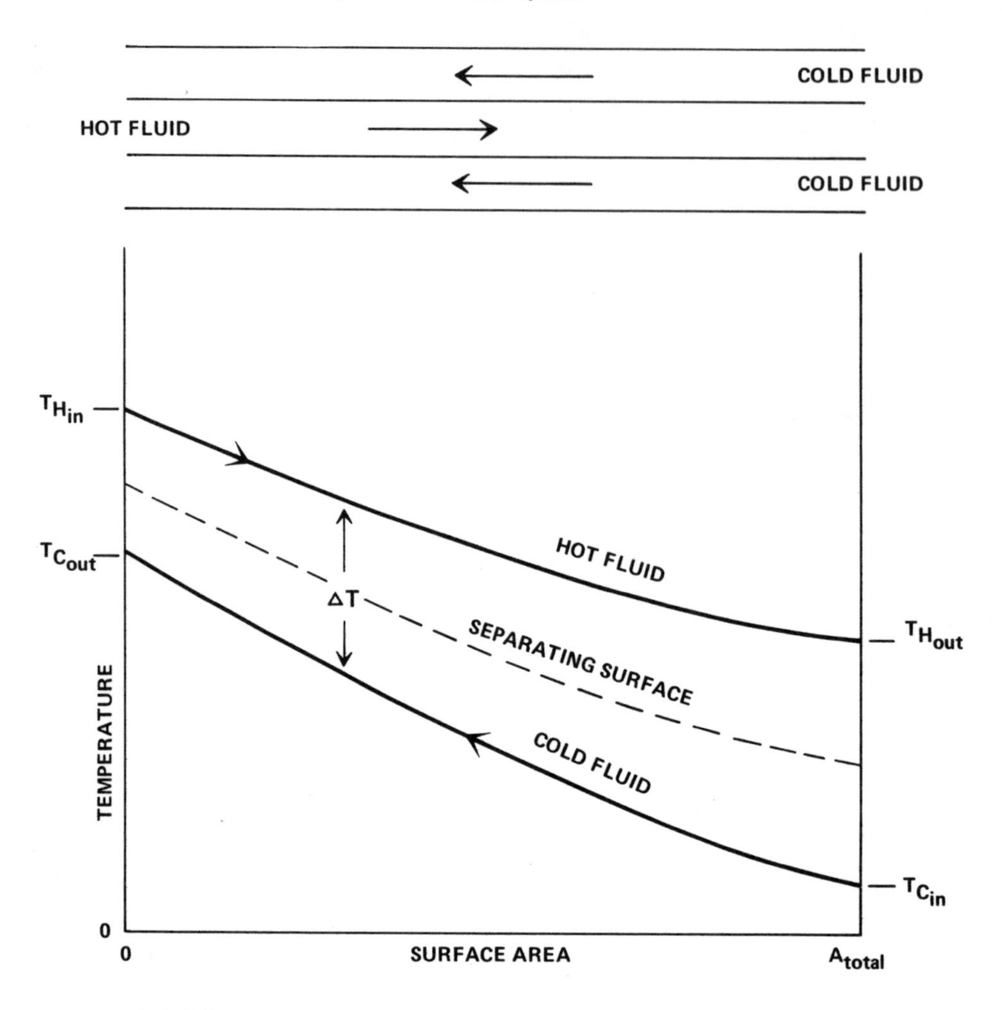

FIGURE 23. Temperature profile in a single-pass counterflow heat exchanger.

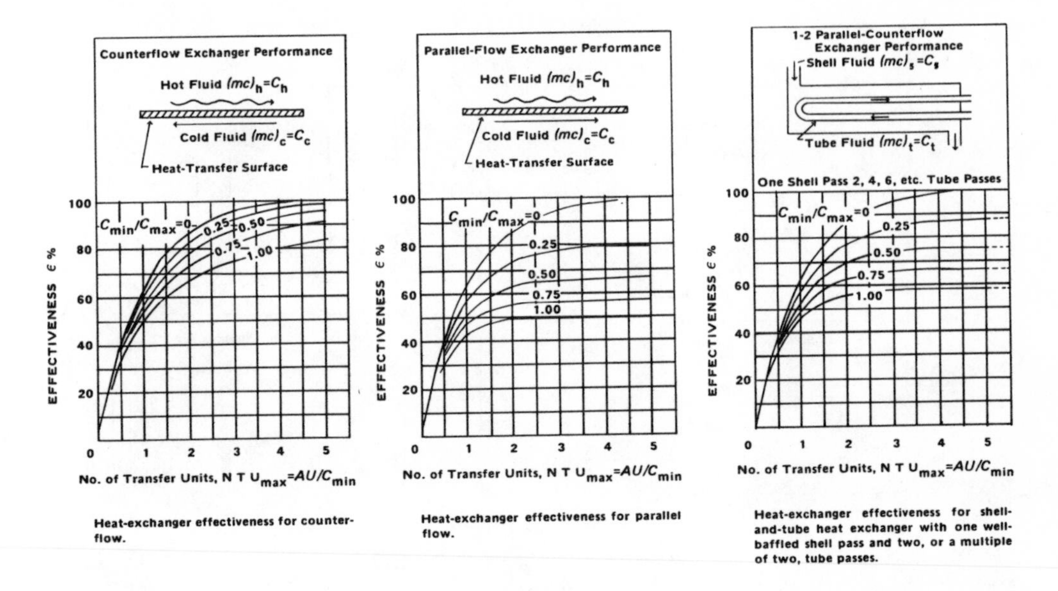

FIGURE 24. Heat exchanger effectiveness. (From Kays, W. and London, A. L., *Compact Heat Exchangers*, 2nd ed., McGraw-Hill, New York, 1964, 24. With permission.)

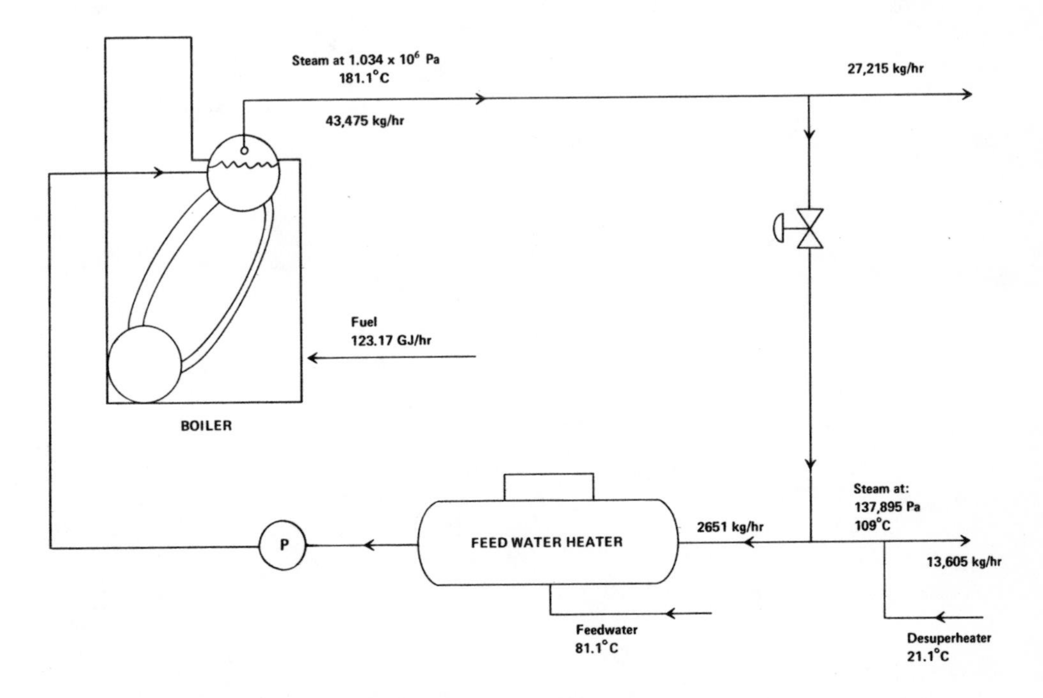

FIGURE 25. Steam plant schematic before adding electrical generation.

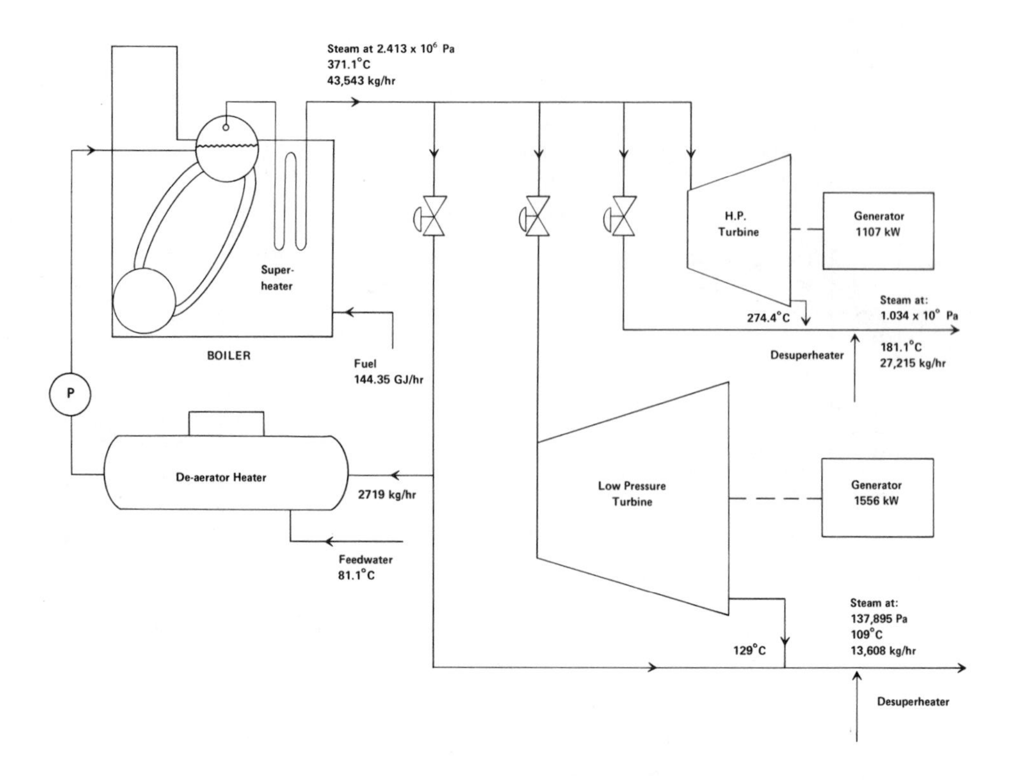

FIGURE 26. Steam plant schematic after installing electrical generation.

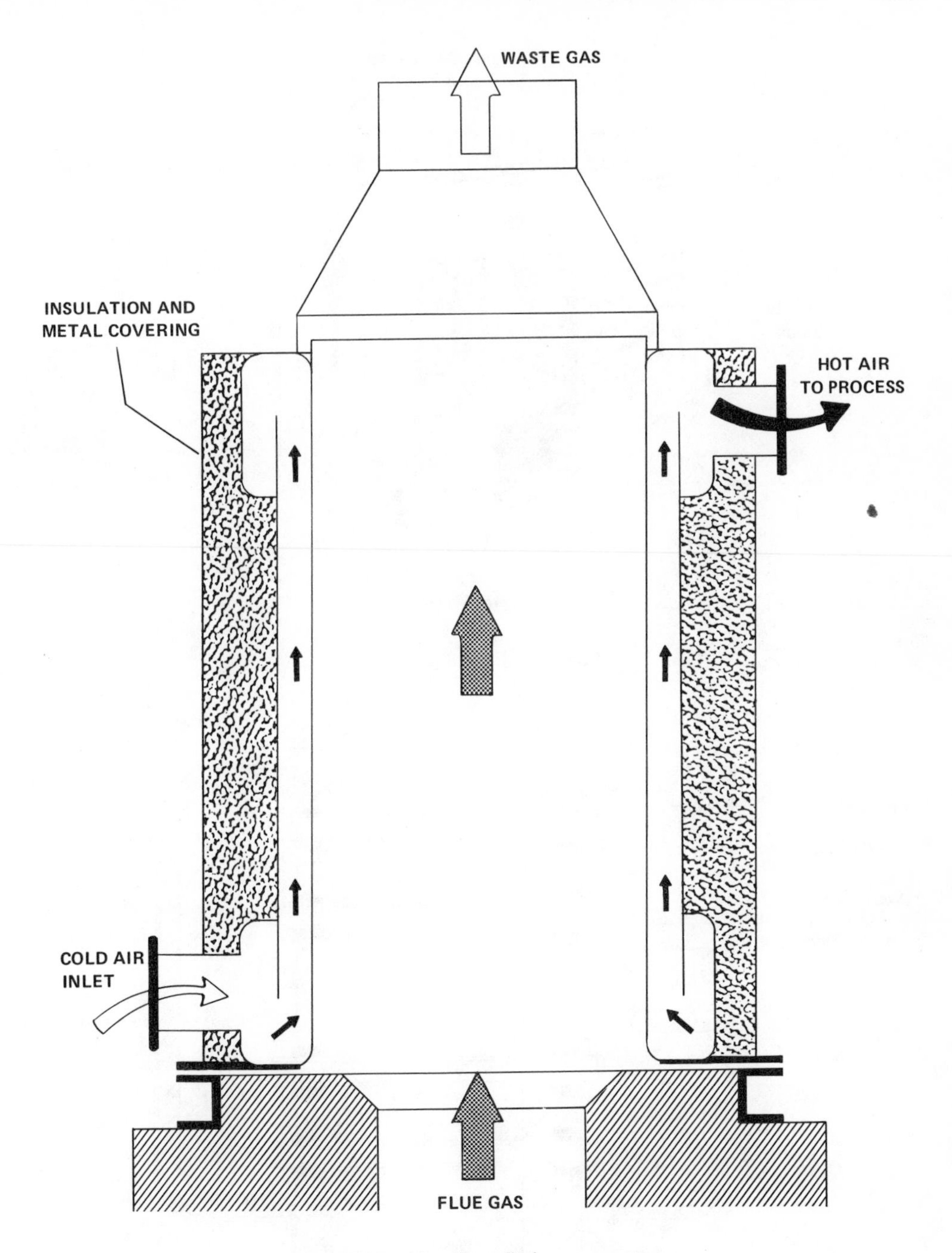

FIGURE 27. Metallic radiation recuperator.

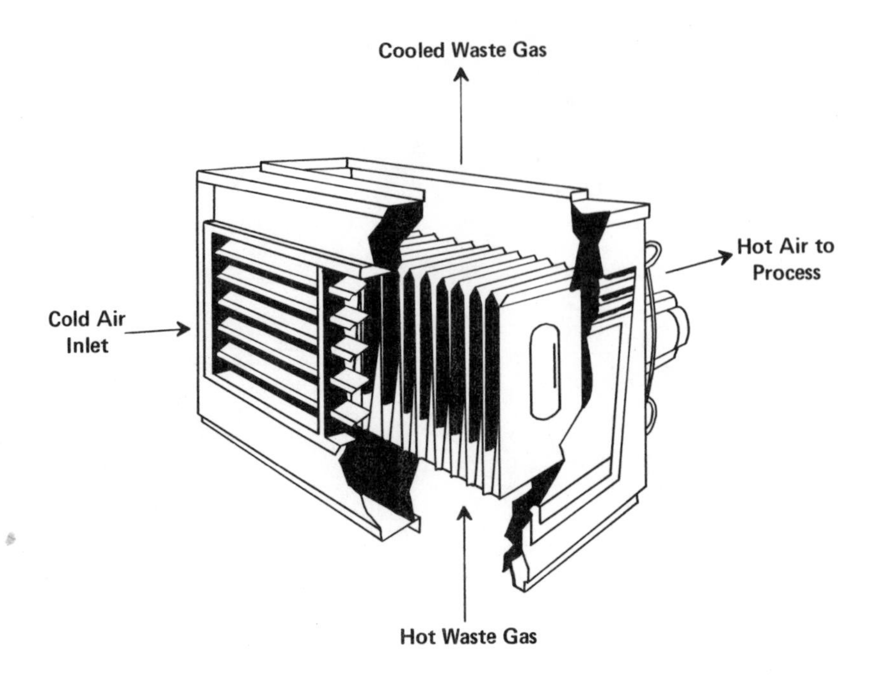

FIGURE 28. Air preheater.

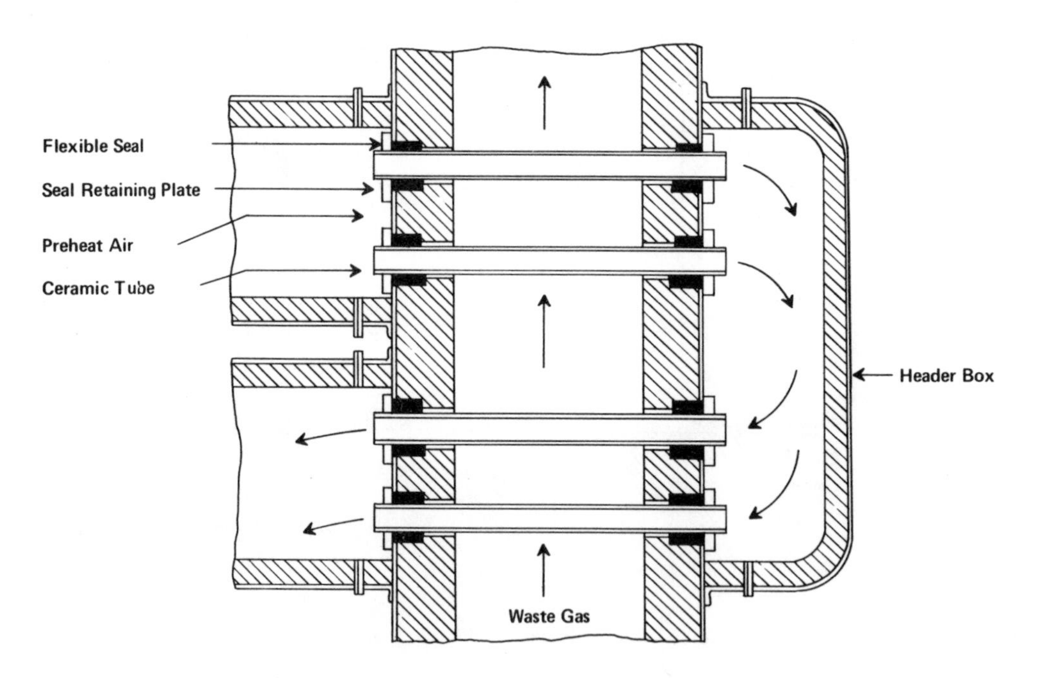

FIGURE 29. Ceramic tube recuperator.

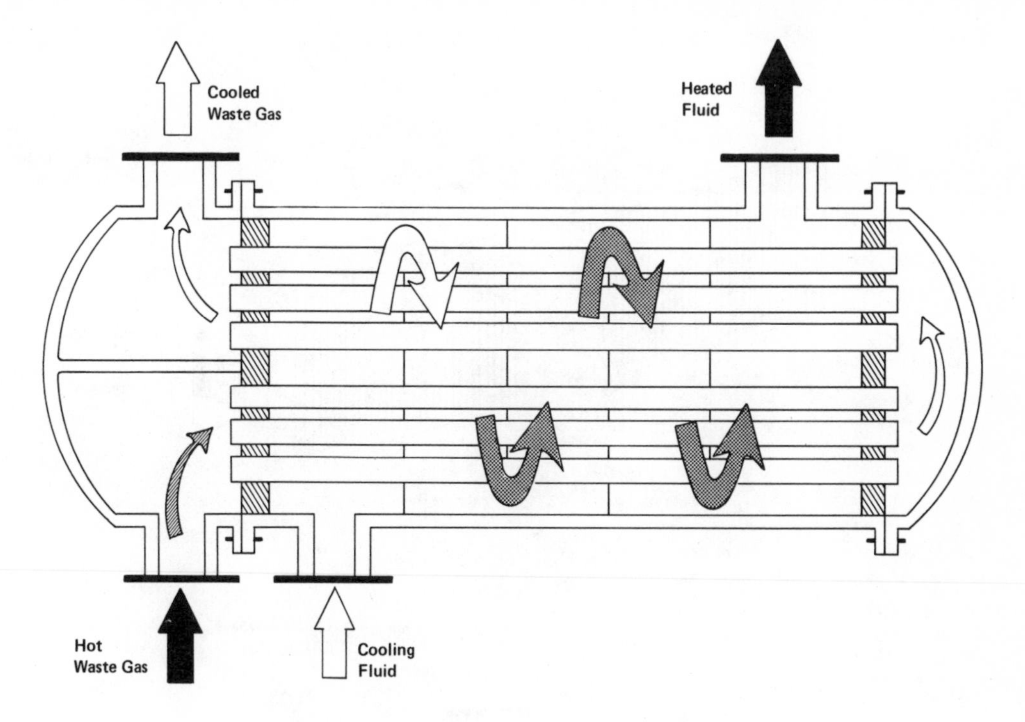

FIGURE 30. Convective tube-type recuperator.

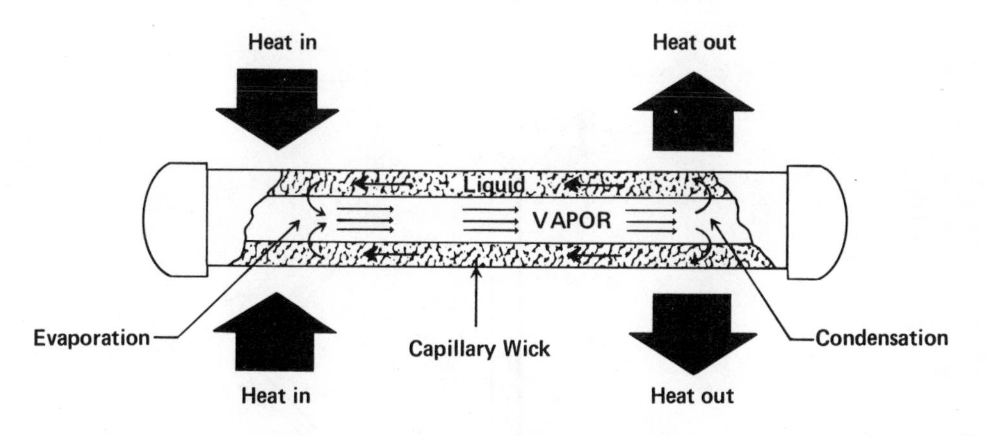

FIGURE 31. Heat pipe.

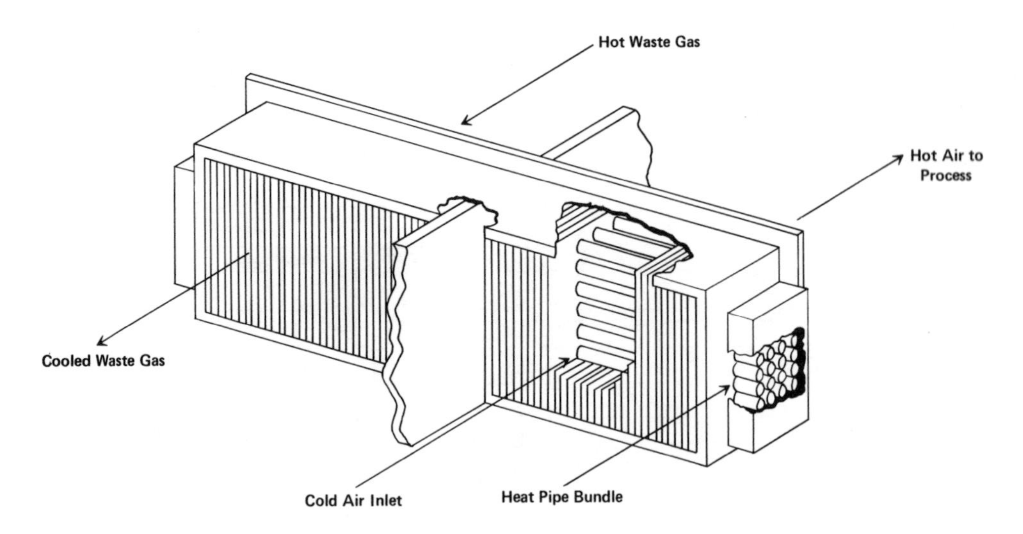

FIGURE 32. Heat pipe recuperator.

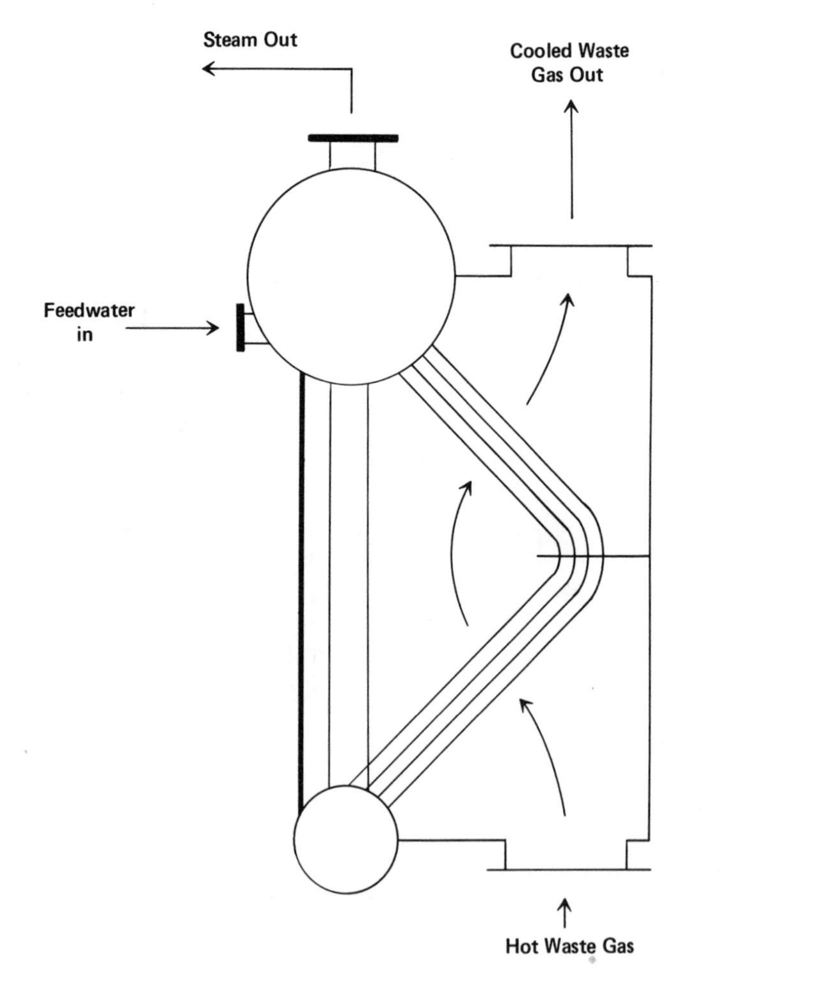

FIGURE 33. Waste heat boiler.

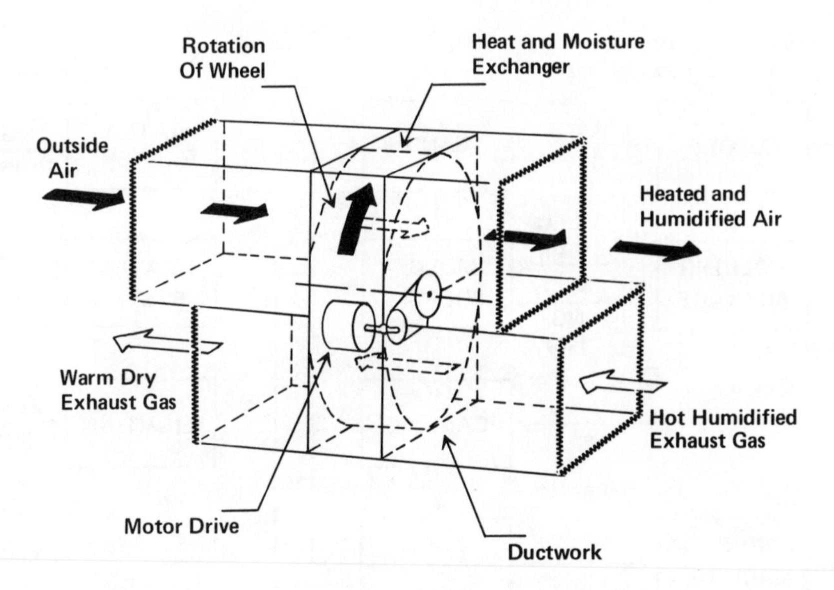

FIGURE 34. Heat wheel.

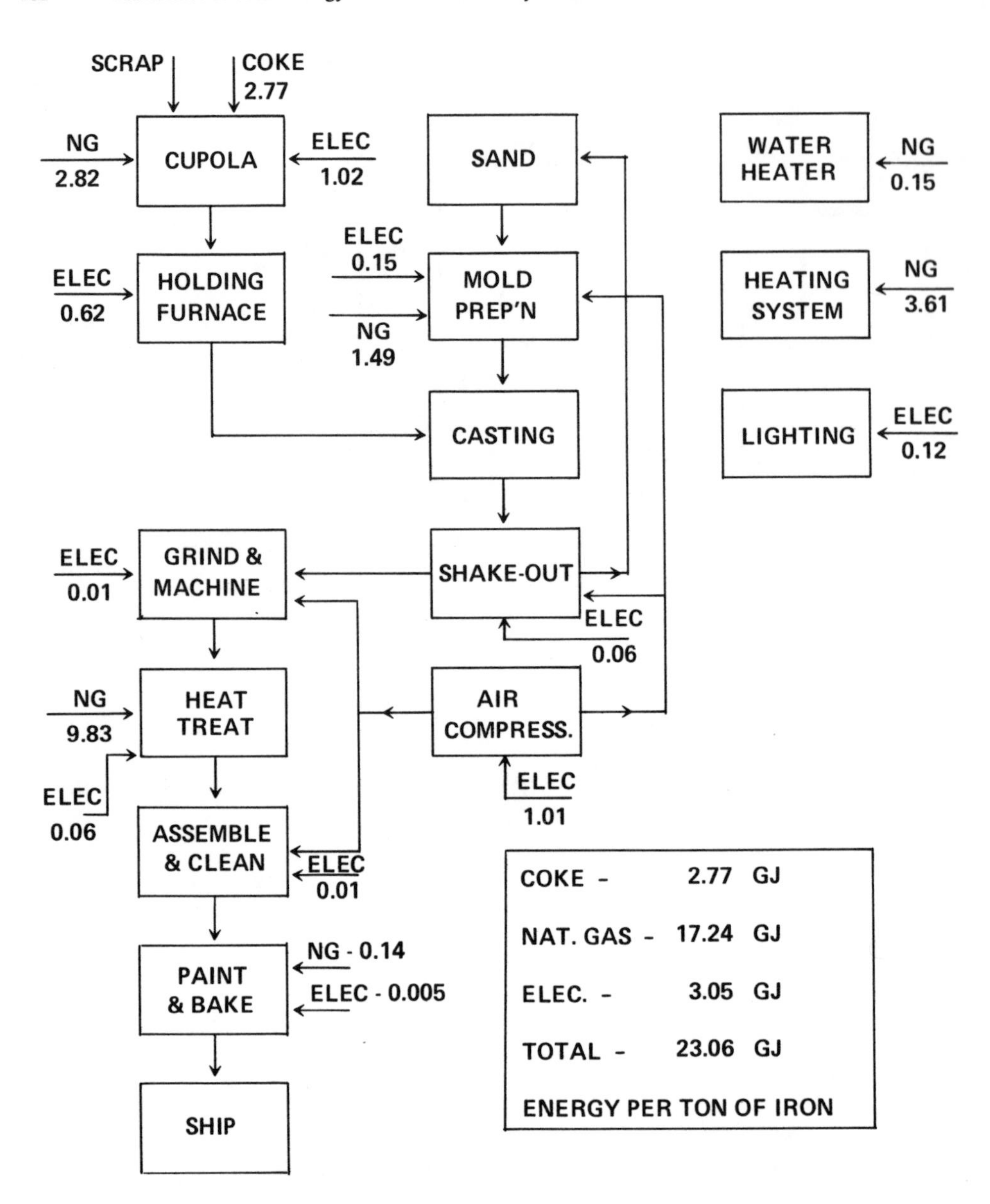

FIGURE 35. Unit energy consumption in GJ per metric ton of product. Numbers preceded by ELEC (electricity) and NG (natural gas) showing inputs to systems have units of GJ/metric ton.

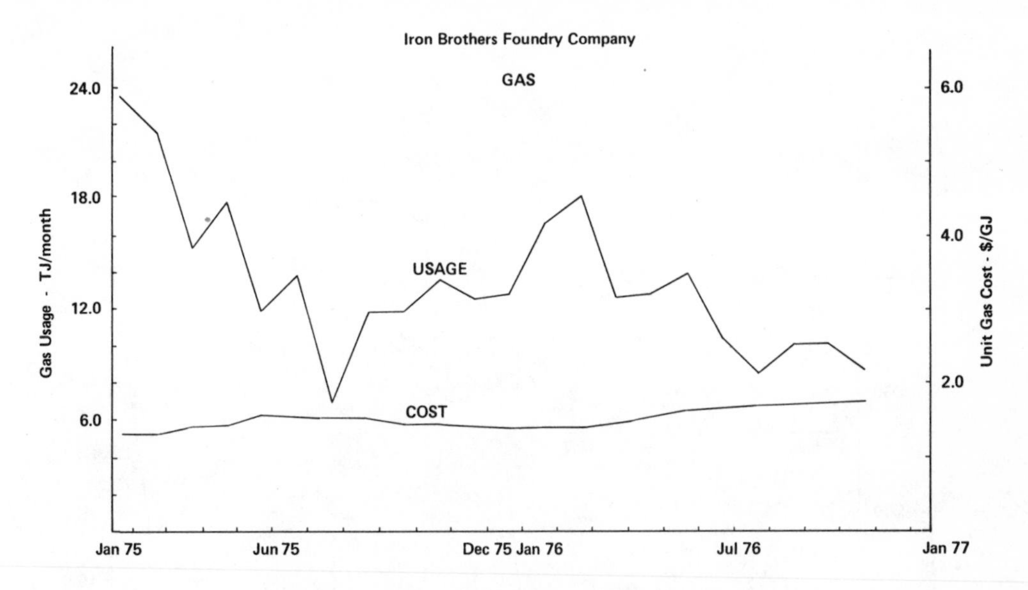

FIGURE 36. Foundry natural gas usage and unit costs.

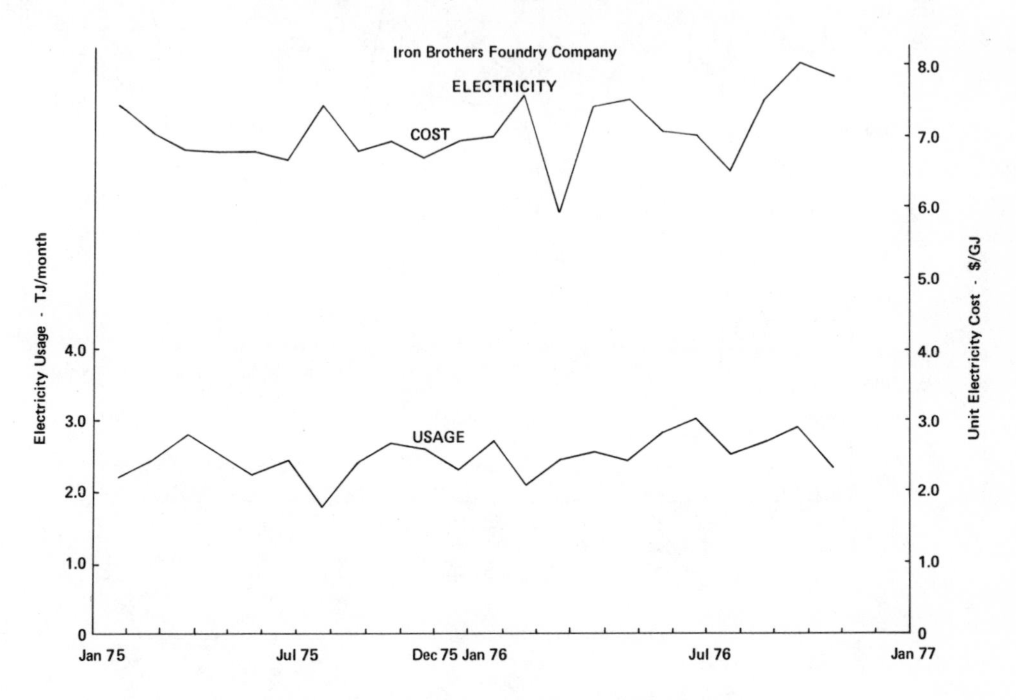

FIGURE 37. Foundry electricity usage and unit costs.

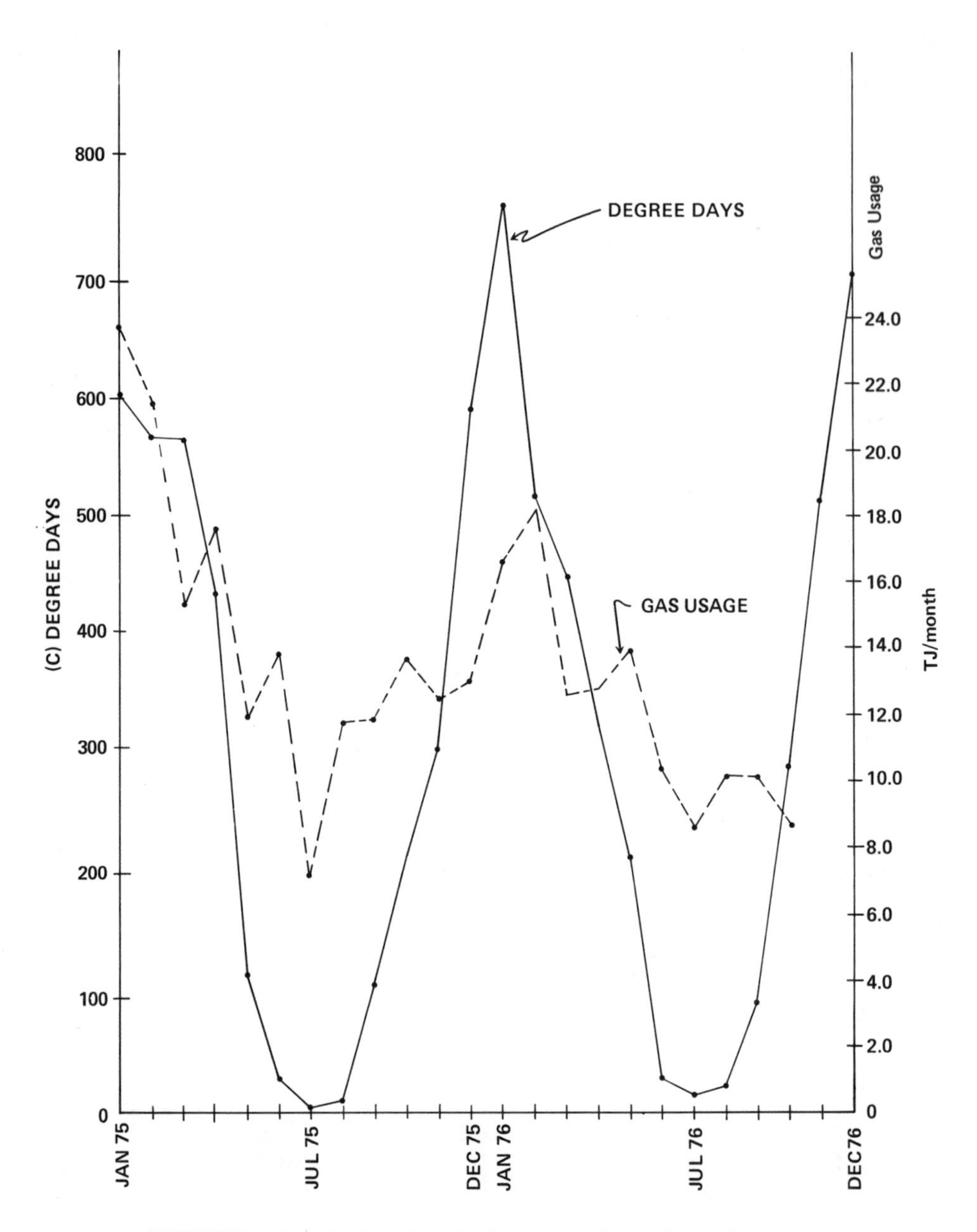

FIGURE 38. Iron Brothers Foundry Company — degree-days and gas usage.

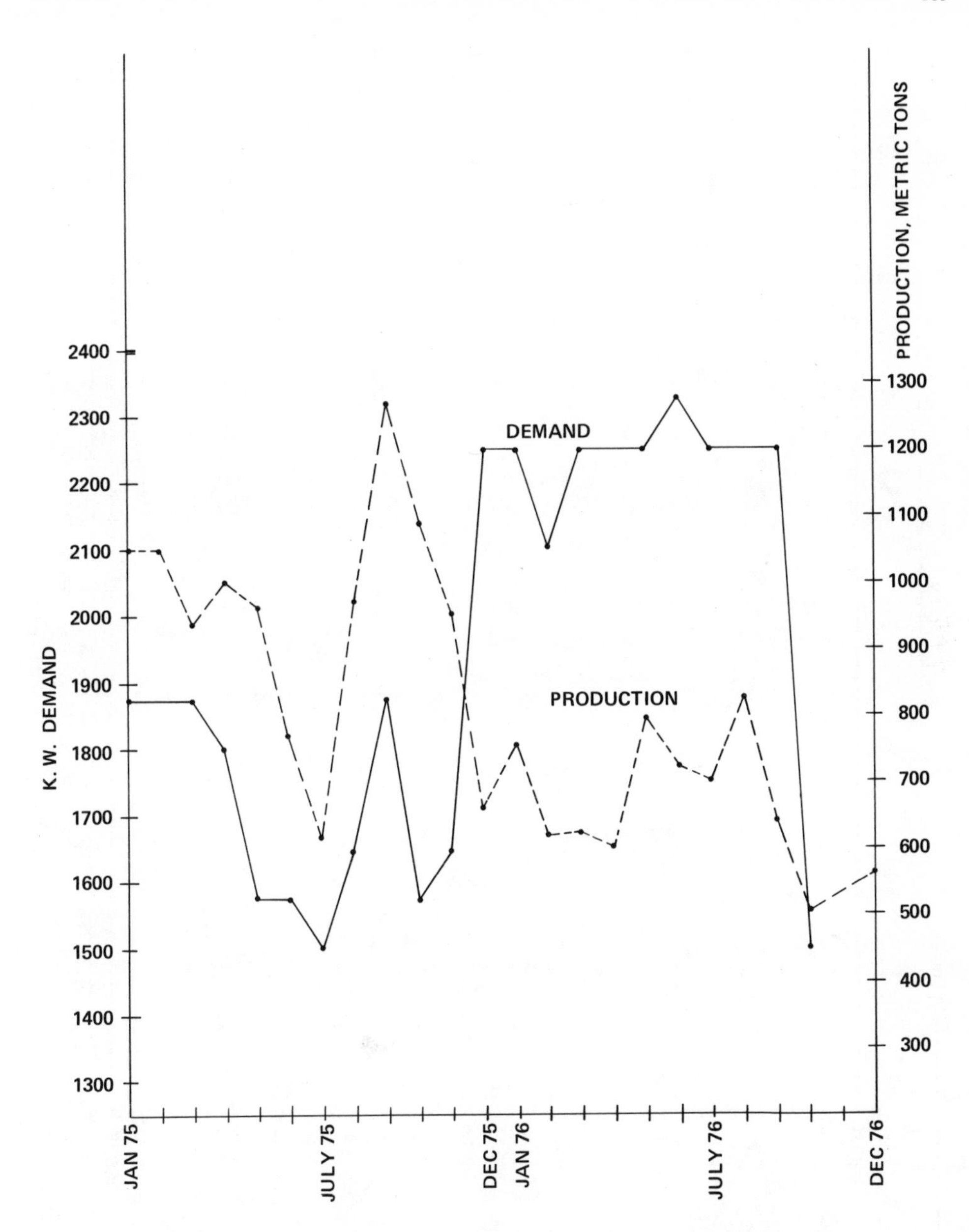

FIGURE 39. Iron Brothers Foundry Company — kW demand and production.

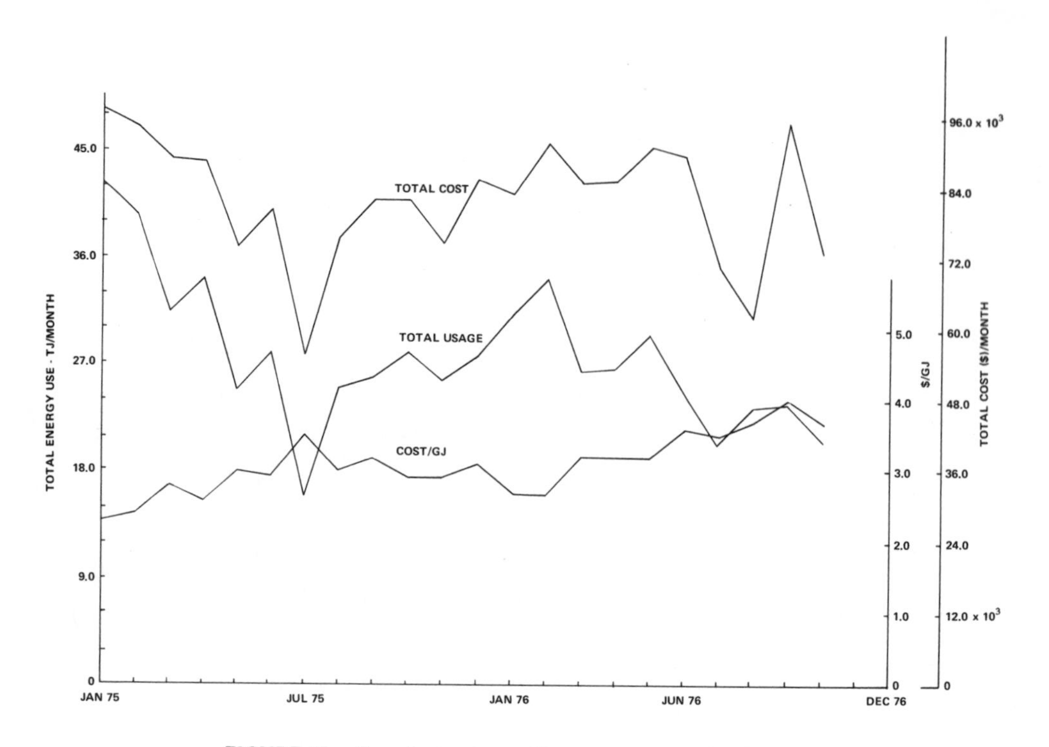

FIGURE 40. Foundry total monthly energy usage and unit costs.

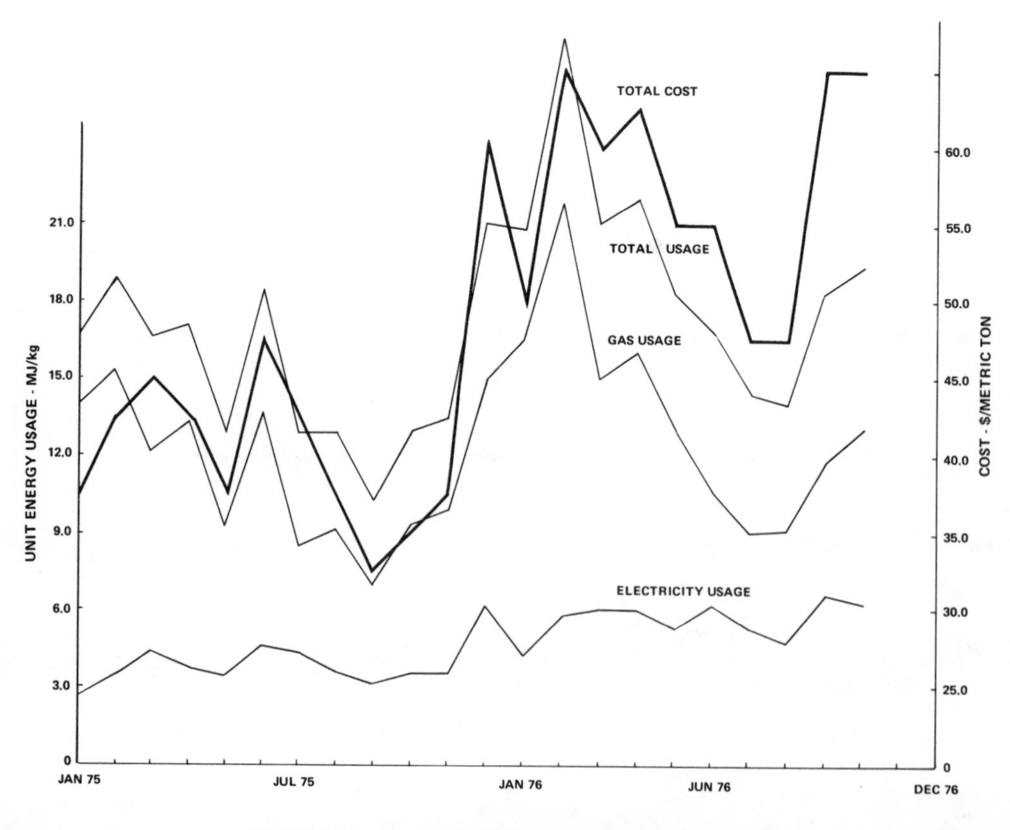

FIGURE 41. Energy use and cost per production unit.

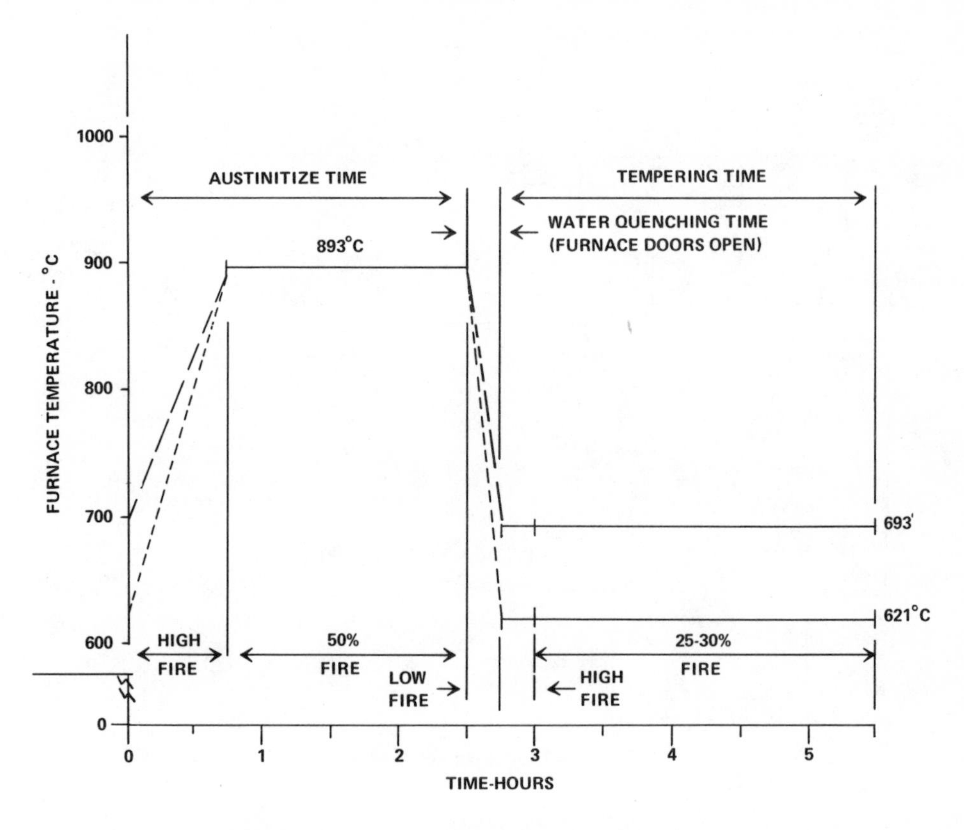

FIGURE 42. Temperature profile — heat treat furnaces.

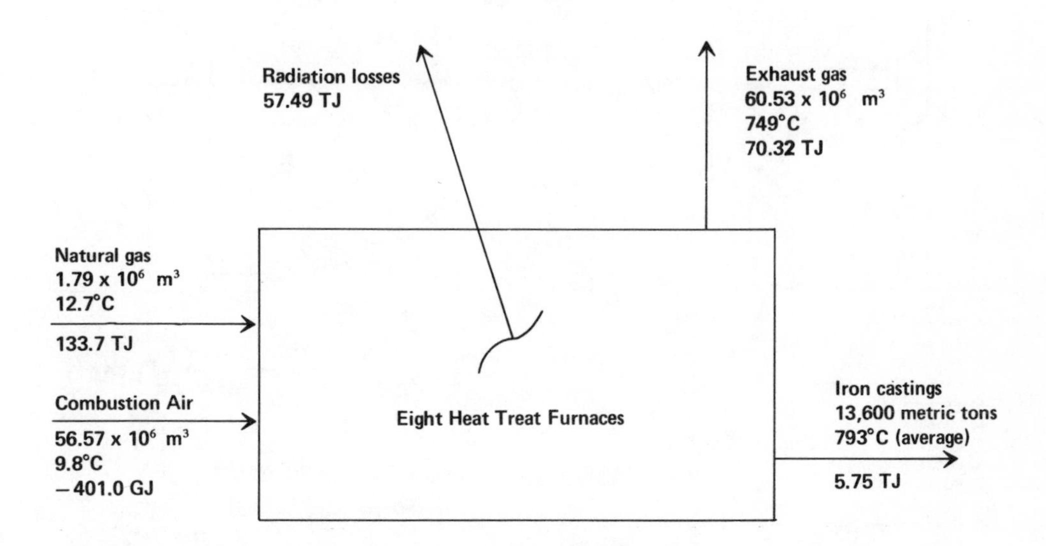

FIGURE 43. Annual heat balance on foundry treatment furnaces — Iron Brothers Foundry Company.

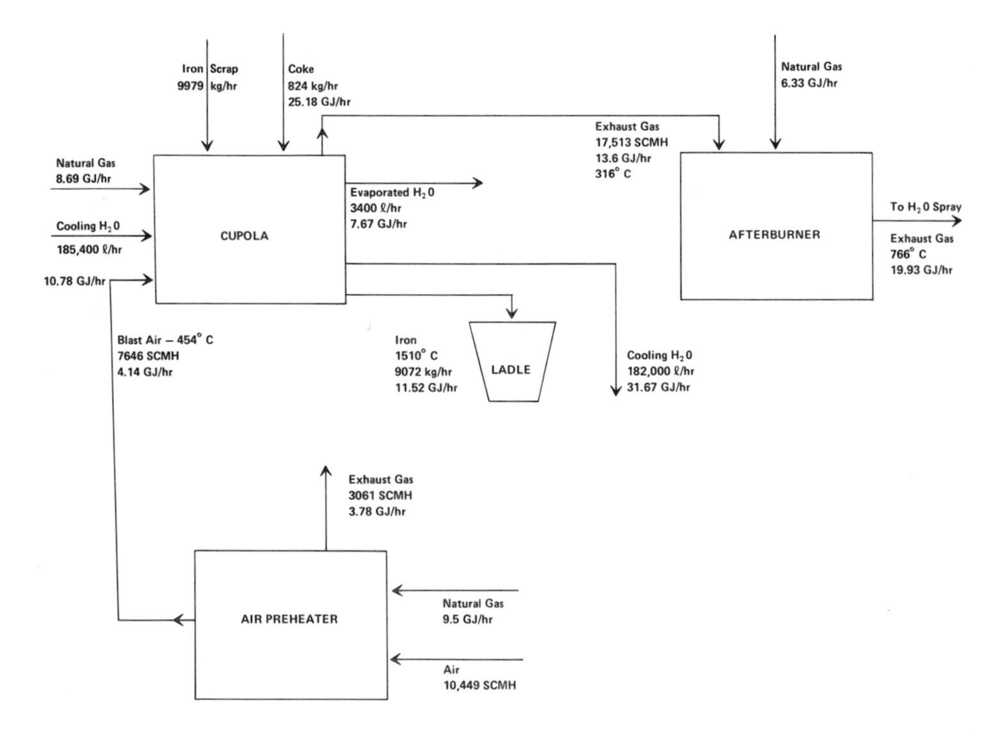

FIGURE 44. Block diagram of cupola.

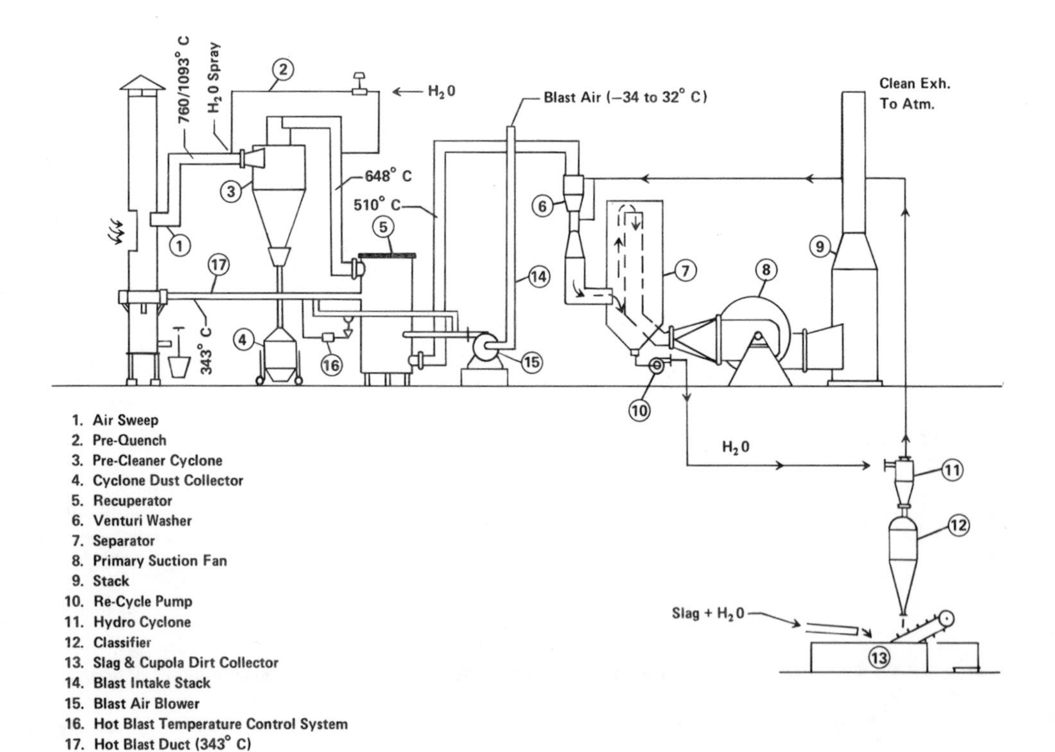

FIGURE 45. Cupola with recuperator installed. (Courtesy of Clean Air Engineering, Inc., Palatine, Ill. With permission.)

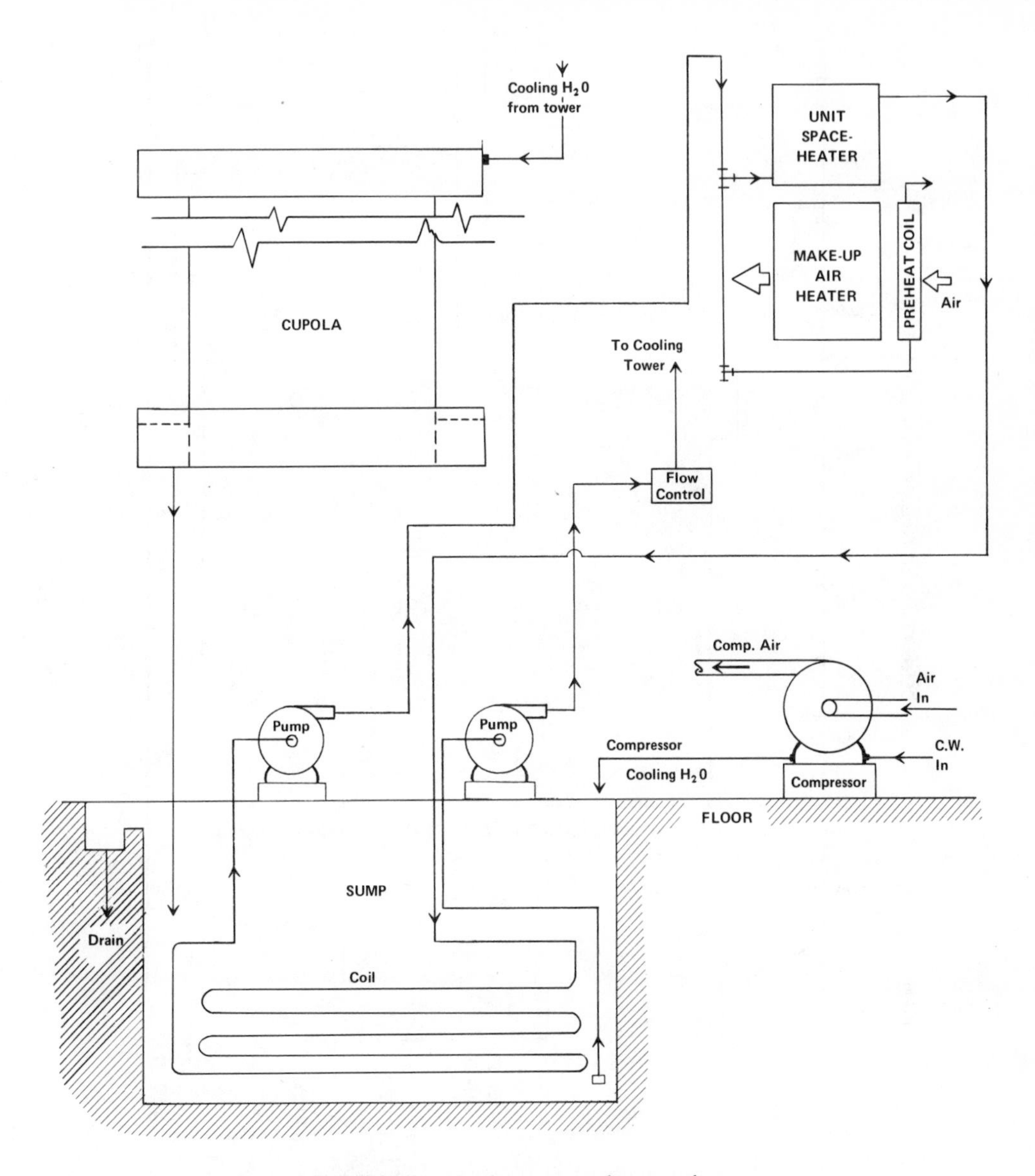

FIGURE 46. Heating system using waste heat.

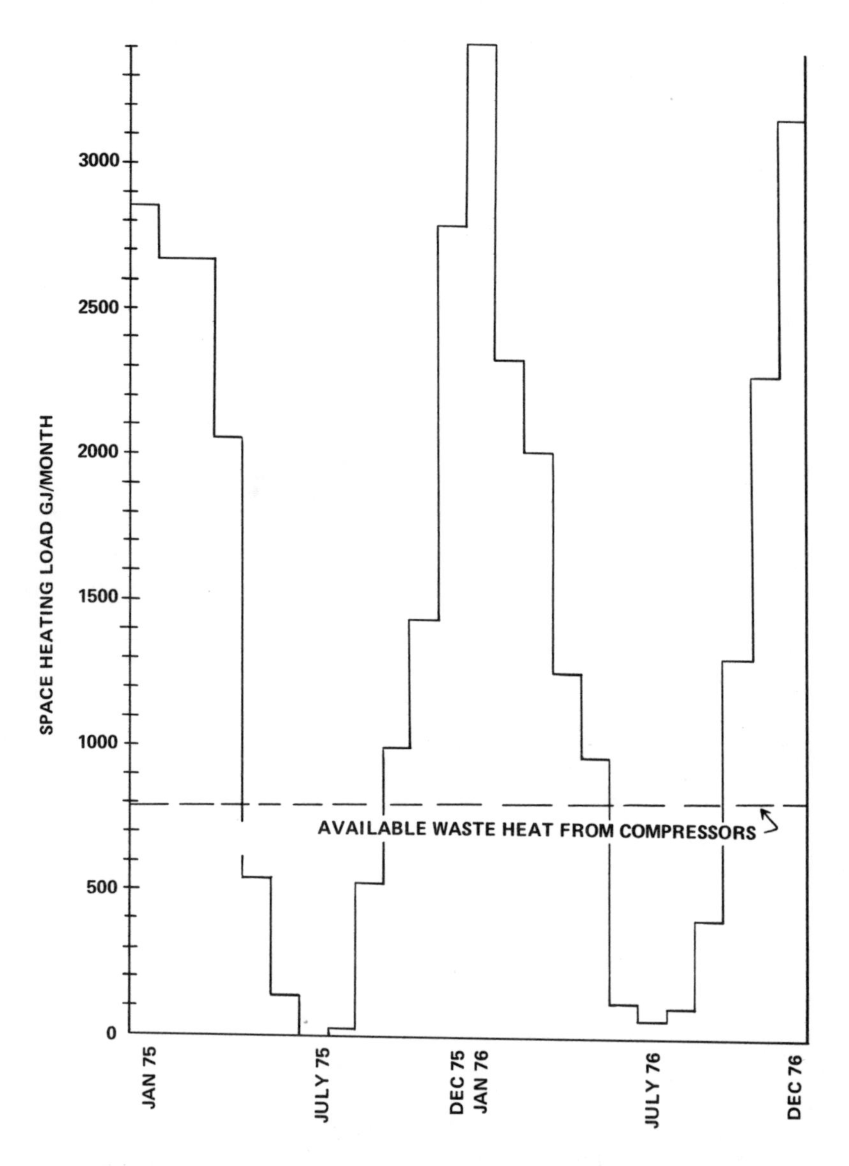

FIGURE 47. Space heating load — Iron Brothers Foundry Company.

REFERENCES

1. **Ross, M. H. and Williams, R. H.,** The potential for fuel conservation, *Technol. Rev.*, 79(4), 49, 1977.
2. **Gatts, R., Massey, R., and Robertson, J.,** Energy Conservation Program Guide for Industry and Commerce, NBS Handbook 115, U.S. Superintendent of Documents, Government Printing Office, Washington, D.C., 1974.
3. The Data Base, The Potential for Energy Conservation in Nine Selected Industries, NTIS PB-243-611/AS, Federal Energy Administration, Washington, D.C., 1975.
4. Evaluation of Building Characteristics Relative to Energy Consumption in Office Buildings, NTIS PB-248-774/2ST, Federal Energy Administration, Washington, D.C., 1976.
5. Energy Audit Procedures, Federal Energy Administration, Federal Register, Part IV, June 29, 1977.
6. **Culbert, M., Looney, Q., Rohrer, W., Rudoy, W., and Skoglund, T.,** Energy Analysis and Information System Workbook, to be published through Office of Energy Programs, U.S. Department of Commerce, Washington, D.C., 20230.
7. **Anon.,** ASHRAE Handbook of Fundamentals, American Society of Heating, Refrigerating and Air-Conditioning Engineers, New York, 1977.
8. Maintenance and Adjustment Manual for Natural Gas and No. 2 Fuel Oil Burners — Technical Information Center, U.S. Department of Energy, TID 27600, 1977.
9. Industrial Boiler Users' Manual, Vol. 2, NTIS PB 262-577, Federal Energy Administration, Washington, D.C., January 1977.
10. Measuring and Improving the Efficiency of Boilers — A Manual for Determining Energy Conservation in Steam Generating Power Plants, NTIS PB 265-713, Federal Energy Administration, Washington, D.C., November 1976.
11. Assessment of the Potential for Energy Conservation Through Improved Boiler Efficiency, Volume I, NTIS PB-262-576/AS, Federal Energy Administration, Washington, D.C., 1977.
12. **Kays, W. and London, A. L.,** *Compact Heat Exchangers,* 2nd ed., McGraw-Hill, New York, 1964.
13. Industrial Ventilation: A Manual of Recommended Practice, Committee on Industrial Ventilation, American Conference of Government Industrial Hygienists, Ann Arbor, Michigan, 1970.
14. A Study of Inplant Electric Power Generation in the Chemical, Petroleum Refining, and Paper and Pulp Industries. Executive Summary, NTIS PB-255-658, Federal Energy Administration, Washington, D.C., 1977.
15. **Kreider, K. and McNeil, M., Eds.,** Waste Heat Management Guidebook, NBS Handbook 121, Superintendent of Documents, Government Printing Office, Washington, D.C., 1976.
16. **Dukelow, S. G.,** Nomograph, Bailey Meter Company, Wickliffe, Ohio.

Chapter 4

ELECTRICAL POWER MANAGEMENT IN INDUSTRY

Craig B. Smith

TABLE OF CONTENTS

I. INTRODUCTION

A. Energy Use in Industry

Industrial energy use accounts for about 40% of all energy used in the U.S. (Table 1). The major end uses, in descending order of importance, are space heating and process heat, electric drives, water heating, air conditioning, and lighting (Table 2).

In terms of end use, electricity accounts for 12.5% of industrial energy. In terms of primary fuel input, electricity is much more significant, accounting for 30.7% of industrial energy. It is significant to note that in 1975, for the first time, the electric generating industry's fuel use surpassed that of industry. Thus electricity generation has now become the nation's number one user of fuels.

B. Industrial Electricity Use

Electricity use in industry is primarily for electric drives, electrolysis, electric heat, air conditioning, space heating, lighting, and refrigeration.[1] Tables 3 and 4 list some representative industrial processes and fuel and electricity use for each. The important ones will be discussed in more detail.

II. ELECTRICITY END USES IN INDUSTRY

A. Electric Drives

Of all U.S. electricity use, approximately 60% is for electric drives of one type or another.[1] Examples include electric motors, machine tools, compressors, refrigeration systems, fans, and pumps. Improvements in these applications would have a significant effect on reducing industrial electrical energy.

Motor efficiency can be improved in some cases by retrofit (modifications, better lubrication, improved cooling, heat recovery), but generally requires purchasing of more efficient units. Particularly for small motors (<10 kW), manufacturers today supply a wide range of efficiency. Greater efficiency in a motor requires improved design, more costly materials, and generally greater first cost.

Loss in electric driven systems may be divided into four categories:

	Typical efficiency (%)
Prime mover (motor)	10—95
Coupling (clutches)	80—99
Transmission	70—95
Mechanical load	1—90

Each category must be evaluated to determine energy management possibilities. In many applications the prime mover will be the most efficient element of the system. Table 5 shows typical induction motor data. Note that both efficiency and power factor decrease with partial load operation.

B. Electrolysis

Industrial uses of electrolysis include electrowinning, electroplating, electrochemicals, electrochemical machining, fuel cells, welding, and batteries. An indirect application of electrolysis is corrosion — which results in a substantial loss of materials which embody energy. It has been estimated that corrosion damage amounts to approximately 1% of the U.S. gross national product.[1]

A major use of electrolytic energy is in electrowinning — the electrolytic smelting

TABLE 1

Estimated U.S. Energy Use — 1975

	10^{15} Btu			Totals		
Sector	Used as fuel	Electricity conversion losses	Used as electricity	10^{15} Btu	10^9 GJ	%
Household and commercial	13.48	7.85	3.78	25.1	26.5	35
Industrial	18.95	5.69	2.71	27.4	28.9	39
Transportation	18.49	0.04	0.02	18.6	19.6	26
Miscellaneous	0.07	—	—	0.07	0.07	—
Totals	50.99	13.58	6.51	71.2	75.1	100

Note: The total equivalent fuel input for electrical generation is 13.58 + 6.51 = 20.09×10^{15} Btu.

From U.S. Department of Interior, Bureau of Mines, News Release, April 5, 1976.

TABLE 2

End Uses of Energy in the U.S. — 1975

Sector	(%)
Transportation	26.2
Space heating	17.6
Process steam	16.0
Direct heat	10.4
Electric drive	8.2
Feedstocks	5.1
Water heating	3.8
Air conditioning	3.3
Commercial lighting	2.6
Refrigeration	2.4
Cooking	1.1
Electrolytic processes	1.1
Other	2.2
Total	100

From National Academy of Engineering, U.S. Energy Prospects — An Engineering Viewpoint, Report prepared by the Task Force on Energy, Washington, D.C., 1974.

of primary metals such as aluminum and magnesium. Current methods require on the order of 13 to 15 kWh/kg; efforts are under way to improve electrode performance and reduce this to 10 kWh/kg.

Electroplating and anodizing are two additional uses of electricity of great importance. Electroplating is basically the electrodeposition of an adherent coating upon a base metal. It is used with copper, brass, zinc, and other metals. Anodizing is roughly the reverse of electroplating, with the workpiece (aluminum) serving as the anode. The reaction progresses inward from the surface to form a protective film of aluminum oxide on the surface.

TABLE 3

Estimated 1973 U.S. Electricity Use Patterns

Use	Percent of U.S. total energy use	Percent of all electricity uses
Electric drives	8	30
Space heating and A/C	6 (est)	22
Light	5	19
Refrigeration	2	7
Water heating and cooling	1	4
Electrolysis	1	4
Miscellaneous	4	14
Total	27	100

From Applied Nucleonics, Inc., Proceedings of an EPRI Workshop on Technologies for Conservation and Efficient Utilization of Electric Energy, EPRI EM-313-R, Electric Power Research Institute, Palo Alto, Calif., July 1976. With permission.

TABLE 4

Energy Use in Selected Industrial Processes (Includes Approximately one third of U.S. Industrial Energy)

	10^{15} Btu						
Industrial processes	Used as fuel	Approximate electricity conversion losses	Used as electricity	10^{15} Btu	10^{9} GJ	%	Temperature range (°C)
Thermal							
Drying	1.06	0.16	0.08	—	—	—	100
Distillation and separation	1.07	0.08	0.04	—	—	—	95—290
Cooking	0.19	—	—	—	—	—	120
Evaporation	0.19	—	—	—	—	—	95—150
Washing	0.11	—	—	—	—	—	40—90
Sterilization	0.10	—	—	—	—	—	65—120
Coking	0.99	—	—	—	—	—	1100
Furnace	0.58	0.24	0.12	—	—	—	1000—1550
Bake oven heater	0.21	0.04	0.02	—	—	—	260—1000
Annealing	0.01	—	—	—	—	—	800—1100
Reactors with preheaters	2.41	0.08	0.04	—	—	—	150—1500
Subtotal	6.92	0.60	0.30	7.82	8.26	77	
Electrolytic							
Electrolysis	0	0.58	0.29	—	—	—	N/A
Subtotal	0	0.58	0.29	0.87	0.92	9	—
Mechanical drives							
Mixing, crushing, grinding, separation	0	0.38	0.19	—	—	—	N/A
Compression	0.06	0.12	0.06	—	—	—	N/A
Refrigeration	0.05	0.10	0.05	—	—	—	N/A
Assembly	0	0.12	0.06	—	—	—	N/A
Extrusion and rolling	0	0.10	0.05	—	—	—	N/A
Filtration	0	0.08	0.04	—	—	—	N/A

Subtotal	0.11	0.90	0.45	1.46	1.54	14
Grand totals	7.03	2.08	1.04	10.15	10.72	100

From Electric Power Research Institute, *EPRI Journal*, 2(10), 40, 1977. With permission.

Fuel cells are devices for converting chemical energy to electrical energy directly through electrolytic action. Currently they represent a small use of energy, but current research is directed at developing large systems suitable for use by electric utilities for small dispersed generation plants.

Batteries are another major use of electrolytic energy, ranging in size from small units with energy storage in the joule or fractional joule capacity up to units proposed for electric utility use which will store 18×10^9J (5 MWh).

Electro-forming, etching, and welding are forms of electrolysis used in manufacturing and material shaping. The range of applications for these techniques stretches from microcircuits to aircraft carriers. In some applications, energy for machining is reduced and reduction of scrap also saves energy. Welding has benefits in the repair and salvage of materials and equipment, reducing the need for energy to manufacture replacements.

C. Electric Process Heat

Electricity is widely used as a source of process heat due to ease of control, cleanliness, wide range in unit capacities (watts to megawatts), safety, and low initial cost. Typical heating applications include resistance heaters (metal sheath heaters, ovens, furnaces), electric salt bath furnaces, infrared heaters, induction and high frequency resistance heating, dielectric heating, and direct arc electric furnaces.

Electric arc furnaces in the primary metals industry are a major use of electricity. Typical energy use in a direct arc steel furnace is about 0.6 kWh/kg. Major opportunities for improved efficiency with electric process heat applications include improved heat transfer surfaces, better insulation, heat recovery, and improved controls.

D. HVAC

Heating, ventilating, and air conditioning (HVAC) is an important use of energy in the industrial sector. The environmental needs in an industrial operation can be quite different from residential or commercial operations. In some cases, strict environmental standards must be met for a specific function or process. More often, the environmental requirements for the process itself are not limiting, but space conditioning is a prerequisite for the comfort of production personnel. The reader should refer to Volume III, Chapter 2 for a more complete discussion of energy management opportunities in HVAC systems.

E. Lighting

Industrial lighting needs range from low level requirements for assembly and welding of large structures (such as shipyards) to the high levels needed for manufacture of precision mechanical and electronic components such as integrated circuits.

Lighting uses about 20% of U.S. electrical energy and 5% of all energy. Of all lighting energy, about 20% is industrial, with the balance being residential, commercial, and miscellaneous. Although there are differences in equipment types and sizes of systems, energy management opportunities in industrial lighting systems are similar to those in residential/commercial systems (refer to Volume III, Chapter 2).

TABLE 5

Typical Induction Motor Data

Power (kW)	Weight (kg)	Horsepower	Weight (lb)	Amp.	Power factor percent ½ load	Power factor percent ¾ load	Power factor percent 4/4 load	Efficiency, % ½ load	Efficiency, % ¾ load	Efficiency, % 4/4 load
Squirrel-cage type										
0.75	30	1	65	3.31	60	71	78	71	76	76
1.5	45	2	100	5.70	71	81	86	78	80	80
3.7	72	5	159	13.4	76	84	87	83	84	84
7.5	116	10	255	26.2	81	87	88	85	86	85
14.9	190	20	418	52.2	85	88	89	84	85	84
29.8	365	40	804	98	86	88.5	89.5	89.5	90	89.5
74.6	802	100	1769	238	85	89.5	90.4	89	90.5	91
149.1	1463	200	3225	463	91	93	94	87	90	90
Wound-rotor type										
3.7	100	5	220	14.3	72.5	80	82.5	78	79	79.5
7.5	152	10	336	26.6	69	79	83	83	84.5	85
18.6	262	25	578	62.9	75	83.5	87	84	86	86.5
37.2	450	50	991	118.4	84	89	90	86	88	88
74.6	1187	100	2618	233	88	90.5	89.5	86	88	88
149.1	1770	200	3902	473	89	91	92	87	89	90
Three phase, 2300 v, 60 cycle, 1775 rpm Squirrel-cage type										
224	1451	300	3200	67	87.5	89.3	90.6	90.0	91.8	92.7
522	2359	700	5200	151	90.2	92.0	92.9	91.6	93.0	93.6
745	3493	1000	7700	212	91.2	92.8	93.7	92.2	93.4	94.0
Wound-rotor type										
224	1769	300	3900	67	84.7	89.0	90.0	90.0	91.8	92.7
522	2608	700	5750	151	88.5	91.8	92.6	91.6	93.0	93.6
745	3833	1000	8450	212	90.0	92.8	93.5	92.2	93.4	94.0

From Smith, C. B., **Ed.**, *Efficient Electricity Use*, 2nd ed., Pergamon Press, New York, 1978. With permission.

III. ENERGY AND POWER MANAGEMENT IN INDUSTRY

A. Setting up an Industrial Power Management Program[1]

The effectiveness of energy utilization varies with specific industrial operations because of diversity of the products and in the processes required to manufacture them. The organization of personnel and operations involved also varies. Consequently, an effective conservation program should be custom designed for each company and its plant operations. There are some generalized guidelines, however, for initiating and implementing an energy management program. Many large companies have already instituted energy management programs and have realized substantial savings in fuel utilization and costs. Smaller industries and plants, however, often lack the technical personnel and equipment to institute and carry out effective programs. In these situations, reliance on external consultants may be appropriate to initiate the program. Internal participation, however, is essential for success. A well-planned, organized, and executed energy management program requires a strong commitment by top management.

Assistance also can be obtained from local utilities. Utility participation would include help in getting the customer started on an energy management program, technical guidance, or making information available. Some utilities today have active programs which include training of customer personnel or provision of technical assistance.

Table 6 summarizes the elements of an effective energy management program. These will now be discussed in more detail.

1. Phase 1: Management Commitment

A commitment by the directors of a company to initiate and support a program is essential. An Energy Coordinator is designated and an energy management committee is formed. The committee should include personnel representing major company activities utilizing energy. A plan is formulated to set up the program with a commitment of funds and personnel. Realistic overall goals and guidelines in energy savings should be established based on overall information in the company records, projected activities, and future fuel costs and supply. A formal organization as described above is not an absolute requirement for the program; smaller companies will simply give the Energy Management Coordination task to a staff member.

2. Phase 2: Audit and Analysis

a. Energy Audit of Equipment and Facilities

Historical data for the facility should be collected, reviewed, and analyzed. The review should identify gross energy uses by fuel types, cyclic trends, fiscal year effects, dependence on sales or work load, and minimum and maximum energy use ratios. Historical data are graphed in a form similar to the one shown in Volume III, Chapter 2.

Historical data assist in planning a detailed energy audit and alert the auditors as to the type of fuel and general equipment to expect. A brief facility "walk-through" is recommended to establish the plant layout, major energy uses, and primary processes or functions of the facility.

The energy audit is best performed by an experienced or at least trained team, since visual observation is the principal means of information gathering and operational assessment. A team would have from three to five members, each with a specific assignment for the audit. For example, one auditor would check the lighting, another the HVAC system, another the equipment and processes, another the building struc-

TABLE 6

Elements of an Energy Management Program

Phase 1:	Management commitment
1.1	Commitment by management to an Energy Management Program
1.2	Assignment of an Energy Management Coordinator
1.3	Creation of an Energy Management Committee of major plant and department representatives
Phase 2:	Audit and analysis
2.1	Review of historical patterns of fuel and energy use
2.2	Facility walk-through survey
2.3	Preliminary analyses, review of drawings, data sheets, equipment specifications
2.4	Development of energy audit plans
2.5	Conduct facility energy audit, covering:
	a. Processes
	b. Facilities and equipment
2.6	Calculation of annual energy use based on audit results
2.7	Comparison with historical records
2.8	Analysis and simulation step (engineering calculations, heat and mass balances, theoretical efficiency calculations, computer analysis and simulation) to evaluate energy management options
2.9	Economic analysis of selected energy management options (life cycle costs, rate of return, benefit-cost ratio)
Phase 3:	Implementation
3.1	Establish energy effectiveness goals for the organization and individual plants
3.2	Determine capital investment requirements and priorities
3.3	Establish measurement and reporting procedures, install monitoring and recording instruments as required
3.4	Institute routine reporting procedures (''energy tracking'' charts) for managers and publicize results
3.5	Promote continuing awareness and involvement of personnel
3.6	Provide for periodic review and evaluation of overall energy management program

From Smith, C. B., **Ed.**, *Efficient Electricity Use*, 2nd ed., Pergamon Press, New York, 1978. With permission.

ture (floor space, volume, insulation, age, etc.), and another the occupancy use schedule, administration procedures, and employees' general awareness of energy management.

The objectives of the audit are to determine how, where, when, and how much energy is used in the facility. In addition, the audit helps to identify opportunities to improve the energy use efficiency of the facility and its operations.

Some of the problems encountered during energy audits are determining the rated power of equipment, determining the effective hours of use per year, and determining the effect of seasonal, climatic, or other variable conditions on energy use. Equipment ratings are often obscured by dust or grease (unreadable nameplates). Complex machinery may not have a single nameplate listing the total capacity, but several giving ratings for component equipment. The effect of load is also important because energy use in a machine operating at less than full load may be reduced along with a possible loss in operating efficiency.

The quantitative assessment of fuel and energy use is best determined by actual measurements under typical operational conditions using portable or installed meters and sensing devices. Such devices include light meters, ammeters, thermometers, air flow meters, recorders, etc. In some situations, sophisticated techniques such as infrared scanning or thermography are useful. The degree of measurement and recording sophistication naturally depends on available funds and the potential savings anticipated. For most situations, however, nameplate and catalog information are sufficient to estimate power demand. Useful information can be obtained from operating personnel and their supervisors — particularly as it relates to usage patterns throughout the day. A sample form which can be used for recording audit data is shown in Chapter 2.

The first three columns of the form are self-explanatory. The fourth column is used for the rated capacity of the device, e.g., 5 kW. The seventh column is used if the device is operated at partial load.

Usage hours (column 8) are based on all work shifts, and are corrected to account for the actual operating time of the equipment. The last three columns are used to convert energy units to a common basis (e.g., MJ or Btu).

Data recorded in the field are reduced easily either by hand or computer. The advantage of using a computer is that uniform results and summaries can be obtained easily in a form suitable for review or for further analysis. Computer analysis also provides easy modification of the results to reflect specific management reporting requirements or to present desired comparisons for different energy use, types of equipment, etc.

b. A Special Case: Energy Audit of a Process

In some manufacturing and process industries it is of interest to determine the energy content of a product. This can be done by a variation of the energy audit techniques described above. Since this approach resembles classical financial accounting, it is sometimes called *energy accounting.*

In this procedure the energy content of the raw materials is determined in a consistent set of energy units. Then, the energy required for conversion to a product is accounted for in the same units. The same is done for energy in the waste streams and the by-products. Finally, the net energy content per unit produced is used as a basis for establishing efficiency goals. A guide and procedures for calculating the energy content of a product are shown in Reference 1.

In this approach, all materials used in the product or to produce it are determined. Input raw materials used in any specific period are normally available from plant records. Approximations of specific energy content for some materials can be found in the literature or can be obtained from the U.S. Department of Commerce or other sources. The energy content of a material includes that due to extraction and refinement as well as any inherent heating value it would have as a fuel prior to processing. Consequently, nonfuel type ores in the ground are assigned zero energy, and petroleum products are assigned their alternate value as a fuel prior to processing in addition to the refinement energy. The energy of an input metal stock would include the energy due to extraction, ore refinement to metal, and any milling operations.

Conversion energy is an important aspect of the energy audit since it is under direct control of plant management. All utilities and fuels coming into the plant are accounted for. They are converted to consistent energy units (joules or Btu) using the actual data available on the fuels or using approximate conversions.

Electrical energy is assigned the actual fuel energy required to produce the electricity. This accounts for power conversion efficiencies. A suggested approach is to assume (unless actual values are available from your utility) that 10.8 MJ (10,200 Btu) is used to produce 3.6 MJe (1 kWh), giving a fuel conversion efficiency of $3.6 \div 10.8 = 0.33$ or 33%.

The energy content of process steam includes the total fuel and electrical energy required to operate the boiler as well as line and other losses. Some complexities are introduced when a plant produces both power and steam, since it is necessary to allocate the fuel used to the steam and power produced. One suggested way to make this allocation is to assume that there is a large efficient boiler feeding steam to a totally condensing vacuum turbine. Then, one must determine the amount of extra boiler fuel that would be required to permit the extraction of steam at whatever pressure while maintaining the constant load on the generator. The extra fuel is considered the energy content of the steam being extracted.

Waste disposal energy is that energy required to dispose of or treat the waste products. This includes all the energy required to bring the waste to a satisfactory disposal state. In a case where waste is handled by a contractor or some other utility service, it would include the cost of transportation and treatment energy.

If the plant has by-products or co-products, then an energy credit is allocated to them. A number of criteria can be used. If the by-product must be treated to be utilized or recycled (such as scrap), then the credit would be based on the raw material less the energy expended to treat the by-product for recycle. If the by-product is to be sold, the relative value ratio of the by-product to the primary product can be used to allocate the energy.

c. *Analysis of Audit Results, Identification of Energy Management Opportunities*

Often the energy audit will identify immediate energy management opportunities, such as unoccupied areas which have been inadvertently illuminated 24 hr per day, equipment operating needlessly, etc. Corrective housekeeping and maintenance action can be instituted to achieve short-term savings with little or no capital investment.

An analysis of the audit data is required for a more critical investigation of fuel waste and identification of the potentials for conservation. This includes a detailed energy balance of each process, activity, or facility. Process modifications and alternatives in equipment design should be formulated, based on technical feasibility and economic and environmental impact. Economic studies to determine payback, return on investment, and net savings are essential before making capital investments.

3. Phase 3: Implementation

At this point goals for saving energy can be established more firmly and priorities set on the modifications and alterations to equipment and the process. Effective measurement and monitoring procedures are essential in evaluating progress in the energy management program. Routine reporting procedures between management and operations should be established to accumulate information on plant performance and to inform plant supervisors of the effectiveness of their operations. Time-tracking charts of energy use and costs can be helpful. Involvement of employees and recognizing their contributions facilitate the achievement of objectives. Finally, the program must be continually reviewed and analyzed with regard to established goals and procedures.

B. Electric Load Analysis

The energy audit methodology is a general tool which can be used to analyze energy use in several forms and over a short or long period of time. Another useful technique, particularly for obtaining a short-term view of industrial electricity use, is an analysis based on evaluation of the daily load curve. Normally this analysis uses metering equipment installed by the utility and therefore available at the plant. However, special metering equipment can be installed if necessary to monitor a specific process or building.

For small installations both power and energy use can be determined from the kilowatt hour meter installed by the utility. Energy in kWh is determined by

$$E = \frac{K_h P_t C_t N}{1000} \quad \text{kWh}$$

where E = electric energy used, kWh; k_h = meter constant, watt hours/revolution; P_t = potential transformer ratio; C_t = current transformer ratio; N = number of revolutions of the meter disk. To determine energy use, the meter would be observed during an operation and the number of revolutions of the disk counted. Then the equation can be used to determine E.*

To determine the average load over some period p (hours), determine E as above for time p and then use the relation that:

$$L = \frac{E}{p} \text{ kW}$$

where E is in kWh; p is in hours; L is the load in kW.

Larger installations will have meters with digital outputs or strip charts. Often these will provide a direct indication of kWh and kW as a function of time. Some also indicate the reactive load (kVARs) or the power factor.

The first step is to construct the daily load curve. This is done by obtaining kWh readings each hour using the meter. The readings are then plotted on a graph to show the variation of the load over a 24-hr period. Table 7 shows a set of readings obtained over a 24-hr period in the XYZ manufacturing plant located in Sacramento, Calif. and operating one shift per day. These readings have been plotted in Figure 1.

Several interesting conclusions can be immediately drawn from this figure.

- The greatest demand for electricity occurs at 11:00.
- Through the lunch break the third highest demand occurs.
- The ratio of the greatest demand to the least demand is approximately 3:1.
- Only approximately 50% of the energy used actually goes into making a product (54% on-shift use, 46% off-shift use).

When presented to management, these facts were of sufficient interest that a further study of electricity use was requested.

Additional insight into the operation of a plant (and into the cost of purchase of electricity) can be obtained from the load analysis. Following a brief discussion of electrical load parameters, a load analysis for the XYZ Company will be described.

Any industrial electrical load consists of lighting, motors, chillers, compressors and other types of equipment. The sum of the capacities of this equipment, in kW, is the *connected load.* The actual load at any point in time is normally less than the connected load since every motor is not turned on at the same time; only part of the lights may be on at any one time, etc. Thus the load is said to be *diversified,* and a measure of this can be gotten by calculating a *diversity factor:*

$$DV = \frac{D_{m1} + D_{m2} + D_{m3} + \text{etc.}}{D_{max}}$$

where: D_{m1} D_{m2} etc. = sum of maximum demand of individual loads kW; D_{max} = maximum demand of plant kW. If the individual loads do not occur simultaneously, (usually they do not) the diversity factor will be greater than unity. Typical values for industrial plants are 1.3 to 2.5.

* The value of K_h is usually marked on the meter. P_t and C_t are usually 1.0 for small installations.

If each individual load operated to its maximum extent simultaneously, the maximum *demand* for power would be equal to the connected load and the diversity factor would be 1.0. However, as pointed out above, this does not happen except for special cases.

The demand for power varies over time as loads are added and removed from the system. It is usual practice for the supplying utility to specify a *demand interval* (usually 0.25, 0.5, or 1.0 hr) over which it will calculate the demand and compute the demand charge using the relationship:

$$D = \frac{E}{p} \quad kW$$

where: D = demand, kW; E = kilowatt hours used during p; p = demand interval, hours. The demand calculated in this manner is an average value, being greater than the lowest instantaneous demand during the demand interval, but being less than the maximum demand during the interval.

Utilities are interested in *peak demand* since this determines the capacity of the equipment they must install to meet the customer's power requirements. This is measured by a demand factor, defined as

$$DF = \frac{D_{max}}{CL}$$

where D_{max} = maximum demand, kW; CL = connected load, kW. The demand factor is normally less than unity; typical values range from 0.25 to 0.90.

Since the customer normally pays a premium for the maximum load placed on the utility system, it is of interest to determine how effectively the maximum load is used. The most effective use of the equipment would be to have the peak load occur at the start of the use period, and continue unchanged throughout it. Normally, this does not occur, and a measure of the extent to which the maximum demand is sustained throughout the period (a day, month, or year) is given by the *Hours Use of Demand:*

$$HUOD = \frac{E}{D_{max}} \quad \text{hours}$$

where: HUOD = Hours Use of Demand, hours; E = energy used in period p, kWh; D_{max} = maximum demand during period p, in kW; p = period over which HUOD is determined, e.g., 1 day, 1 month, or 1 year (p is always expressed in *hours*).

The *Load Factor* is another parameter which measures the plant's ability to use electricity efficiently. In effect it measures the ratio of the average load for a given period of time to the maximum load which occurs during the same period. The most effective use results when the load factor is as high as possible once E or HUOD have been minimized (it is always less than one). The load factor is defined as:

$$LF = \frac{E}{(D_{max})(p)}$$

where: LF = load factor (dimensionless); E = energy used in period p, kWh; D_{max} = maximum demand during period p, in kW; p = period over which load factor is determined, (e.g., 1 day, 1 month, or 1 year) in hours. Another way to determine LF is from the relation

$$LF = \frac{HUOD}{p}$$

Still another method is to determine the average load, L = kWh/p during p divided by p and then use the relation:

$$LF = \frac{L}{D_{max}}$$

These relations are summarized for convenience in Table 8.

Returning to the XYZ plant, the various load parameters can now be calculated. Table 9 summarizes the needed data and the results of the calculations.

The most striking thing shown by the calculations is the hours use of demand, equal to 15.6. This is a surprise, since the plant is only operating one shift. The other significant point brought out by the calculations is the low load factor.

An energy audit of the facility was conducted and the major loads were evaluated. The audit results revealed a number of energy management opportunities whereby both loads (kW) and energy use (kWh) could be reduced.

The audit indicated that inefficient lighting (on about 12 hr per day) could be replaced in the parking lot. General office lighting was found to be uniformly at 100 fc; by selective reduction and task lighting the average level could be reduced to 75 fc. The air conditioning load would also be reduced. Improved controls could be installed to automatically shut down lighting during off-shift and weekend hours (the practice had been to leave the lights on). Some walls and ceilings were selected for repainting to improve reflectance and reduced lighting energy. It was found that the air conditioning chillers operated during weekends and off-hours; improved controls would prevent this. Also, the ventilation rates were found to be excessive and could be reduced. In the plant, compressed air system leaks, heat losses from plating tanks, and on-peak operation of the heat treat furnace represented energy and load management opportunities.

The major energy management opportunities were evaluated to have the following potential savings, with a total payback of 5.3 months:

	Savings	
	kW	kWh/yr
More efficient parking lot lighting	16	67,000
Reduce office lighting	111	495,000
Office lighting controls to reduce off-shift use	—	425,000
Air conditioning system controls and smaller fan motor	71	425,000

TABLE 7

Kilowatt Hour Meter Readings for XYZ Manufacturing Company

Time meter read	Elapsed kWh	Notes concerning usage	% of total usage
1:00 (a.m.)	640		
2:00	610		
3:00	570		
4:00	570	7 hr preshift use	
5:00	640		
6:00	770		
7:00	1,120		
Subtotal	4,920		17%
8:00	1,470		
9:00	1,700		
10:00	1,790		
11:00	1,850 ← (PEAK)	9 hr on shift use	
12:00 (noon)	1,830		
13:00 (1 p.m.)	1,790		
14:00 (2 p.m.)	1,790		
15:00 (3 p.m.)	1,760		
16:00 (4 p.m.)	1,690		
Subtotal	15,670		54%
17:00 (5 p.m.)	1,470		
18:00 (6 p.m.)	1,310		
19:00 (7 p.m.)	1,210		
20:00 (8 p.m.)	1,090	8 hr postshift use	
21:00 (9 p.m.)	960		
22:00 (10 p.m.)	800		
23:00 (11 p.m.)	730		
24:00 (12 p.m.)	640		
Subtotal	8,210		29%
Grand totals	28,800		100%

Compressed air system repairs and reduction of heat losses from plating tanks	—	200,000
Shift heat treat oven off-peak	37	—
Totals	235	1,612,000

The average daily savings of electricity amounted to approximately 4400 kWh/day. This led to savings of $33,000 per year, with the cost of the modifications being $14,500.

The revised electricity use was found to be:

	kWh	%
Pre-shift	4,400	18
On shift	14,400	59
Post-shift	5,600	23
Totals	24,400	100

This can be compared to the original situation (Table 7). See also Figure 2, which shows the daily load curve after the changes have been made.

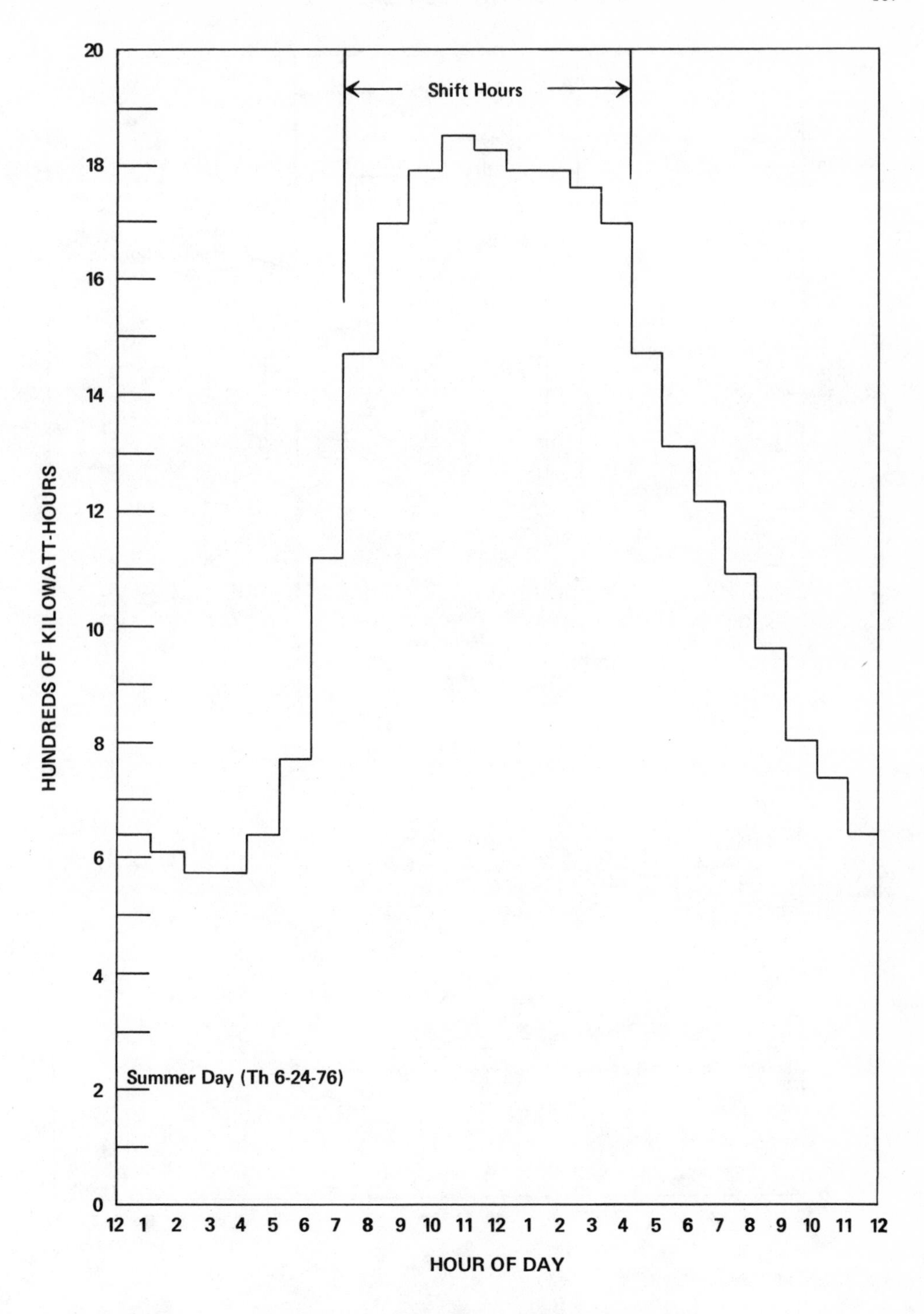

FIGURE 1. Daily load curve XYZ company.

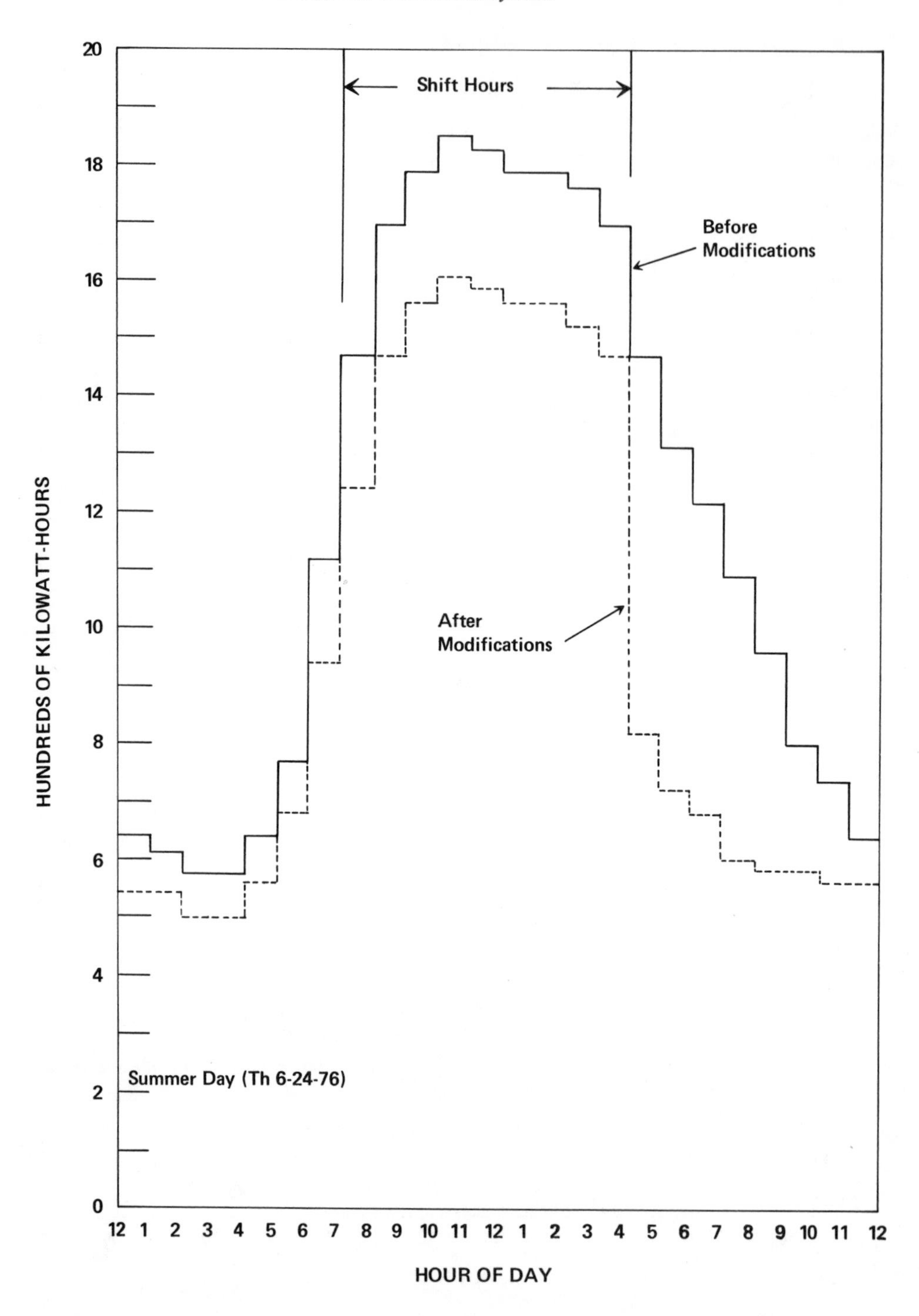

FIGURE 2. Daily load curve for XYZ company after modifications.

TABLE 8

Summary of Load Analysis Parameters

Formulae

$$E = \frac{K_h P_t C_t N}{1000}$$

$$L = \frac{E}{P}$$

$$DV = \frac{D_{m1} + D_{m2} + D_{m3}}{D_{max}}$$

$$D = \frac{E}{p} \quad D_{max} = \frac{E_{max}}{p}$$

$$DF = \frac{D_{max}}{CL}$$

$$HUOD = \frac{E}{D_{max}}$$

$$LF = \frac{E}{(D_{max})(p)} = \frac{HUOD}{p} = \frac{L}{D_{max}}$$

Definitions

E = Electric energy used in period p, kWh

E_{max} = Maximum energy used during period p, kWh

K_h = Meter constant, watt hours/revolution

P_t = Potential transformer ratio

C_t = Current transformer ratio

N = Number of revolutions of the meter disk

L = Average load, kW

p = Period of time used to determine load, demand, electricity use, etc., normally 1 hour, day, month, or year; measured in hours

DV = Diversity factor, dimensionless

D_{max} = Maximum demand in period, p, kW

D_{m1}, D_{m2}, etc. = Maximum demand of individual load, kW

D = Demand during period p, kW

DF = Demand factor for period p, dimensionless

CL = Connected load, kW

HUOD = Hours use of demand during period p, hours

LF = Load factor during period p, dimensionless

TABLE 9A

Data for Load Analysis of XYZ Plant

P = 24 hr

E = 28800 kWh/day

E_{max} = 1850 kWh

CL = 2792 kW

$D_{m1,\,2,\,3,\,etc.}$ = 53, 62, 144, 80, 700, 1420 kW

TABLE 9B

Sample Calculations for XYZ Plant

1. $D_{max} = \dfrac{E_{max}}{p} = \dfrac{1850 \text{ kWh}}{1 \text{ hr}} = 1850 \text{ kW}$

2. $DV = \dfrac{D_{m1} + D_{m2} + D_{m3} \ldots}{D_{max}} = \dfrac{2459}{1850} = 1.33$

3. $DF = \dfrac{D_{max}}{CL} = \dfrac{1850}{2792} = 0.66$

4. HUOD (daily) $= \dfrac{E}{D_{max}} = \dfrac{28800 \text{ kWh/day}}{1850 \text{ kW}} = 15.6 \text{ hr/day}$

5. LF (daily) $= \dfrac{HUOD}{p} = \dfrac{15.6}{24.0} = 0.65$

The load parameters after the changes were made can be found:

$$D_{max} = \frac{1850 - 235}{1 \text{ hr}} = 1615 \text{ kW}$$

$$HUOD = \frac{24{,}400 \text{ kWh/day}}{1615} = 15.1 \text{ hrs/day}$$

$$LF = \frac{15.1}{24} = 0.63$$

The percentage of use on shift is now higher. Note that D_{max} has been improved significantly (reduced by 13%); the HUOD has improved slightly (about 3% lower now); and the LF is slightly lower.

Further improvements are undoubtedly still possible in this facility; they should be directed first at reducing nonessential uses, thereby reducing HUOD.

So far the discussion has dealt entirely with power and has neglected the reactive component of the load. In the most general case the apparent power in kVA which must be supplied to the load is the vector sum of the active power in kW and the *reactive power* in KVAR:*

$$|S| = \sqrt{P^2 + Q^2}$$

where: S = apparent power, kVA; P = active power, kW; Q = reactive power,

* The reader who is unfamiliar with these terms should refer to a basic electrical engineering text.

KVAR. In this notation the apparent power can be considered a vector of magnitude S and angle θ, where θ is commonly referred to as the phase angle and is given as

$$\theta = \tan^{-1}\left(\frac{\text{KVAR}}{\text{kW}}\right)$$

Another useful parameter is the *power factor* given by

$$\text{pf} = \cos\theta$$

The power factor is also given by

$$\text{pf} = \frac{|P|}{|S|}$$

The power factor is always less than or equal to unity. A high value is desirable since it implies a small reactive component to the load. A low value means the reactive component is large.

The importance of the power factor is related to the reactive component of the load. Even though the reactive component does not dissipate power (it is stored in magnetic or electric fields), the switch gear and distribution system must be sized to handle the current required by the apparent power, or vector sum of the active and reactive components. This results in a greater capital and operating expense. The operating expense is increased due to the standby losses which occur in supplying the reactive component of the load.

Power factor can be improved by adding capacitors to the load to compensate for part of the inductive reactance. The benefit of this approach depends on the economics of each specific case and generally requires a careful review or analysis.

These points can be clarified with an example. Consider the distribution system shown in Figure 3. Four loads are supplied by a 600 A bus. Load A is a distant load which has a large reactive component and a low power factor (pf = 0.6). To supply the active power requirement of 75 kW, an apparent power of 125 kVA must be provided and a current of 150 A is required.

The size of the wire to supply the load is dictated by the current to be carried and voltage drop considerations. In this case, # 3/0 wire which weighs 508 lbs per 1000 ft and has a resistance of 0.062 Ω per 1000 ft is used. Since the current flowing in this conductor is 150 A, the power dissipated in the resistance of the conductor is:

$$P = \sqrt{3}\, i^2 r = (1.732)\ (150^2)\ (0.124)\ \text{W}$$

$$P = 4.8\ \text{kW}$$

Similar calculations can be made for load B, which uses #1 wire, at 253 lbs per 1000 ft and 0.12 Ω/1000 ft.

Now the effect of the power factor is visible. Although the active power is the same for both load A and load B, load A requires 150 A vs. 90 A for load B. The i^2r standby

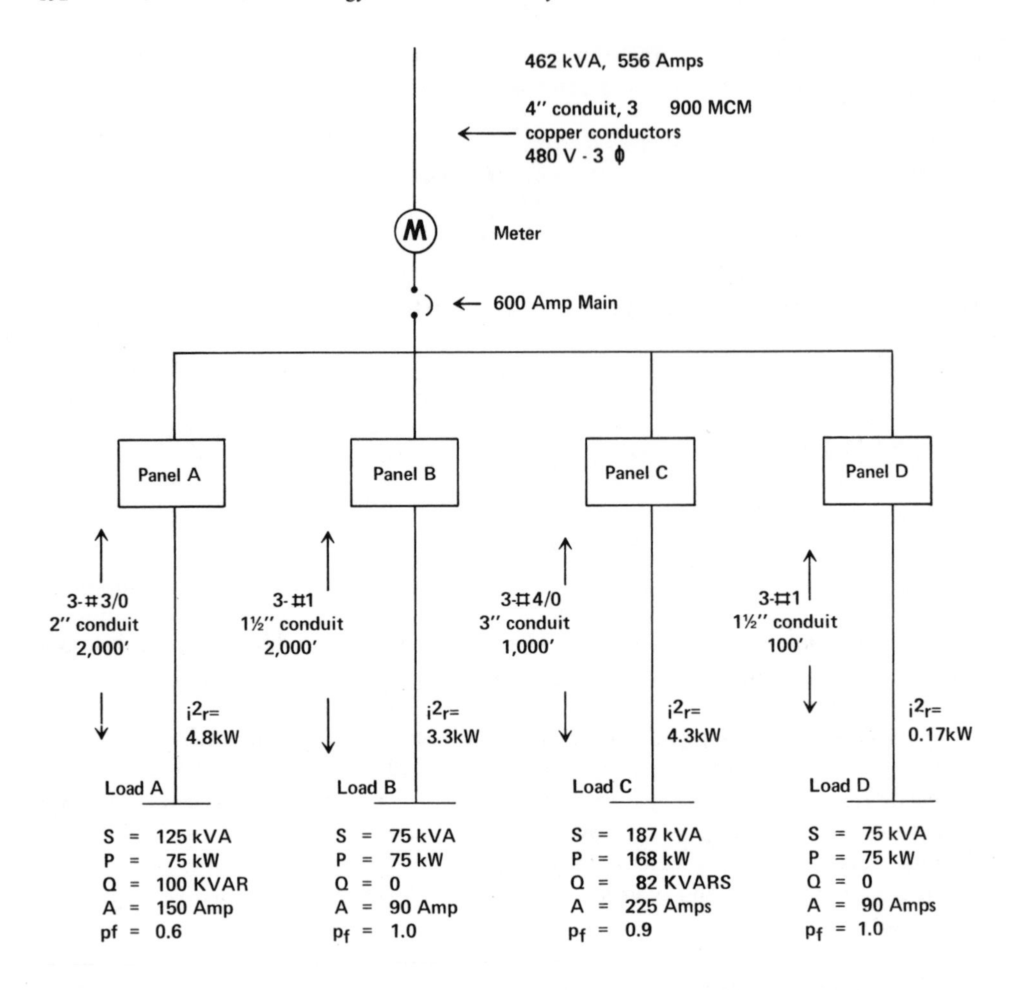

FIGURE 3. Electrical load diagram Building 201, XYZ company.

losses are also higher for load A as opposed to load B. The installation cost to service load B is roughly one half that of load A, due to the smaller wire, switches, and conduit sizes possible.

Note the greater power dissipation due to the long conduit run of load B compared to load D. For large loads which are served over long distances and which operate continuously, consideration should be given to using larger wire sizes to reduce standby losses.

An estimate of the annual cost of power dissipated as heat in these conduit runs can be made if the typical operating hours of each load are known:

Load	Line losses in kW	Operating hours/year	kWh/year
A	4.8	2,000	9,600
B	3.3	2,000	6,600
C	4.3	4,000	17,200
D	0.17	2,000	340
Total			33,740

At an average cost of 3¢/kWh (includes demand and energy costs), the losses in the distribution system alone are $1011 per year. Over the life of the facility, this is a major expense for a totally unproductive use of energy.

C. Energy Management Strategies for Industry*

Energy management strategies for industry can be grouped into three categories:

- Operational and maintenance strategies
- Retrofit or modification strategies
- New design strategies

The order in which these are listed corresponds approximately to increased capital investment and increased implementation times. Immediate savings at little or no capital cost can generally be achieved by improved operations and better maintenance. Once these "easy" savings have been realized, additional improvements in efficiency will require capital investments.

1. Electric Drives and Electrically Driven Machinery

About 80% of industrial electricity use is for electrical motor driven equipment.[6] Integral horsepower (> 0.75 kW) motors account for more than 90% of industrial electricity use, even though fractional horsepower (<0.75 kW) motors are more numerous. Major industrial electricity loads are[6]

	Percent of industrial electricity (1973)
Pumps	24
Electrolysis, electric heat, HVAC, lighting, miscellaneous	23
Compressors	14
Blowers and fans	12
Miscellaneous integral motor applications	9
DC drives	8
Machine tools	7
Fractional hp applications	3
Total	100%

The typical industrial motor is a polyphase motor rated at 11.2 kW (15 hp) and having a life of about 40,000 hr. The efficiencies of electric motors under 15 hp have degraded over the past 15 to 20 years due in part to low energy costs. Larger motors are about as efficient as previously.[6] Higher efficiency motors can be found on the market today, but usually at a premium price.

Most efficient use of motors requires that attention be given to the following:

Optimum power — Motors operate most efficiently at rated voltage. Three phase power supplies should be balanced; an unbalance of 3% can increase losses 25%.[1]

Good motor maintenance — Provide adequate cooling, keep heat transfer surfaces and vents clean, provide adequate lubrication. Improved lubrication alone can increase efficiency a few percentage points.

Equipment scheduling — Turn equipment off when not in use; schedule large motor operation to minimize demand peaks.

Size equipment properly — Match the motor to the load and to the duty cycle. Motors operate most efficiently at rated load.

* This section draws extensively on material in Reference 1, particularly Chapter 2 of Reference 1. I would like to acknowledge my debt to various authors, and note my appreciation to Pergamon Press, New York, for granting permission to use this material.

Evaluate continuous vs. batch processes — Sometimes a smaller motor operating continuously will be more economical.

Power factor — Correct if economics dictate savings. Motors have the best power factor at rated load.

Retrofit or new designs permit use of more efficient motors. For motors up to about 10 to 15 kW (15 to 20 hp) there are wide variations in efficiency. Select the most efficient motor for the job. Check to verify that the additional cost (if any) will be repaid by the savings which will accrue over the life of the installation.

In addition to reviewing the electric drive system, consider the power train and the load. Friction results in energy dissipation in the form of heat. Bearings, gears, and belt drives all have certain losses, as do clutches. Proper operation and maintenance can reduce energy wastage in these systems and improve overall efficiency.

Material shaping and forming, such as is accomplished with machine tools, requires that electrical energy be transformed into various forms of mechanical energy. The energy expenditure is related to the material and to the depth and speed of the cut. By experimenting with a specific process, it is possible to establish cutting rates which are optimum for the levels of production required and which are most efficient in terms of energy use. Motors are not the only part of the electric drive system which sustain losses. Other losses occur in the electric power systems which supply the motor.[1] Electric power systems include substations, transformers, switchgear, distribution systems, feeders, power and lighting panels, and related equipment. Possibilities for energy management include:

Use highest voltages which are practical — For a given application, doubling the voltage cuts the required current in half and reduces the i^2r losses by a factor of four.

Eliminate unnecessary transformers — They waste energy. Proper selection of equipment and facility voltages can reduce the number of transformers required and cut transformer losses. Remember, the customer pays for losses when the transformers are on his side of the meter. Example: it is generally better to order equipment with motors of the correct voltage, even if this costs more, than to install special transformers.

Energy losses are an inherent part of electric power distribution systems — This is primarily due to i^2r losses and transformers. The end use conversion systems for electrical energy used in the process also contribute to energy waste. Proper design and operation of an electrical system can minimize energy losses and contribute to the reduction of electricity bills. Where long feeder runs are operated at near-maximum capacities, check to see if larger wire sizes would permit savings and be economically justifiable.

The overall power factor of electrical systems should be checked for low power factor — This could increase energy losses and the cost of electrical service, in addition to excessive voltage drops and increased penalty charges by the utility. Electrical systems studies should be made and consideration should be given to power factor correction capacitors. In certain applications as much as 10 to 15% savings can be achieved in a poorly operating plant.

Check load factors — This is another parameter which measures the plant's ability to use electrical power efficiently. It is defined as the ratio of the actual kWh used to the maximum demand in kW times the total hours in the time period. A reduction in demand to bring this ratio closer to unity without decreasing plant output means more economical operation. For example, if the maximum demand for a given month (200 hr) is 30,000 kWe and the actual kWh is 3.6×10^6 kWh, the load factor is 60%. Proper management of operations during high demand periods which may extend only 15 to 20 min can reduce the demand during that time without curtailing production. For example, if the 30,000 kWe could be reduced to 20,000 kWe, it would increase the load factor to about 90%. Such a reduction could amount to a $15,000 to $25,000 reduction in the electricity bill.

Reduce peak loads wherever possible — Many nonessential loads can be shed during the demand peak without interrupting production. These loads would include such items as air compressors, heaters, coolers, and air conditioners. Manual monitoring and control is possible but is often impractical because of the short periods of time that are normally involved and the lack of centralized control systems. Automatic power demand control systems are available.

Provide improved monitoring or metering capability, submeters, or demand recorders — While it is true that meters alone will not save energy, plant managers need feedback to determine if their energy management programs are taking effect. Often the installation of meters on individual processes or buildings leads to immediate savings of 5 to 10% by virtue of the ability to see how much energy is being used and to test the effectiveness of corrective measures.

2. Fans, Blowers, Pumps

Simple control changes are the first thing to consider with these types of equipment. Switches, timeclocks, or other devices can insure that they do not operate except when needed by the process.

Heat removal or process mass flow requirements will determine the size of fans and pumps. Often there is excess capacity, ether as a result of design conservatism, or because of process changes subsequent to the installation of equipment. The required capacity should be checked, since excess capacity leads to unnecessary demand charges and decreased efficiency.

For fans, the volume rate of air flow Q varies in proportion to the speed of the impeller:

$$Q = c_f N \quad m^3/sec$$

where: Q = air flow in m^3/sec; c_f = a constant with units m^3/r; and N = fan speed, r/sec. The pressure developed by the fan varies as the square of the impeller speed. The important rule, however, is that the power needed to drive the fan varies as the cube of the speed:

$$P = p_c N^3 \quad W$$

where: P = input power in watts, and p_c = a constant with units $W \cdot sec^3/r^3$.

The cubic law of pumping power indicates that if the air flow is to be doubled, eight (2^3) times as much power must be supplied. Conversely, if the air flow is to be cut in half, only one eighth ($\frac{1}{2}^3$) as much power must be supplied.

Air flow (and hence power) can be reduced by changing pulleys or installing smaller motors. Figure 5 in Chapter 2 shows the relative efficiency of various approaches.

Pumps follow laws similar to fans, the key being the cubic relationship of power to the volume pumped through a given system. Small decreases in flow rate such as might be obtained with a smaller pump or gotten by trimming the impeller, can save significant amounts of energy.

3. Air Compressors[1]

Compressed air is a major energy use in many manufacturing operations. Electricity used to compress air is converted into heat and potential energy in the compressed air stream. Efficient operation of compressed air systems therefore requires the recovery of excess heat where possible, as well as the maximum recovery of the stored potential energy.

Efficient operation is achieved by:

Selecting the appropriate type and size of equipment for the duty cycle required — Process requirements vary, depending on flow rates, pressure, and demand of the system. Energy savings can be achieved by selecting the most appropriate equipment for the job. The rotary compressor is more popular for industrial operations in the range of 20 to 200 kW, even though it is somewhat less efficient than the reciprocal compressor. This has been due to lower initial cost and reduced maintenance. When operated at partial load, reciprocating units can be as much as 25% more efficient than rotary units. However, newer rotary units incorporate a valve that alters displacement under partial load conditions and improves efficiency. Selection of an air-cooled vs. a water-cooled unit would be influenced by whether water or air was the preferred medium for heat recovery.

Proper operation of compressed air systems can also lead to improved energy utilization — Obviously, air leaks in lines and valves should be eliminated. The pressure of the compressed air should be reduced to a minimum. The percentage saving in power required to drive the compressor at a reduced pressure can be estimated from the fan laws described previously. For example, suppose the pressure was reduced to one half the initial value. Since pressure varies as the square of the speed, this implies the speed would be 70.7% of the initial value. Since power varies as the cube of the speed, the power would now be $0.707^3 = 35\%$ of the initial value. Of course, this is the theoretical limit; actual compressors would not do as well, and the reduction would depend on the type of compressor. Measurements indicate that actual savings would be about one half of the theoretical limit; reducing pressure 50% would reduce brake horsepower about 30%. To illustrate this point further, for a compressor operating at 6.89×10^5 N/m^2 (100 psi) and a reduction of the discharge pressure to 6.20×10^5 N/m^2 (90 psi), a 5% decrease in brake horsepower would result. For a 373 kW (500 hp) motor operating for 1 year, the 150,000 kWh savings per year would result in about $5,000 per year at current electric power costs.

The intake line for the air compressor should be at the lowest temperature available — This normally means outside air. The reduced temperature of air intake results in a smaller volume of air to be compressed. The percentage horsepower saving relative to a 21°C (70°F) intake is shown in Table 10.

Leakage is the greatest efficiency offender in compressed air systems — The amount of leakage should be determined and measures taken to reduce it. If air leakage in a plant is more than 10% of the plant demand, a poor condition exists. The amount of leakage can be determined by a simple test during off production hours by noting the time that the compressor operates under load compared with the total cycle. This indicates the percentage of the compressor's capacity which is used to supply the plant air leakage. Thus if the load cycle compared with the total cycle were 60 sec compared with 180 sec, the efficiency would be 33%, or 33% of the compressor capacity is the amount of air leaking in m^3/min (ft^3/min).

Recover heat where feasible — There are sometimes situations where water-cooled or air-cooled compressors are a convenient source of heat for hot water, space heating, or process applications. As a rough rule of thumb, about 300 J/m^{3}min of air compressed (~10 Btu/ft^3 min) can be recovered from an air-cooled rotary compressor.

TABLE 10

Power Requirements and Compressor Air Inlet Temperatures

Temperature of air intake °C	°F	Intake volume in m^3 required to deliver 1000 m^3 of free air at 21°C (70°F)	% kW Saving or increase relative to 21°C (70°F) intake
−1	30	925	7.5 Saving
5	40	943	5.7 Saving
10	50	962	3.8 Saving
16	60	981	1.9 Saving
21	70	1000	0
27	80	1020	1.9 Increase
32	90	1040	3.8 Increase
37	100	1060	5.7 Increase
43	110	1080	7.6 Increase
49	120	1100	9.5 Increase

From Smith, C. B., **Ed.**, *Efficient Electricity Use,* 2nd ed., Pergamon Press, New York, 1978. With permission.

Substitute electric motors for air motor (pneumatic) drives — Electric motors are far more efficient. Typical vaned air motors range in size from 0.15 to 6.0 kW (0.2 to 8 hp), cost $200 to $1200, and produce 1.4 to 27 N·m (1 to 20 ft lbs) of torque at 620 kN/m² (90 psi) air pressure. These are used in manufacturing operations where electric motors would be hazardous, or where light weight and high power are essential. Inefficiency results from air system leaks and the need (compared to electric motors) to generate compressed air as an intermediate step in converting electrical to mechanical energy.

Review air usage in paint spray booths — In paint spray booths and exhaust hoods, air is circulated through the hoods to control dangerous vapors. Makeup air is constantly required for dilution purposes. This represents a point of energy rejection through the exhaust air.

Examination should be made of the volumes of air required in an attempt to reduce flow and unnecessary operation. Possible mechanisms for heat recovery from the exhaust gases should be explored using recovery systems.

4. Electrolytic Operations

Electrolysis is an important industrial use of electricity, particularly in the primary metals industry, where it is used in the extraction process for several important metals. Energy management opportunities include:

Improved design and materials for electrodes — Evaluate loss mechanisms for the purpose of improving efficiency.

Examine electrolysis and plating operations for savings — Review rectifier performance, heat loss from tanks, and the condition of conductors and connections.

Welding is another electrolytic process. Alternating current welders are generally preferable when they can be used, since they have a better power factor, better demand characteristics, and more economical operation.

Welding operations can also be made more efficient by the use of automated systems which require 50% less energy than manual welding. Manual welders deposit a bead only 15 to 30% of the time the machine is running. Automated processes, however, reduce the no-load time to 40% or less. Different welding processes should be compared in order to determine the most efficient process. Electro-slag welding is suited only for metals over 1 cm (0.5 in.) thick but is more efficient than other processes.

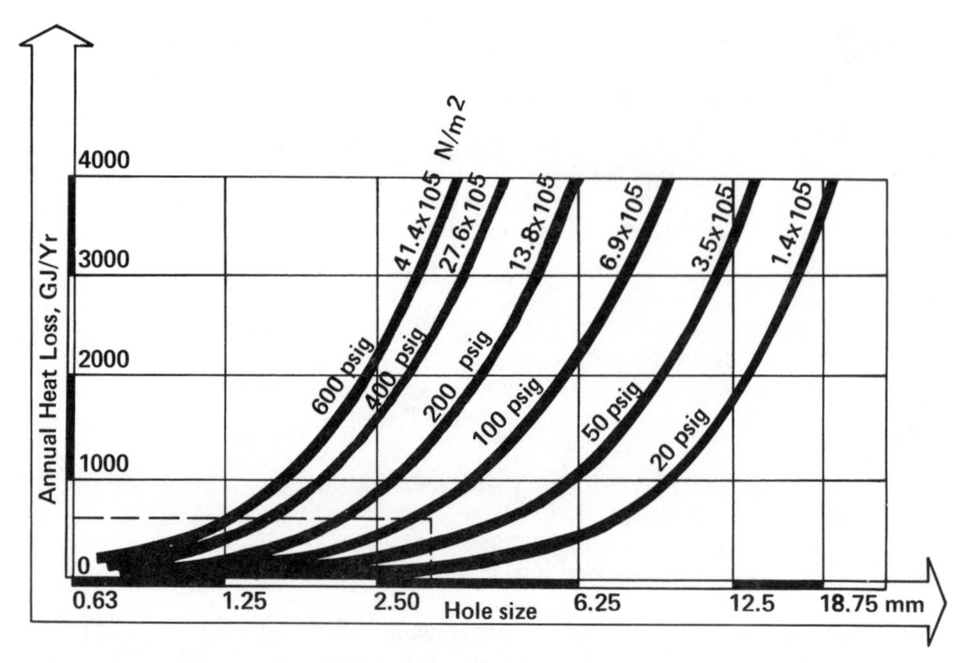

FIGURE 4. Heat loss from steam leaks. (From Smith, C. B., Ed., *Efficient Electricity Use*, 2nd ed., Pergamon Press, New York, 1978. With permission.)

Two other significant applications of electrolysis of concern to industry are batteries and corrosion. Batteries are used for standby power, transportation, and other applications. Proper battery maintenance, and improved battery design contribute to efficient energy use.

Corrosion is responsible for a large loss of energy-intensive metals every year and thus indirectly contributes to energy wastage. Corrosion can be prevented and important economies realized, by use of protective films, cathodic protection, and electroplating or anodizing.

5. Electric Process Heat and Steam Systems

Inasmuch as approximately 40% of the energy utilized in industry goes toward the production of process steam, it presents a large potential for energy misuse and fuel waste from improper maintenance and operation. Even though electrically generated steam and hot water is a small percentage of total industrial steam and hot water, the electrical fraction is likely to increase as other fuels increase in price. This makes increased efficiency even more important. For example:

Steam leaks from lines and faulty valves result in considerable losses — These losses depend on the size of the opening and the pressure of the steam, as illustrated by Figure 4.

Steam traps are major contributors to energy losses when not functioning properly — A large process industry might have thousands of steam traps which could result in large costs if they are not operating correctly. Steam traps are intended to remove condensate and noncondensable gases while trapping or preventing the loss of steam. If they stick open, orifices as large as 6 mm (0.25 in) can allow steam to escape. Such a trap would allow 5.7 GJ/year (6 MBtu/year) of heat to be rejected to the atmosphere on a 6.89×10^5 N/m (100 psi) pressure steam line. Many steam traps are improperly sized, contributing to an inefficient operation. Routine inspection, testing, and a correction program for steam valves and traps are essential in any energy program and can contribute to cost savings.

Poor practice and design of steam distribution systems can be the source of heat waste up to 10% or more — It is not uncommon to find an efficient boiler or process plant joined to an inadequate steam distribution system. Modernization of plants results from modified steam requirements. The old distribution systems are still intact, however, and can be the source of major heat losses. Large steam lines intended to supply units no longer present in the plant are sometimes used for minor needs such as space heating and cleaning operations which would be better accomplished with other heat sources.

Steam distribution systems operating on an intermittent basis require a start-up warming time to bring the distribution system into proper operation — This can extend up to 2 or 3 hr which puts a demand on fuel needs. Not allowing for proper ventilating of air can also extend the start-up time. In addition, condensate return can be facilitated if it is allowed to drain by gravity into a tank or receiver and is then pumped into the boiler feed tank.

Proper management of condensate return — Proper management can lead to great savings. Lost feed water must be made up and heated. For example, every 0.45 kg (1 lb) of steam which must be generated from 15°C feedwater instead of 70°C feedwater requires an additional 1.056×10^5 J (100 Btu) more than the 1.12 MJ (1063 Btu) required or a 10% savings of fuel. A rule of thumb is that a 1% fuel saving results for every 5°C increase in feedwater temperature. Maximizing condensate recovery is an important fuel saving procedure.

Poorly insulated lines and valves due either to poor initial design or a deteriorated condition — Heat losses from a poorly insulated pipe can be estimated using Figure 5. This curve shows that a poorly insulated line can lose ~1000 GJ/year (10^9 Btu/year) or more per 30 m (100 ft) of pipe. At steam costs of $2.00/GJ, this translates to $2,000 savings per year.

Improper operation and maintenance of tracing systems — Steam tracing is used to protect piping and equipment from cold weather freezing. The proper operation and maintenance of tracing systems will not only insure the protection of traced piping but also saves fuel. Occasionally these systems are operating when not required. Steam is often used in tracing systems and many of the deficiencies mentioned above apply (e.g., poorly operating valves, insulation, leaks).

Reduce losses in process hot water systems — Electrically heated hot water systems are used in many industrial processes for cleaning, pickling, coating, or etching components. Hot or cold water systems can dissipate energy. Leaks and poor insulation should be repaired.

6. Electrical Process Heat

Industrial process heat applications can be divided into four categories: direct-fired, indirect-fired, fuel, or electric. Here we shall consider electric direct-fired installations (ovens, furnaces) and indirect-fired (electric water heaters and boilers) applications.

Electrical installations use metal sheath resistance heaters, resistance ovens or furnaces, electric salt bath furnaces, infrared heaters, induction and high frequency resistance heaters, dielectric heaters, and direct arc furnaces.

From the housekeeping and maintenance point of view, typical opportunities would include:

Repair or improve insulation — Operational and standby losses can be considerable, especially in larger units. Remember that insulation may degrade with time or may have been optimized to different economic criteria.

Provide finer controls — Excessive temperatures in process equipment waste energy. Run tests to determine the minimum temperatures which are acceptable, then test in-

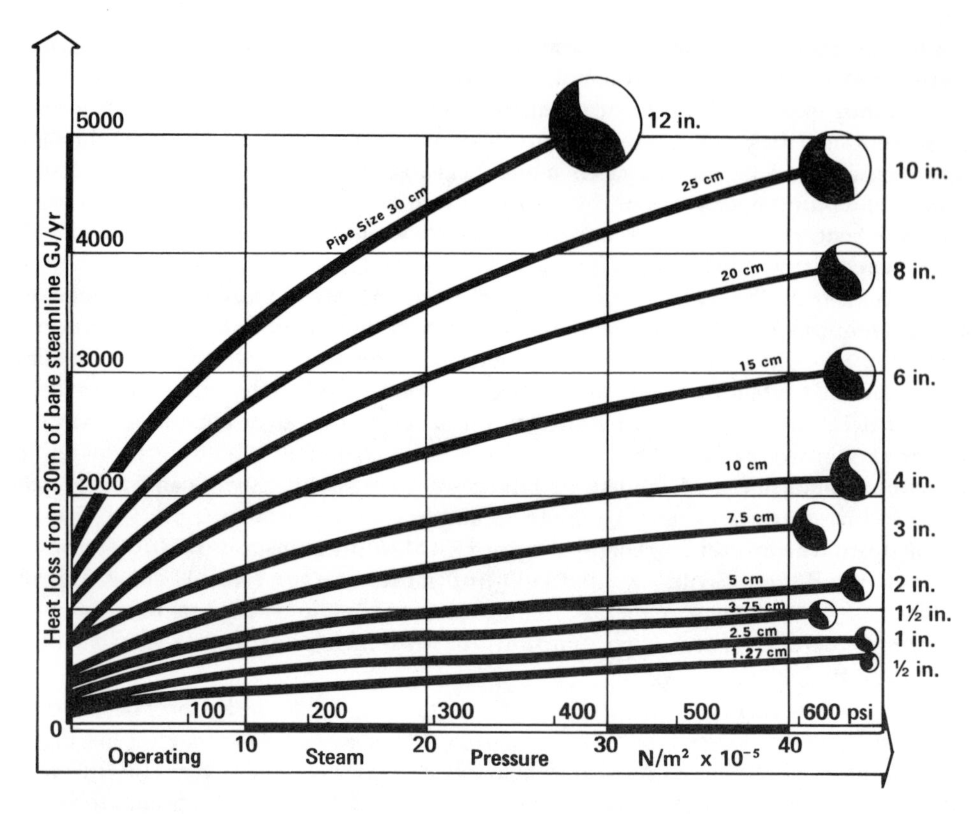

FIGURE 5. Heat loss from bare lines. (From Smith, C. B., Ed., *Efficient Electricity Use,* 2nd ed., Pergamon Press, New York, 1978. With permission.)

strumentation to verify that it can provide accurate process control and regulation.

Practice heat recovery — This is an important method, applicable to many industrial processes as well as HVAC systems, etc. It is described in more detail in the next section.

7. Heat Recovery

Exhaust gases from electric ovens and furnaces provide excellent opportunities for heat recovery. Depending on the exhaust gas temperature, exhaust heat can be used to raise steam or to preheat air or feedstocks.

Another potential source of waste heat recovery is the exhaust air which must be rejected from industrial operations in order to maintain health and ventilation safety standards. If the reject air has been subjected to heating and cooling processes, it represents an energy loss inasmuch as the makeup air must be modified to meet the interior conditions. One way to reduce this waste is through the use of heat wheels or similar heat exchange systems.

Energy in the form of heat is available at a variety of sources in industrial operations, many of which are not normally derived from primary heat sources. Such sources include electric motors, crushing and grinding operations, air compressors, and drying processes. These units require cooling in order to maintain proper operation. The heat from these systems can be collected and transferred to some appropriate use such as space heating or water heating.

The heat pipe is gaining wider acceptance for specialized and demanding heat transfer applications. The transfer of energy between incoming and outgoing air can be accomplished by banks of these devices. A refrigerant and a capillary wick are permanently sealed inside a metal tube setting up a liquid-to-vapor circulation path. Thermal energy applied to either end of the pipe causes the refrigerant to vaporize. The refrigerant vapor then travels to the other end of the pipe where thermal energy is removed. This causes the vapor to condense into liquid again and the condensed liquid then flows back to the opposite end through the capillary wick.

Industrial operations involving fluid flow systems which transport heat such as in chemical and refinery operations offer many opportunities for heat recovery. With proper design and sequencing of heat exchangers, the incoming product can be heated with various process steams. For example, proper heat exchanger sequence in preheating the feedstock to a distillation column can reduce the energy utilized in the process.

Many process and air conditioning systems reject heat to the atmosphere by means of wet cooling towers. Poor operation can contribute to increased power requirements.

Water flow and air flow should be examined to see that they are not excessive — The cooling tower outlet temperature is fixed by atmospheric conditions if operating at design capacity. Increasing the water flow rate or the air flow will not lower the outlet temperature.

The possibility of utilizing heat which is rejected to the cooling tower for other purposes should be investigated — This includes preheating feedwater, heating hot water systems, space heating, or other low temperature applications. If there is a source of building exhaust air with a lower wet bulb temperature, it may be efficient to supply this to a cooling tower.

8. Power Recovery

Power recovery concepts are an extension of the heat recovery concept described above. Many industrial processes today have pressurized liquid and gaseous streams at 150 to 375°C (300 to 700°F) that present excellent opportunities for power recovery. In many cases high pressure process stream energy is lost by throttling across a control valve.

The extraction of work from high-pressure liquid streams can be accomplished by means of hydraulic turbines (essentially diffuser-type or volute-type pumps running backwards). These pumps can be either single or multistage. Power recovery ranges from 170 to 1340 kW (230 to 1800 hp). The lower limit of power recovery approaches the minimum economically justified for capital expenditures at present power costs.

9. Heating, Ventilating, Air Conditioning Operation

The environmental needs in an industrial operation can be quite different from those in a residential or commercial structure. In some cases strict environmental standards must be met for a specific function or process. More often the environmental requirements for the process itself are not severe; however, conditioning of the space is necessary for the comfort of operating personnel, and thus large volumes of air must be processed. Quite often opportunities exist in the industrial operation where surplus energy can be utilized in environmental conditioning. A few suggestions follow:

Review HVAC controls — Building heating and cooling controls should be examined and preset.

Ventilation, air, and building exhaust requirements should be examined — A reduction of air flow will result in a savings of electrical energy to motor drives and additionally reduce the energy requirements for space heating and cooling. Due to the fact that pumping power varies as the cube of the air flow rate, substantial savings can be achieved by reducing air flows where possible.

Do not condition spaces needlessly — Review air conditioning and heating operations, seal off sections of plant operations which do not require environmental conditioning, and use air conditioning equipment only when needed. During nonworking hours the environmental control equipment should be shut down or reduced. Automatic timers can be effective.

Provide proper equipment maintenance — Insure that all equipment is operating efficiently. (Filters, fan belts, and bearings should be in good condition.)

Use only equipment capacity needed — When multiple units are available, examine the operating efficiency of each unit and put their operation in sequence in order to maximize overall efficiency.

Recirculate conditioned (heated or cooled) air where feasible — If this can not be done, perhaps exhaust air can be used as supply air to certain processes (e.g., a paint spray booth) to reduce the volume of air which must be conditioned.

For additional energy management opportunities in HVAC systems, refer to Volume III, Chapter 2.

10. Lighting

Industrial lighting needs range from low level requirements for assembly and welding of large structures (such as shipyards) to the high levels needed for manufacture of precision mechanical and electronic components (e.g., integrated circuits).

There are four basic housekeeping checks which should be made:

Is a more efficient lighting application possible — Remove excessive or unnecessary lamps.

Is relamping possible? — Install lower wattage lamps during routine maintenance.

Will cleaning improve light output? — Fixtures, lamps, and lenses should be cleaned periodically.

Can better controls be devised? — Eliminate turning on more lamps than necessary.

For modification, retrofit, or new design, consideration should be given to the spectrum of high efficiency lamps and luminaires which are available. For example, high pressure sodium lamps are finding increasing acceptance for industrial use with savings of nearly a factor of five compared to incandescent lamps. Refer to Volume III, Chapter 2 for additional details.

11. General Industrial Processes

The variety of industrial processes is so great that detailed specific recommendations are outside the scope of this chapter. Useful sources of information are found in trade journals, vendor technical bulletins, and manufacturers' association journals. These suggestions are intended to be representative, but by no means do they cover all possibilities.

In machining operations, eliminate unnecessary operations and reduce scrap — This is so fundamental from a purely economical point of view that it will not be possible to find significant improvements in many situations. The point is that each additional operation and each increment of scrap also represent a needless use of energy. Machining itself is not particularly energy-intensive. Even so, there are alternate technologies which can not only save energy but reduce material wastage as well. For example, powder metallurgy generates less scrap, and is efficient if done in induction-type furnaces.

Use stretch forming — Forming operations are more efficient if stretch forming is used. In this process sheet metal or extrusions are stretched 2 to 3% prior to forming, which makes the material more ductile so that less energy is required to form the product. The finished part is etronger and therefore thinner sheets and lighter extrusions can be used.

Use alternate heat treat methods — Conventional heat treat methods such as carburizing are energy intensive. Alternate approaches are possible. For example, a hard surface can be produced by induction heating, which is a more efficient energy process. Plating, metalizing, flame spraying, or cladding can substitute for carburizing although they do not duplicate the fatigue strengthening compressive skin of carburization or induction hardening.

Use alternative painting methods — Conventional techniques using solvent-based paints require drying and curing at elevated temperature. Powder coating is a substitute process in which no solvents are used. Powder particles are electrostatically charged and attracted to the part being painted so that only a small amount of paint leaving the spray gun misses the part and the overspray is recoverable. The parts can be cured rapidly in infrared ovens which require less energy than standard hot air systems. Water-based paints and high solids coatings are also being used and are less costly than solvent-based paints. They use essentially the same equipment as the conventional solvent paint spray systems so that the conversion can be made at minimum costs. New water-based emulsion paints contain only 5% organic solvent and require no afterburning. High solids coatings are already in use commercially for shelves, household fixtures, furniture, and beverage cans and require no afterburning. They can be as durable as conventional finishes and are cured by either conventional baking or ultraviolet exposure.

Substitute for energy intensive processes such as hot forging — Hot forging may require a part to go through several heat treatments. Cold forging with easily wrought alloys may offer a replacement. Lowering the preheat temperatures may also be an opportunity for savings. Squeeze forging is a relatively new process in which molten metal is poured into the forging dye. The process is nearly scrap free, requires less press power, and promises to contribute to more efficient energy utilization.

Movement of materials through the plant creates opportunities for saving energy. Material transport energy can be reduced by:

Combine processes or relocate machinery to reduce transport energy — Sometimes merely relocating equipment can reduce the need to haul materials.

Turn off conveyors and other transport equipment when not needed — Look for opportunities where controls can be modified to permit shutting down of equipment not in use.

Use gravity feeds wherever possible — Avoid unnecessary lifting and lowering of products.

D. Demand Management

The cost of electrical energy for medium to large industrial and commercial customers generally consists of two components. One component is the *energy charge*, which is based on the cost of fuel to the utility, the average cost of amortizing the utility generating plant, and on the operating and maintenance costs experienced by the utility. Energy costs for industrial users in the U.S. are typically in the range of 0.01 to 0.05 $/kWh.

The second component is the *demand charge*, which reflects the investment cost the utility must make to serve the customer. Besides the installed generating capacity needed, the utility also provides distribution lines, transformers, substations, and switches whose cost depends on the size of the load being served. This cost is recovered in a demand charge which typically is1 to 4 $/kW · month.

Demand charges typically account for 10 to 50% of the bill, although wide variations are possible depending on the type of installation. Arc welders, for example, have relatively high demand charges, since the installed capacity is great (10 to 30 kW for a typical industrial machine) while the energy use is low.

From the utility point of view it is advantageous to have its installed generating capacity operating at full load as much of the time as possible. This requires a high capacity factor. In the U.S. today capacity factors are typically in the range of 50 to 60%. To follow load variations conveniently, the utility operates its largest and most economical generating units continuously to meet its base load, and then brings smaller (and generally more expensive) generating units on line to meet peak load needs.

Today consideration is being given to "time-of-day" or *peak load pricing* as a means of assigning the cost of operating peak generating capacity to those loads which require it. See Volume 1, Chapter 5 for a discussion of utility pricing strategies.

From the viewpoint of the utility, *load management* implies maintaining a high capacity factor and minimizing peak load demands. From the customer's viewpoint, *demand management* means minimizing electrical demands (both on and off peak) so as to minimize overall electricity costs.

Utilities are experimenting with several techniques for load management. Besides rate schedules which encourage the most effective use of power, some utilities have installed remotely operated switches which permit the utility to disconnect nonessential parts of the customers' load when demand is excessive. These switches are actuated by a radio signal, through the telephone lines, or over the power grid itself through a harmonic signal ("ripple frequency") which is introduced into the grid.

Customers can control the demand of their loads by any of several methods:

- Manually switching off loads ("load shedding")
- Use of timers and interlocks to prevent several large loads from operating simultaneously
- Use of controllers and computers to control loads and minimize peak demand by scheduling equipment operation
- Energy storage, e.g., producing hot or chilled water during off-peak hours and storing it for use on peak

Demand can be monitored manually (by reading a meter) or automatically using utility-installed equipment or customer-owned equipment installed in parallel with the utility meter. For automatic monitoring, the basic approach involves pulse counting.

The demand meter produces electronic pulses, the number of which is proportional to demand in kW. Demand is usually averaged over some interval, e.g., 15 min, for calculating cost. By monitoring the pulse rate electronically, a computer can project what the demand will be during the next demand measurement interval, and can then follow a preestablished plan for shedding loads if the demand set point is likely to be exceeded.

Computer control can assist in the dispatching of power supply to the fluctuating demands of plant facilities. Large, electrically based facilities are capable of forcing large power demands during peak times which exceed the limits contracted with the utility or cause penalties in increased costs. Computer control can "even out" the load by shaving peaks and "filling in" the valleys, thus minimizing power costs. In times of emergency or fuel curtailment, operation of the plant can be programmed to provide optimum production and operating performance under prevailing conditions. Furthermore, computer monitoring and control provide accurate and continuous records of plant performance.

It should be stressed here that many of these same functions can be carried out by manual controls, time clocks, microprocessors, or other inexpensive devices. Selection of a computer system must be justified economically on the basis of the number of parameters to be controlled and the level of sophistication required. Many of the benefits described above can be obtained in some types of operations without the expense of a computer.

ACKNOWLEDGMENTS

I acknowledge many stimulating discussions with Lou Adrian, Max Good, and Rocco Fazzolare. Their ideas and comments are reflected in this chapter.

REFERENCES

1. **Smith, C. B., Ed.,** *Efficient Electricity Use,* 2nd ed., Pergamon Press, New York, 1978.
2. U.S. Department of Interior, Bureau of Mines, News Release, April 5, 1976.
3. National Academy of Engineering, U.S. Energy Prospects — An Engineering Viewpoint, Report prepared by the Task Force on Energy, Washington, D.C., 1974.
4. Applied Nucleonics Co., Inc., Proceedings of an EPRI Workshop on Technologies for Conservation and Efficient Utilization of Electric Energy, EPRI EM-313-SR, Electric Power Research Institute, Palo Alto, Calif., July 1976.
5. Electric Power Research Institute, *EPRI Journal,* 2(10), 40, 1977.
6. A. D. Little, Inc., Energy Efficiency and Electric Motors, FEA Conservation Paper No. 58, Federal Energy Administration, Washington, D.C., August 1976.

Chapter 5

ENERGY STORAGE

Stanley W. Angrist and Charles E. Wyman

TABLE OF CONTENTS

I. THE USES OF ENERGY STORAGE

Energy storage is not a new branch of technology — indeed its roots reach back to antiquity. It is clear that several methods of mechanical energy storage were employed long before Newton formalized that subject in his treatise on the laws of motion. The flywheel, one of the devices to be considered in some detail here, plays an essential role in the potter's wheel which is mentioned in the Old Testament.

Thermal energy storage was also used long ago. Homes with thick adobe walls were constructed by the Pueblo Indians in the American Southwest. The walls would absorb energy during the hot days, and the interior would remain cool. The stored heat would warm the living quarters against the cold desert nights.

Due to the declining availability of inexpensive fossil fuels, energy storage is again assuming an important role. Whenever the availability of an energy resource and the load requirements do not match, it is necessary to store energy if the energy source is to service noncoincident portions of the load. The examples of the ancient flywheel and adobe walls provide just this ability to extend periods of energy excess into times of deficiency. Modern mechanical means include sophisticated flywheels, pumped hydroelectric, and compressed air storage; thermal energy may be accumulated as sensible heat changes, latent heat changes, or as changes in chemical bond energy. In addition, batteries and other electrochemical processes may be used to store electrical energy in chemical forms, and fuels such as hydrogen may be produced. Current applications for storage include solar energy processes, utility peak shaving, and electric load leveling. Storage may also become important in such areas as medium-to-lightweight vehicles and industrial waste heat utilization, although these applications will not be discussed here.

A. Energy Storage In Solar Applications

Since solar radiation is an inherently time dependent energy resource, storage of energy is essential if solar is to meet energy needs at night or during daytime periods of cloud cover and make a significant contribution to total energy needs. Since radiant energy can be converted into a variety of forms, energy may be stored as thermal, chemical, kinetic, or potential energy. Generally, the choice of the storage media is related to the end use of the energy and the process employed to meet that application. For thermal conversion processes, storage as thermal energy itself is often most cost-effective. For photochemical or photovoltaic processes, storage is more appropriate in chemical form. A device which produces mechanical energy directly stores energy easily as kinetic or potential energy.

The location and type of energy storage in the overall system is often not very well defined. For example, in a solar thermal electric plant, steam is produced by concentrating solar collectors and is used to run a turbine. The turbine powers a generator to produce electricity. As shown for this process in Figure 1, energy may be stored thermally between the collector and turbine, mechanically between the turbine and generator, or chemically by a battery between the generator and the user. Since the turbine and generator are not 100% efficient, more energy must be stored for a given final output of electricity the closer the storage is located to the solar source, assuming that the storage units are all of the same efficiency. On the other hand, locating storage near the source reduces the required capacity of all the subsequent units since they are used for longer periods of time; and better conversion efficiencies, lower capital costs, and higher utilization of equipment result. Ultimately the decision as to where to locate storage in the system depends on the cost of the energy delivered as well as the reliability of the device.

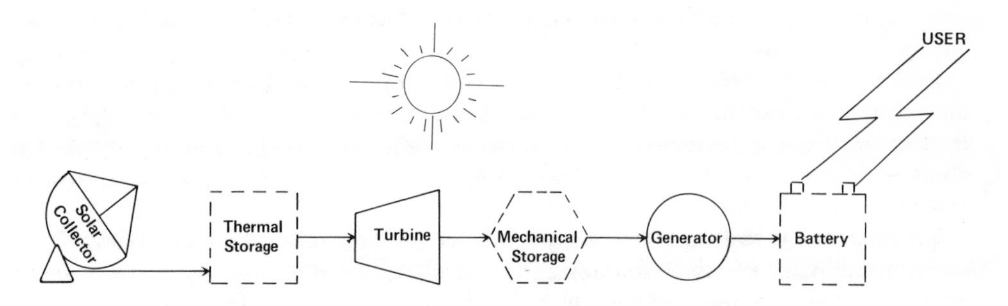

FIGURE 1. Possible locations for storage in a solar thermal electric plant.

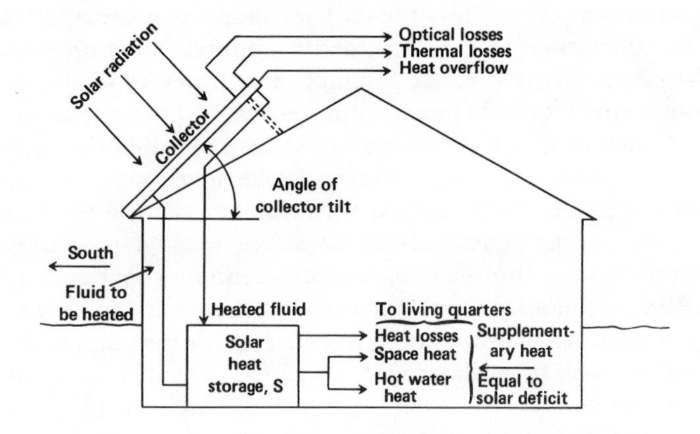

FIGURE 2. A residential solar heating system. (From Lof, G. O. G. and Tybout, R. A., *Sol. Energy,* 14, 253, 1973. With permission.)

The optimum capacity of the storage device for a given solar process depends on the time dependence of the solar availability, the nature of the load, the cost of auxiliary energy, and the price of the process components. These factors must all be weighed carefully for a particular application to arrive at the system design (including storage size) which minimizes the final cost of delivered energy. Such a cost optimization has been performed in a classic study of solar home heating by Löf and Tybout.[1,2] The following discussion will focus on some storage aspects of their work to illustrate the generally accepted role of storage current in solar home heating systems.

The system studied consists of flat-plate solar collectors tilted at some angle from the horizontal with necessary pumps or blowers to transfer heat from the collectors to storage. All the solar energy provided goes through the storage unit first as shown in Figure 2. Storage may be in a tank of water or a bin of dry crushed rock depending on whether water or air, respectively, is the heat transfer fluid. The stored energy could be used for both heating the living space and hot water. Since the storage unit is located inside the house, heat losses from storage are to the house itself and do not penalize the solar heating efficiency. A full sized backup heating unit is included since it is generally considered costly per unit of energy provided to meet the total heating demands by solar energy alone. Both the conventional heat source and the solar heating unit use the same heat distribution system to serve the house heating needs. (see Volume II, Chapter 8 for details.)

Relationships were incorporated into a computer program to describe the performance of the home heating system. To make the results adaptable to different system requirements, all the design equations were written on the basis of one square foot of collector area. The program requires that the collector temperature exceed the storage temperature by 6° C (10°F) before fluid is circulated between the storage device and the collectors to overcome heat losses in transfer and justify the electricity cost to operate the pump or blower. Since the collector temperature must exceed that in storage before fluid can be circulated profitably, it is more difficult for the collector to get to a high enough temperature to provide additional energy to a storage device which is already charged to high temperatures than to add heat to a discharged storage unit.

The objective is to find the solar system component sizes and the mixture of solar and conventional fuel which minimize the cost of delivered energy. At the time of the referenced study, commercial solar heating systems were not widely used; and the price information was somewhat sketchy. In addition, the costs employed in the analysis were for 1961 to 1962 and since they are no longer applicable, they will not be discussed in detail here. However, the authors found that changes in price levels did not significantly affect the conclusions of the study; and more recent work has substantiated the storage requirements.[3] Therefore, the findings of Löf and Tybout still have merit in matching storage requirements to home heating needs.

Weather data from eight different sites within the U.S. were employed in conjunction with the developed model to determine solar heating system performance. These locations, shown in Table 1, were judged representative of a range of world climates. Cities in parentheses in the table are locations where actual data was taken if the city identified in the analysis is different from the actual monitoring station by a few miles. A full year of hour-by-hour data of horizontal surface solar radiation, atmospheric temperature, cloud cover, and wind speed was selected for the year the Weather Bureau advised to be most typical for each location.

Each of the seven design and demand parameters was varied independently while maintaining all the others constant at levels believed to be nearest their optimum to keep the number of computer runs reasonable. Each variable was optimized for a 28.5 MJ per °C-day (15,000 Btu/°F day) and 47.5 MJ per °C-day (25,000 Btu/°F day) house. The influence of storage capacity on energy cost is shown in Figure 3 for three of the cities studied. The percent heating by solar energy with these designs is shown along the lines drawn. The absolute costs shown on the ordinate for fixed collector size do not represent general least cost values; but they were judged adequate for the relative comparison made in the study.

One observation from Figure 3 is the occurrence of the optimum at 0.2 to 0.3 MJ of storage capacity per °C temperature rise and m^2 of collector area (10 to 15 Btu/°F-ft^2) for each of the sites shown, with a range of 0.2 to 0.4 MJ/°C-m^2 (10 to 20 Btu/°F-ft^2) for all sites studied. Thus storage capacity varies quite closely with collector area. However, Lof and Tybout found that the optimum collector area changes significantly with sites (details are available in their references[1,2]). The somewhat fixed storage capacity to collector area ratio means that the larger collector areas required in colder climates must be accompanied by more storage to meet the higher fluctuation in solar insolation. Overall, the increase in both storage and collector capacities causes the energy delivery cost to be higher for colder cities such as Boston in Figure 3.

Another important conclusion from this study is the moderate storage capacity required. If the storage device could swing through the full 64°C (115°F) temperature range allowed between the low heating limit of 29°C (85°F) and ceiling for water of 93°C (200°F), only about one winter day's heat should be stored at Albuquerque in the optimum sized device. In Boston about 2 days heat should be stored in the optimum case. In practice, even less heat could be stored since a temperature swing of less than 64°C (115°F) is expected in the winter. In any event, storage durations on the order of one week are not currently considered cost-effective as shown in Figure 3 for the fixed collector size.

TABLE 1

Eight Cities Used in Löf and Tybout Study

Year	Site	Climate classification (Trewartha, including alphabetic code)	
1955	Miami	Aw:	Tropical savannah
1959	Albuquerque	BS:	Tropical and subtropical steppe
1956	Phoenix	BW:	Tropical and subtropical desert
1955	Santa Maria	Cs:	Mediterranean or dry summer subtropical
1955	Charleston	Ca:	Humid subtropical
1960	Seattle-Tacoma	Cb:	Marine west coast
1959	Omaha (North Omaha)	Da:	Humid continental, warm summer
1958	Boston (Blue Hill)	Db:	Humid continental, cool summer

From Löf, G. O. G. and Tybout, R. A., *Solar Energy*, 14, 253, 1973. With permission.

B. Energy Storage In Central-Station Power Plants

Most electric utilities that serve the public experience a variable daily, weekly, and seasonal demand for their product — electric power.[4,5] Utilities are required by their charters to find economical ways to generate power over large swings of electric load; they must also have sufficient generating capacity to satisfy maximum demand plus, for reliability, a sizeable system reserve.

In order to satisfy these conflicting requirements, electric utilities use different kinds of generating equipment in their systems. So-called "base-load plants" are used to service that part ofthe system load that continues 24 hr a day every day of the year. Base-load plants are designed to operate with high efficiency on the least expensive fuels available. These two requirements generally lead to high plant capital or first costs, but the low fuel costs and high load factor more than compensate for the high initial cost and produce the lowest cost power in the system. Base-load plants today are primarily coal- or nuclear-powered. (Of course, if a utility has hydropower available, it will supply as much of its base-load power as water-level conditions would permit; hydropower has essentially zero fuel cost.)

The intermediate load which represents most of the daily demand swing is served by several different types of equipment. The equipment which is generally shut down at night is typically made up of older less efficient fossil fuel plants and, more recently, gas turbines. The peak load which may persist for only a few hours is usually met by the oldest fossil fuel equipment and gas turbine units.

Historically this generation mix has worked quite well for electric utilities, but as fossil fuel costs have steadily increased, the old mix has become more costly. At present, utilities must have considerable capacity that is used only a few hours a day to provide peaking- and intermediate-load power. However, if large-scale energy storage were available, relatively efficient and economical base-load power generation equipment could be employed at higher load factors with the excess over off-peak demand being used to charge the energy storage system. During periods of peak demand, the storage system would supply power, thereby reducing the need for fuel-burning peaking equipment and at the same time reducing expensive fuel consumption. Furthermore, the increased base-load capacity would replace a part of the intermediate generation, producing additional cost and fossil fuel savings, especially if the added base-load plants were to use nonfossil primary fuels. Currently, utilities which are located in favorable geographic regions handle peak loads and sometimes intermediate loads

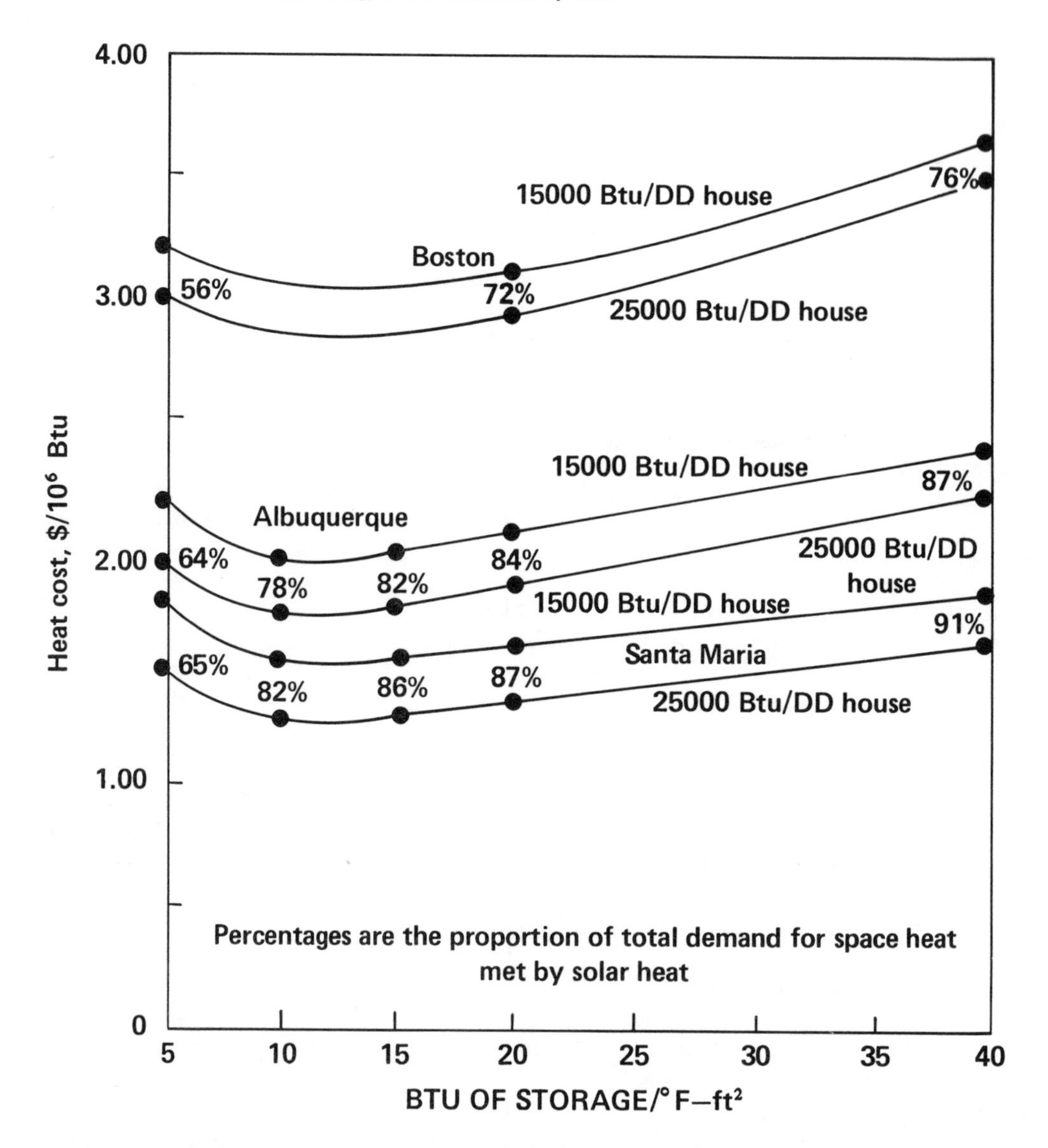

FIGURE 3. The influence of storage capacity on the cost of solar heat. (From Lof, G. O. G. and Tybout, R. A., *Sol. Energy*, 14, 253, 1973. With permission.)

with pumped storage units, if available. Using energy from a storage system to generate peaking power is termed "peak shaving;" the term "load leveling" describes the use of storage to eliminate conventional intermediate load cycling equipment.

It should be understood that the advantages of energy storage are not obtained without cost. The cost arises from the fact that it is impossible to get back every kilowatt-hour that is stored in the storage devices. Thus the real significance of energy storage is not a net saving of energy but a shifting of demand from inefficient units using costly fuels to more efficient primary units using less expensive fuels.

If storage technologies can be developed to meet the criteria for technical and economic feasibility that electric utilities demand, energy storage will certainly see widespread use in utility systems. Technical feasibility means that such systems must meet utility standards for operating life, reliability, safety, and compatibility with existing equipment. Economic feasibility implies that the total annual cost of electric energy delivered from energy storage systems must be equal to or less than the cost of energy from nonstorage equipment used for peaking and intermediate power generation.

The capital cost of an energy storage system is expressed to a first approximation as:

$$C_T = C_P + tC_s \quad (1)$$

where C_T is the total capital cost in dollars per kilowatt, C_P is that portion of the capital cost proportional to the power rating, C_s is that portion of the capital cost proportional to the system's energy storage capacity in dollars per kilowatt-hour, and t is the number of hours per day during which energy can be delivered from storage. Thus the rate at which energy is delivered and the size of the storage device influence the total system cost.

Several preliminary analyses have been carried out with the goal of establishing probable ranges for technical and economic feasibility criteria. Table 2 gives the results of one such study.[5] In this analysis, distributed energy storage costs are assumed to include a credit of $60/kW to reflect the fact that transmission costs will be considerably less for distributed storage systems than for central storage systems.

Kalhammer and Zygielbaum[5] come to the following conclusions about energy storage systems:

1. Energy storage systems with what appear to be attainable technical and economic characteristics and charged with power from modern base-load plants, promise to be more economical than gas turbines and coal gas-fired combined-cycle machines for generation of peak and intermediate power up to approximately 2500 hr of annual operation.
2. For operating periods between about 3000 and 5000 hr per year, coal gas-fired combined cycle power plants promise to have more favorable economics than energy storage systems.
3. Those energy storage devices that are economical in relatively small sizes (such as batteries or flywheels) have particularly favorable economics for peak-power generation (less than 1000 hr of annual operation), because credits can be claimed for transmission and distribution capital cost savings as a result of siting close to the load, and capital costs are largely proportional to energy storage capacity and, thus, are low for short periods of daily operation.
4. Depending on annual operating time and local conditions, combinations of different energy storage methods might be used to achieve the lowest-cost peak- and intermediate-power generation in future electric utility systems.

Figure 4 summarizes these results.[5] Specific methods of storing energy and their usefulness to electric utilities will be cited in subsequent sections.

C. Energy Storage In Residences And Commercial Buildings

Base-load power plants can also be used to provide a larger fraction of the load if the electrical demand is nearly uniform over extended periods. If lower electricity rates are offered during periods of low demand, customers will tend to adjust their use pattern to take advantage of the savings. Thus electric clothes dryers, dishwashers, and other electrical appliances with sizeable electrical demands will be employed more often during the off-peak time periods. As a result, the peak demand will be lowered while the off-peak demand will rise.

TABLE 2

Feasibility Criteria for Utility Energy Storage

			Capital cost for central storage		Capital cost for distributed storage	
Energy storage application	Efficiency (%)	Life (Years)	C_p ($/kW)	C_s ($/kWhr)	C_p ($/kW)	C_s ($/kWhr)
Peak shaving	≥60	≥20	40—90	7—20	60—150	7—25
Load leveling	≥70	≥30	50—110	5—15	50—170	5—18

From Kalhammer, F. R. and Zygielbaum, P. S., Paper No. 74-WA/Ener-9, American Society of Mechanical Engineers, New York, 1974. With permission.

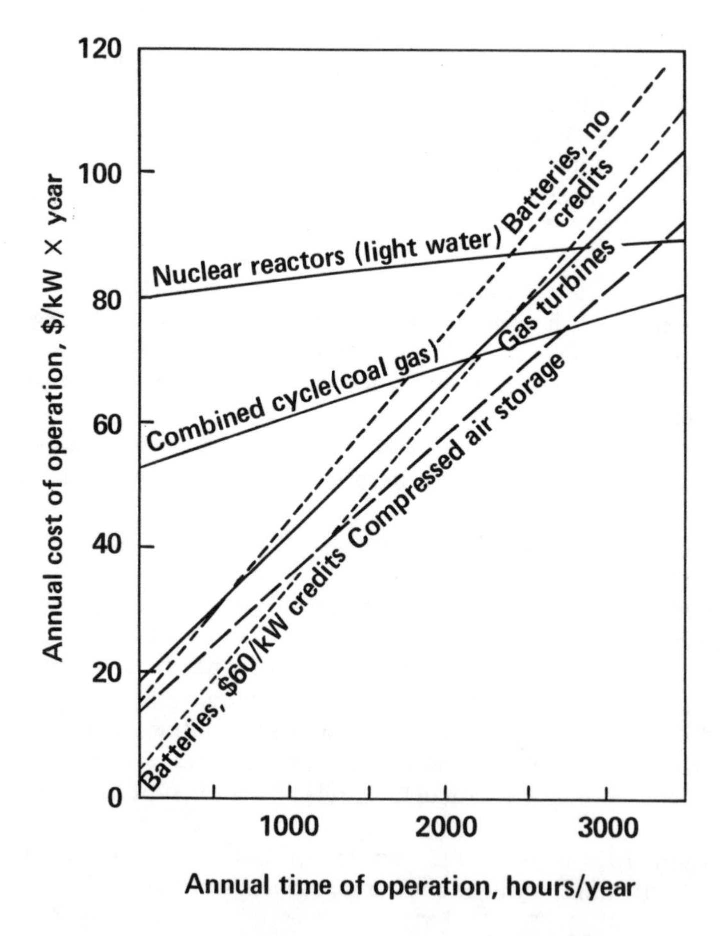

FIGURE 4. The economics of peak- and intermediate-cycling power generation as a function of annual hours of operation. The lower dashed line for batteries gives a $69/kW credit because of savings in transmission and distribution costs. (From Kalhammer, F. R. and Zygielbaum, P. S., Paper No. 74-WA/Ener 9, American Society of Mechanical Engineers, New York. With permission.)

Although many major electrical appliances can be used in the off-peak electrical periods, electrical home heating systems must be able to supply heat according to the demand. Since the latter is related to unpredictable weather patterns for reasonable comfort limits, it is not possible to shift the electrical requirement for conventional home heating systems. However, if a thermal storage medium is heated up electrically during off-peak hours, the energy can be released from storage during peak-load hours. Thus the daily load curve is smoothed, and the growth of winter peaks is retarded by displacing loads into the off-peak valley of the utility's load curve.

Utilities in the U.S. do not offer off-peak rates as a rule, and electric storage heating has consequently not been commercially successful. Utilities and their regulators apparently do not feel that electric storage heating is a cost-effective method of providing space heating in the U.S. On the other hand, favorable off-peak rate structures are offered in a number of European countries; and customer installation of storage heating systems has been successfully carried out in West Germany, Britain, Switzerland, Austria, Belgium, France, and Ireland. One of the major differences between the European and U.S. electric demand experience is that most U.S. utilities face peak demands in the summer while the Europeans see heavier loads in the winter. However, it is anticipated that the continuation of the current pace of electric heating system installation in the U.S. will shift the demand to winter peaking. Such is already the case for many northern states.[7]

Electric storage heating offers significant benefits for winter peaking utilities:[7]

- A slow down in the growth rate of winter peak loads with a consequent saving in transmission and generating capacity over that required otherwise
- Substitution of base-load generating plants for peaking and intermediate equipment due to the smoother daily demand curve
- A cost reduction in the electricity supplied enabling a greater market penetration for electricity

Of course, a favorable rate structure must be offered the customer to allow for a cost savings after the customer has paid for the purchase or rental of the storage equipment. Sufficiently large savings have been provided in the European countries previously mentioned to result in a significant penetration of storage heating systems. An example of the experience in England and Wales will be summarized in the following discussion based on Reference 6. More details on this example as well as the findings for West Germany can be found in that reference.

Under floor heating for air-raid shelters in World War II was the first important storage heater for Britain. Storage radiators such as in Figure 5 were introduced for the residential market in 1961 with later versions incorporating thermostatic and possibly weather-monitor controllers. A further refinement was the introduction of a fan into the unit to produce the storage fan heater shown in Figure 6. These latter two free standing concepts are used in existing housing in Britain, but the growth of fuel fired forced air systems in the 1960s led to the development of the central "Electricaire" heating system of Figure 7. The unit stores energy in a central unit of cast iron or refractory bricks with distribution throughout the building by warm air ducts. In 1971, a device was introduced called a "Centralac" unit which is similar to the Electricaire but uses an air-to-water heat exchanger to provide forced hot water heating. Electric domestic hot water storage units are also heated in off-peak hours to supply the daily needs of an average family.

In England and Wales, the Central Electricity Generating Board (CEGB) generates electricity and transmits it to 12 Area Boards who in turn distribute it to the customers. These Area Boards may set the retail rates for electricity within established guidelines. In the 1950s, the Area Boards began offering special off-peak rates to encourage use of floor warming in buildings; but not until the CEGB began basing capacity charges on contributions to the peak in the early 1960s and the storage radiator became avail-

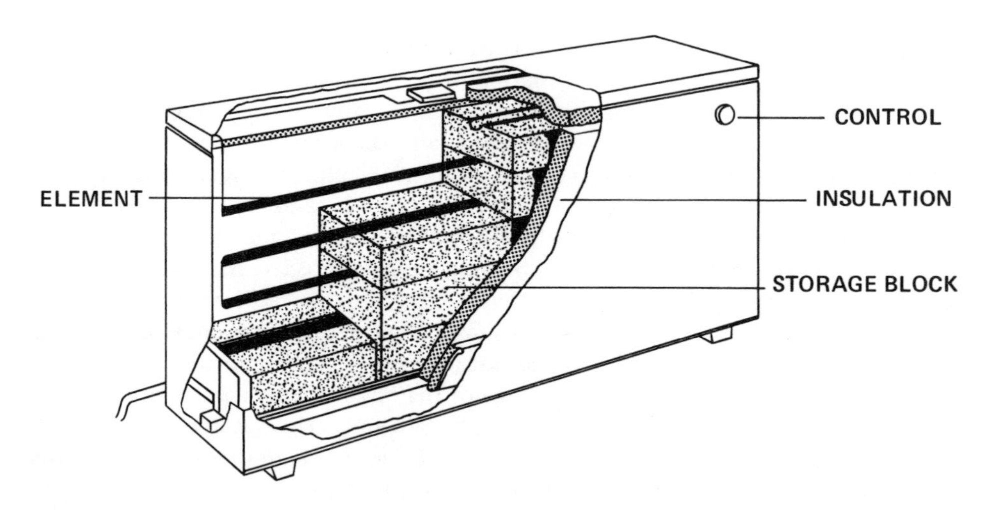

FIGURE 5. An electric storage radiator. (From Asbury, J. G. and Kouvalis, A., Rep. ANL/ES-50, Argonne National Laboratory, Argonne, Ill., May 1976.)

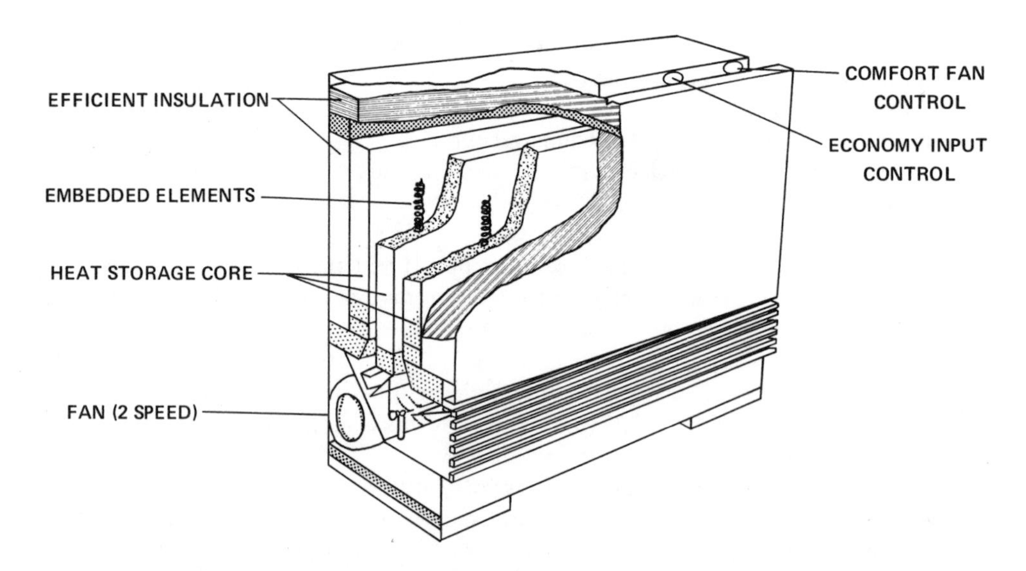

FIGURE 6. An electric storage fan heater. (From Asbury, J. G. and Kouvalis, A., Rep. ANL/ES-50, Argonne National Laboratory, Argonne, Ill., May 1976.)

able, did off-peak electricity rates become actively promoted. Representative rates for storage heaters and floor warmers are shown in Table 3 for the South Western Electricity Board (SWEB). The afternoon rate was required to boost the device since capacity was not enough to last the entire day.

In the period from 1962 to 1966, the rapid growth in storage heater installations and the low rate afternoon boost period caused a shift in the load curve from a midday valley to a midday peak. As a result, the off-peak tariffs for new customers were shifted to only the night period of generally eight hours at so-called White Meter rates by several Area Boards. This change allowed a daytime boost at regular rates if required, but the higher daytime rates encouraged night consumption of electricity if possible. Improvements in heater design and increases in storage capacity resulted in devices that could operate with only the nighttime charge. Figure 8 compares the

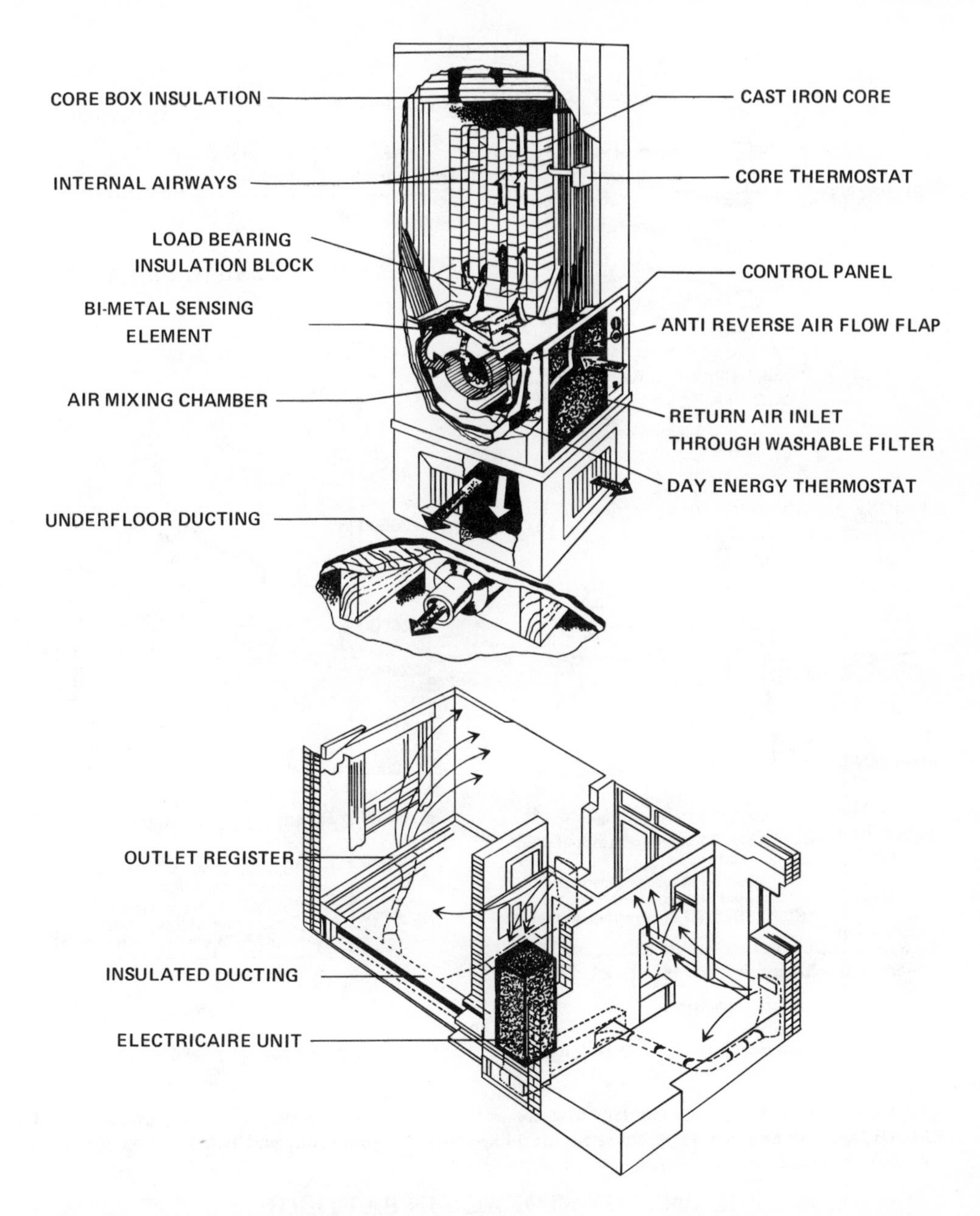

FIGURE 7. Electricaire storage heating system. (From Asbury, J. G. and Kouvalis, A., Rep. ANL/ES-50, Argonne National Laboratory, Argonne, Ill., May 1976.)

CEGB annual average daily load curve and the average weekday daily winter load curve for 1972/1973 with the 1960/1961 curves normalized to the same constant daily load line, and the effect of storage is obvious.

Table 4 summarizes the average electric rate structure used from 1973 to 1976 for England and Wales. Figure 9 presents the growth in the major types of storage heating units in Great Britain over the period in which off-peak rates were introduced. About 20% of the electricity for residential use in England and Wales was sold under off-peak or White Meter night rates in 1975.

TABLE 3

SWEB 1962 Off-peak Electricity Rates

Rate	Hours of availability	Total hours	Rate, pence/kWh	Ratio to normal domestic rate
A	23:00—07:30 and 11:00—16:00	11	0.85	0.62
B	19:00—07:30 and 13:00—16:00	15.5	1.0	0.73

From Asbury, J. G. and Kouvalis, A., Rep. ANL/ES-50, Argonne National Laboratory, Argonne, Ill., May 1976.

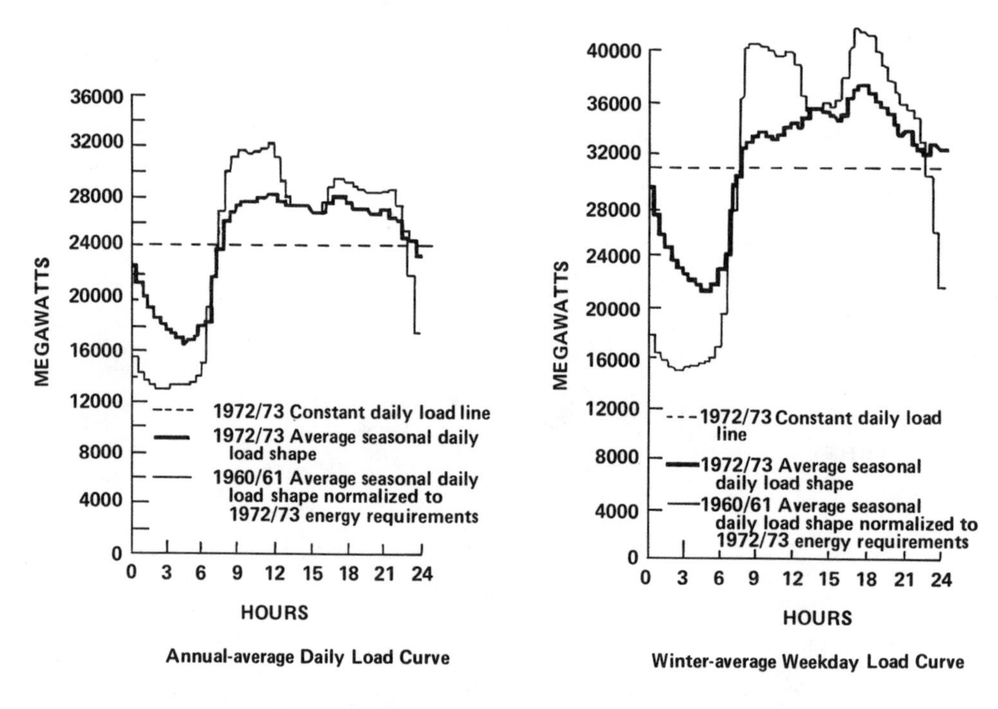

FIGURE 8. The 1972/1973 CEGB daily load curves versus those of 1960/1961. (From Asbury, J. G. and Kouvalis, A., Rep. ANL/ES-50, Argonne National Laboratory, Argonne, Ill., May 1976.)

II. ENERGY STORAGE IN BATTERIES

There is now renewed interest in the possible use of rechargeable batteries for bulk energy storage in utility systems and in vehicles because of: (1) the specific advantages of batteries (which include distributed storage of energy for central station plants with its significant economic and siting benefits, as well as the rapid installation and demand-responsive capacity for growth of essentially modular storage units); and (2) the emergence of new battery concepts for both central station and transportation applications that appear to offer promise of meeting the stringent cost and life requirements of these applications of energy storage.

The major challenge in battery energy storage systems for either central station power or transportation use is to develop a battery that has a significant cycle life and can be mass produced at very low cost. Several different approaches toward these goals

TABLE 4

Average Domestic Tariff Prices in England and Wales. 1973—1976

Year (at January)	Normal domestic		White meter			Preserved OP (8 + 3 hr)
	pence/kWh	FC	Day	Night	FC	
1973	0.950	8.20	1.019	0.409	12.20	0.466
1974	1.246	8.64	1.315	0.536	12.88	0.611
1975	1.311	8.64	1.380	0.564	12.88	0.643
1976	1.944	8.64	2.040	0.914	12.88	1.022

Notes: 1. Average tariff prices relate to the 12 Electricity Boards in England and Wales.
2. Prices are given in average pence per kWh with the annual fixed charges (FC).
3. Preserved off-peak (OP) tariffs (with midday boost) apply to existing customers who chose to remain on these tariffs after they were withdrawn for new customers. Off-peak tariffs of this type are still offered by six Electriciy Boards.

From Asbury, J. G. and Kouvalis, A., Rep. ANL/ES-50, Argonne National Laboratory, Argonne, Ill., May 1976. With permission.

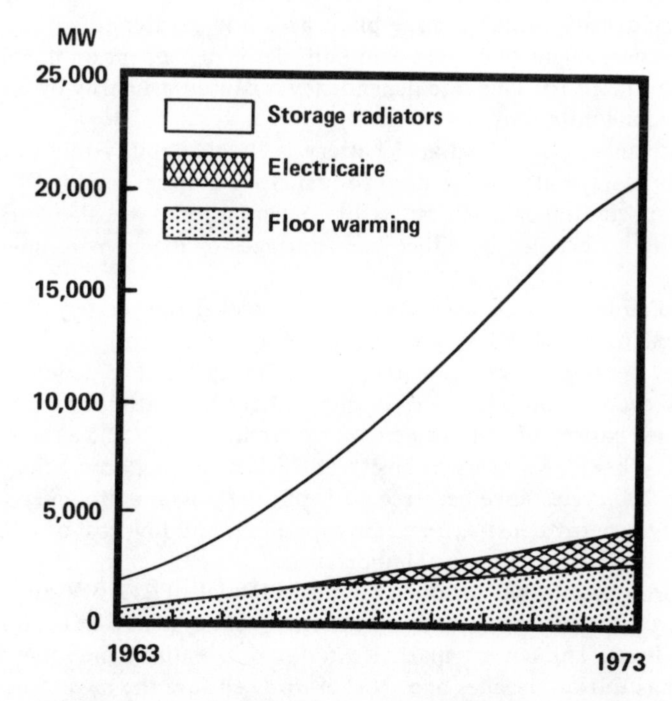

FIGURE 9. The growth in electric storage heating in Great Britain from 1963 to 1973. (From Asbury, J. G. and Kouvalis, A., Rep. ANL/ES-50, Argonne National Laboratory, Argonne, Ill., May 1976.)

are being pursued in current battery development programs. One approach is to develop batteries of modest or low energy density around relatively inexpensive aqueous electrochemical systems. To achieve the low capital cost targets, low cost containment materials, cell designs, and production techniques must be used or developed. Reviews of various battery systems may be found in References 5, 7, and 8. The batteries discussed here will be divided into four categories: aqueous electrolyte, metal-air, high temperature, and organic electrolyte.

A. Aqueous Electrolyte Batteries

Most of the aqueous electrolyte batteries are well developed and some have been used in the operation of commercial and research vehicles. An experimental electric vehicle operated with lead-acid batteries has obtained a 145-mile range at steady speed and a 65-mile range in stop-and-go traffic;[7] acceleration was unacceptably low, however.

The lead-acid battery (actually lead-lead dioxide) is rugged and reliable. It has been a battery standard for nearly a century. Its cost now appears to be too high for widespread appeal to the electric utility industry. A completely installed system (including such items as cooling and AC conversion) costs about $400/kW for 3 hr of storage, $525/kW for 5 hr, and $1000/kW for 10 hr — equivalent to $133, $105, and $100/kWh, respectively.[9] Utilities view these batteries as having a relatively short life, about 10 years.

The lead-acid battery is characterized by a relatively low energy density. (Lead is cheap at about 55¢/kg but the battery requires about 25 kg for each kWh of storage. The low energy density means a large plant area and greater foundation supports.) The attainable energy density of lead-acid batteries is dependent on the discharge rate as illustrated in Figure 10. This rate dependency is caused primarily by mass transport and ionic diffusion limitations.

The main advantage of a lead-acid battery is its availability today for situations where, for example, the alternative need for extensive underground transmission facilities produces substantial capital cost credits. Such batteries are also attractive if new peaking capacity is curtailed by either fuel shortages or local environmental prohibitions.

An alternate aqueous electrolyte system is the nickel-zinc battery. It has an open circuit potential of 1.706 V, a theoretical energy density of 322 Wh/kg, and an achieved energy density of over 66 Wh/kg. Over 300 cycles at 65% depth of discharge and up to 1500 cycles at 50% depth of discharge have been attained in the laboratory. The zinc-bromine battery has an open circuit potential of 1.8 V and a theoretical energy density of 432 Wh/kg. An energy density of 48 Wh/kg has been achieved on small batteries. Over 200 cycles have been reached, but the possible lifetime could be much greater. The high energy density, high rate capability, and low cost of materials might cause this system to have commercial importance.[7]

The nickel-iron system has an open circuit potential of 1.370 V and a theoretical energy density of 267 Wh/kg. This is an extremely old system that is both low in cost and very long-lived. The battery marketed today is essentially the same design developed by Thomas Edison decades ago. Nickel-iron cells are the most rugged in service today with actual lives in excess of 20 years.

The nickel-cadmium battery permits a very high power drain, has outstanding cycle life, and can achieve energy densities better than the lead-acid system. The system has a theoretical energy density of 220 Wh/kg and an open-circuit potential of approximately 1.35 V depending on state-of-charge. The main drawback to widespread use of nickel-cadmium batteries is the limited world supply of cadmium and consequently its high cost.

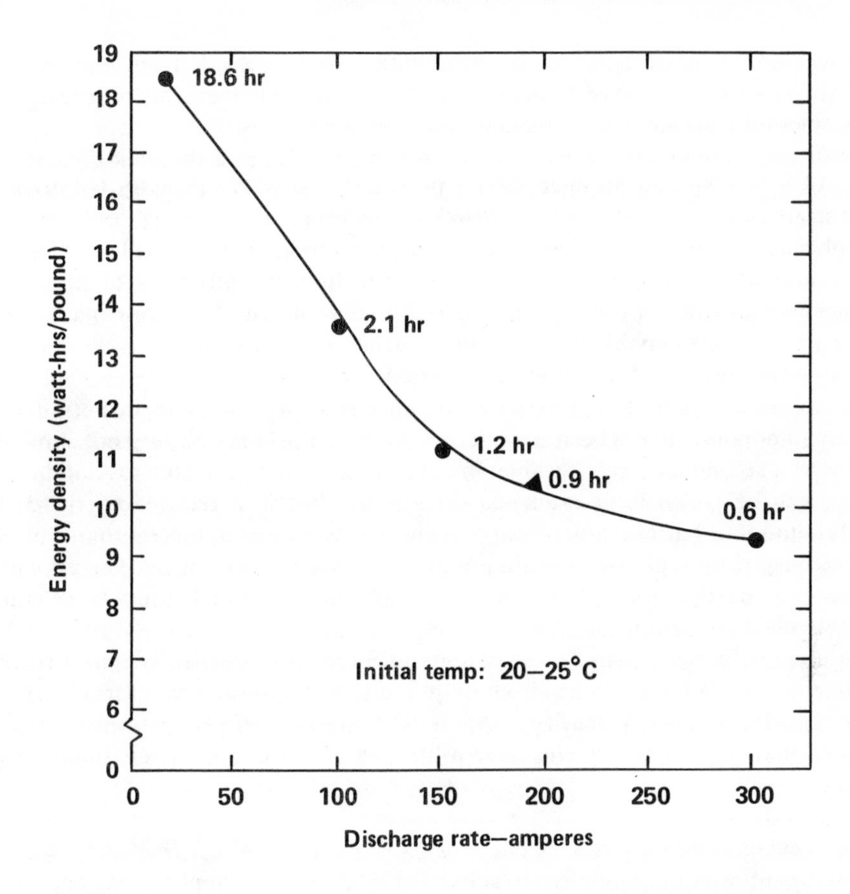

FIGURE 10. The effect of discharge rate and time of discharge on energy density. (From Gross, S., Proc. Battery Council Int. Golden Anniversary Symp., London, May 1974. With permission.)

The nickel-hydrogen cell is a relatively recent development in the area of alkaline storage batteries. This system combines the best electrode from the nickel-cadmium system with the best electrode from the hydrogen-oxygen fuel cell system. The system has an open circuit potential of 1.358 V and a theoretical energy density of 390 Wh/kg. An energy density of 55 Wh/kg has been achieved on prototype cells, and design studies reveal that 88 Wh/kg should be attainable. Power densities of 88 W/kg have been realized, and it is believed that the power density might be raised to 440 W/kg in an optimized design. A possible improvement to the nickel-hydrogen battery would be the development of a means to store hydrogen in a solid — perhaps as a metal hydride.

The zinc-chlorine battery is potentially an important system even though it is the most complex because of the use of a flowing electrolyte and the external storage of chlorine. The couple has an operating potential of 1.9 V and a theoretical energy density of more than 440 Wh/kg. The system has an edge in development and has already performed well in sizes larger than any other advanced battery that has been built. Because the battery uses two relatively cheap materials, it could offer relatively inexpensive storage after development.

B. Metal-Air Batteries

In this battery system, a metal forms the negative electrode, and a gas electrode using oxygen from the air is the positive electrode. Such systems are potentially very flexible, but practical problems such as the development of improved high-rate air electrodes with nonnoble metal catalysts remain to be solved. High energy density metal-air batteries have severe thermal problems, as well: they can overheat, or for high temperature design, the problem of initial heating arises.[7]

The zinc-air system has received the most attention. It has a theoretical energy density of 1350 Wh/kg and an open circuit potential of 1.65 V. Primary batteries have demonstrated energy densities of 330 Wh/kg. However, some investigators believe that the problems encountered in zinc-air batteries are so basic that they will not likely yield with additional development.[7] Such problems include the difficulty of producing a compact enough zinc deposit during charge to avoid interelectrode shorting, achieving a good air electrode capable of high current densities at low gas pressure, and loss of water in the air exhausted from the air electrodes.

The aluminum-air battery appears to be an attractive battery from both a weight and cost standpoint. It has been used successfully as a primary battery but shows little promise as a secondary (rechargeable) material in aqueous electrolytes. Aluminum apparently can be cycled in nonaqueous electrolytes, but with reduced energy density. One aluminum-air battery under development is a two-kilowatt system that is mechanically recharged by replacing the aluminum and alkaline electrolyte. The air cathode uses platinum catalyst, though it is believed that nonnoble metals could be substituted in commercial applications.

Considerable development has been done in Sweden on iron-air batteries where an experimental 30 kWh battery has been built and used to power a small truck achieving 66 Wh/kg with an energy density of 99 Wh/kg expected in production models. It is estimated that these batteries will have a lifetime of about 500 cycles, limited by the cathode.

C. High Temperature Batteries

These batteries are potentially attractive for both traction applications and central-station work. High power capability is achieved by use of low resistance electrolyte materials such as fused salts and by operating at elevated temperatures to increase the charge current density. The benefits of high temperature operation are not achieved without cost — especially troublesome are material and seal problems. The large increase in solute (up to 25%) when electrolyte salts melt poses a design problem. On the other hand, high temperature cells use electrode materials in a liquid state, thus avoiding morphological changes that occur with solid electrodes and thereby offering at least the promise of long life.

The sodium-sulfur cell has received the most publicity of all the high temperature battery systems. It has a theoretical energy density of 790 Wh/kg and an open circuit potential of about 1.8 V, depending on state of charge. The operating temperature of this cell is approximately 300 to 350°C with all reactants and products in the liquid state. Standby temperatures cannot be lower than 230°C. The use of sodium-sulfur cells for bulk energy storage in central-station power plants along with test results on a series of tubular cells is presented by Mitoff and Bush.[10]

The lithium-metal sulfide ($LiSi/FeS_2$) design has a theoretical energy density of 950 Wh/kg. The operating potential is around 1.4 V with an operating temperature of 400 to 450°C. These cells have a very high specific energy but show a rapid degradation of capacity with time. The high operating temperature of the lithium-metal sulfide battery raises serious problems of lifetime. The question of lithium availability also must be answered. Nevertheless, this battery has demonstrated an ability to be cycled.

A lithium tellurium tetrachloride system has achieved the elusive goal of a cycle life in excess of 2000 cycles without lithium dendrites forming or other problems.[7] An open circuit potential of 3.1 V and a theoretical energy density of 1120 Wh/kg are possible. The electrolyte is a molten eutectic of lithium chloride and potassium chloride operating at about 400°C. The active lithium negative materials, rather than being liquid, are alloyed with aluminum to form a solid and then encapsulated in a screen. This system has attained an energy density of 84 Wh/kg and a projected level of 130 Wh/kg.

Another high temperature battery which might find application in central-station plants and for motive uses is the aluminum-chlorine battery which has a theoretical energy density of 1430 Wh/kg and an open circuit potential of 2.1 V. The electrolyte in this system is molten $AlCl_3$-KCl-NaCl, the latter two constituents being a binary eutectic and the amount of $AlCl_3$ varying with cell state of charge. It has been found that the electrolyte can melt as low as 70°C, but a realistic operating temperature is 150 to 250°C. It is believed that operating above 200°C and the use of ceramic separators would help to solve problems of aluminum dendrite formation and excessive blockage of the aluminum by the $AlCl_3$ discharge product. Since overall electrical efficiencies of 87% have been observed with these cells, they should appear especially attractive as storage units in central-station power systems.

A battery with an operating temperature near that of the aluminum-chlorine battery uses sodium-antimony trichloride ($Na/SbCl_3$). It has a theoretical energy density of 770 Wh/kg and an operating potential near 2.6 V. However, the current density per unit area of electrolyte is about one third that of the sodium-sulfur battery. The cost and availability of antimony raise questions about large-scale use of this battery.

D. Organic Electrolyte Batteries

Much effort has been expended on cells that use lithium as a high energy negative electrode in organic electrolyte batteries. Such systems typically suffer from low discharge rates and low charge rates; nevertheless they do appear to hold some promise because they possess wide operating temperature ranges.

One cell that has received some attention is the lithium-sulfur dioxide battery which has a theoretical energy density of 1090 Wh/kg and open circuit potential of 2.95 V. A small primary cell has delivered 265 Wh/kg. This particular system seems to be capable of operating at high discharge rates. At present, this system is used in a primary battery; but it is capable of being produced as a secondary battery.

Another lithium-based battery is the lithium-lamellar dichalcogenide battery which is a new class of rechargeable lithium systems utilizing lamellar transition metal dichalcogenides such as niobium diselenide as host structures for cathodic nonmetals such as iodine and sulfur. Open circuit potentials of three volts have been achieved. Batteries have operated for more than 1100 cycles at low current densities and ambient temperature. Propylene carbonate was the electrolyte.[7]

Another cell utilizing propylene carbonate as the electrolyte is the lithium-bromine battery which has an open circuit potential of 4.05 V and a theoretical energy density of 1110 Wh/kg. An experimental cell was cycled 1785 times, although current densities were only 30% of the initial value at the end of cycling. It appears that the long life of this system is due in part to a bromine shuttle mechanism which limits self-discharge.[11] This system with further development might find application where high energy density is required.

E. Evaluation of Selected Candidate Secondary Batteries

The characteristics required of a battery to be used in central-station work or for motive power are in some ways different. However, both applications will require that the battery be manufactured at low cost. Long battery life is important, but there is a limit to the increase in first cost that will be acceptable in order to achieve long-lived systems. Table 5 lists five promising battery candidates for load-leveling purposes and their characteristics.

TABLE 5

Load-leveling Batteries: Candidates and Characteristics

	Operating temperature (°C)	Theoretical cell energy density (Wh/kg)	Design cell energy density (Wh/kg)	Design modular volumetric energy density (Wh/in^3)	Depth of discharge (%)	Density—10-hr rate (mA/cm^2)	Active material cost ($/kWh)	Operating potential (V)	Demonstrated cell size (kWh)	Demonstrated cell life (cycles)	Critical materials
Lead-acid (Pb/PbO_2)	20—30	240	20	0.75	25	10—15	8.50	1.9	>20	>2000	Lead
Sodium-sulfur (Na/S)	300—350	790	150	2.5	85	75	0.49	1.8	0.5	400	None
Sodium-antimony trichloride ($Na/SbCl_3$)	200	770	110	2.0	80—90	25	2.35	2.6	0.02	175	Antimony
Lithium-metal sulfide ($LiSi/FeS_2$)	400—450	950	190	3.5	80	30	4.27	1.4	1.0	1000	Lithium
Zinc-chlorine (Zn/Cl_2)	50	460	55	0.7	100	40—50	0.74	1.9	1.7	100	Ruthernium (catalyst)

From *EPRI J.*, No. 8, October 1976. With permission.

III. THERMAL STORAGE OF ENERGY

A. Sensible Heat Storage

Energy is stored as sensible heat by raising the temperature of a solid or liquid. This is the simplest way to store thermal energy, and current technology is generally adequate for good system design. Most thermal storage devices now in operation, including those for electrical storage heating or solar heating discussed in the examples, utilize sensible heat storage. The amount of energy stored, Q, is equal to the integral of the specific heat, C_p, between the peak and minimum temperatures (temperature swing) experienced by the storage medium:

$$Q = \int_{T_1}^{T_2} C_P \, dT \tag{2}$$

The temperatures T_1 and T_2 can be any values useful for the application provided the properties of the medium are not altered over the storage temperature range. This latter condition, however, requires that the material chosen must be thermally stable and undergo no phase change between the temperature extremes. To be economically attractive, the substances should also be inexpensive, have a high heat capacity, high density, and acceptably low vapor pressure.

The energy storage capacities of a number of liquids suitable for sensible heat storage are shown in Figure 11 as a function of the temperature swing of the storage media. A constant specific heat has been used to plot the data. Figure 12 shows the storage capacity per 1975 dollar of storage material cost on the same basis. Water appears to be the best sensible heat storage liquid since it is inexpensive and has a high specific heat. However, an antifreeze must be added to water if the fluid temperature can drop below 0°C, and this adds significantly to the system costs. In addition, above 100°C, the storage tank must be able to contain the vapor pressure of water; and the storage tank cost rises sharply with temperature beyond this point. From this viewpoint, organic oils, molten salts, and liquid metals are more desirable for high temperature operation since they circumvent the vapor pressure problems; but significant limitations in handling, containment, storage capacities, cost, and useful temperature range are evident for each as shown in Table 6.[13,14,15]

The vapor pressure difficulties associated with water can also be avoided by storing thermal energy as sensible heat in solids. In addition, many inorganic solids are chemically inert even at high temperatures. To store the same quantity of energy, larger storage vessels are needed than for water since the heat capacity of the solids is less (Figure 13). The amount of energy stored per 1975 dollar invested in storage media only, although not as high as water, is still acceptable as shown in Figure 14. In fact, the cost of water and many solids is so low that the storage costs are influenced more by the price of containers and heat exchangers than by the storage materials. Direct contact between the solid storage medium and a heat transfer fluid is vital to minimize the cost of heat exchange in a solid storage medium, and the storage volume must be increased by up to 50% to allow for fluid passage. While air is generally acceptable as a heat transfer fluid for low temperature home heating systems, other fluids such as high pressure helium or heat transfer oils are generally required in high temperature installations to provide adequate heat transfer capability.[15,16] The heat transfer fluid must be carefully chosen to be compatible with the solids as well. The problems asso-

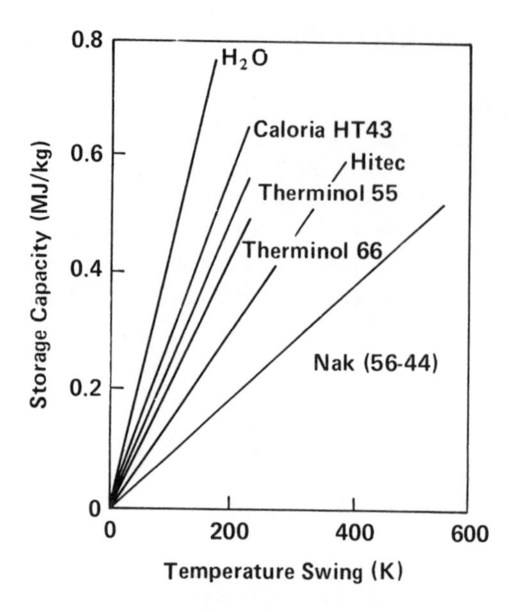

FIGURE 11. The storage capacity temperature swing for selected liquid sensible heat storage media. (From Bramlette, T. T., Green, R. M., Bartel, J. J., Ottesen, D. K., Schafer, C. T., and Brumleve, T. D., Rep. SAND 75-8063, Sandia Laboratories, Albuquerque, New Mexico, March 1976.)

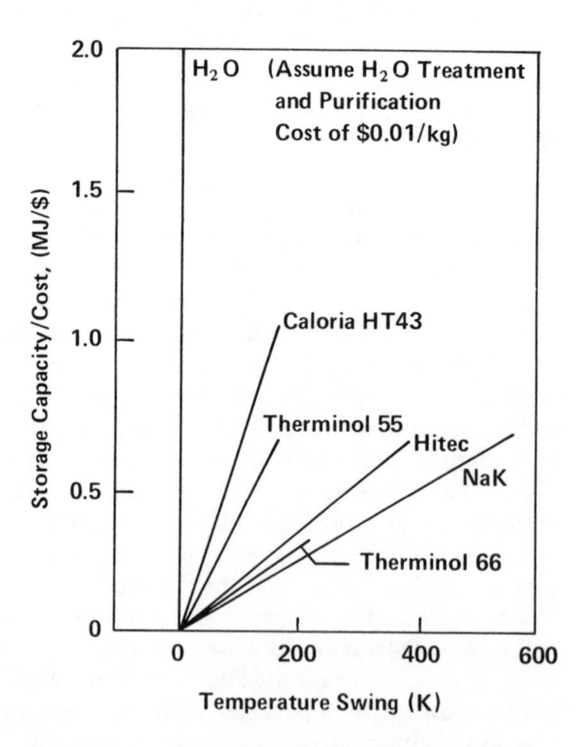

FIGURE 12. The energy stored per unit cost for selected liquid sensible heat storage media. (From Bramlette, T. T., Green, R. M., Bartel, J. J., Ottesen, D. K., Schafer, C. T., and Brumleve, T. D., Rep. SAND 75-8063, Sandia Laboratories, Albuquerque, New Mexico, March 1976.)

TABLE 6

Liquid Sensible Heat Storage Media

Medium	Fluid type	Cost ($/kg)	Temp. range (°C)	Heat capacity ($Jkg^{-1} K^{-1}$)	Comments
Caloria HT43	Oil	0.30	−9 to 310	2300	Non-oxidizing environment required at high temperatures. Cracking occurs at high temperatures and volatile materials may be formed, lowering the flash point. May polymerize at high temperatures to increase viscosity.
Therminol 55	Oil	0.60	−18 to 316	2500	
Therminol 66	Oil	2.03	−9 to 343	2100	
Hitec	Molten Salt	0.60	150 to 590	1550	Long-term stability unknown above 550°C. Stainless steel or other expensive containers probably required above 450°C. Inert atmosphere required at high temperatures. Heated lines required to prevent freezing.
Draw Salt	Molten Salt	0.44	250 to 590	1560	
Sodium					Stainless steel or suitable alternate containers required. Requires sealed system. Reacts violently with water, oxygen, and other oxidizing materials.
Sodium-	Liquid Metal	0.90	125 to 760	1300	
Potassium	Liquid Metal	—	49 to 760	1050	

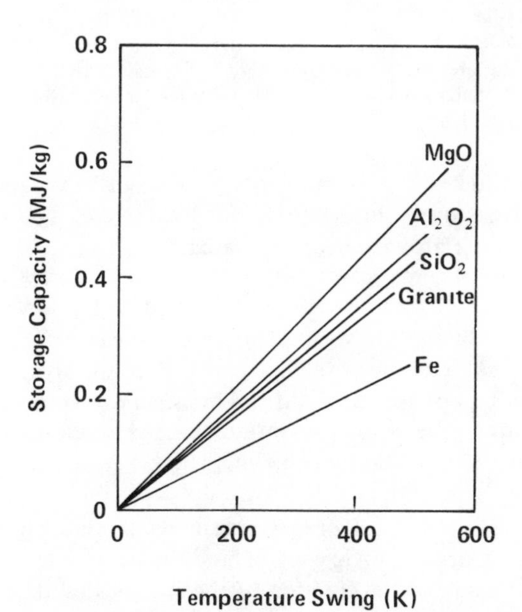

FIGURE 13. The storage capacity vs. temperature swing for selected solid sensible heat storage media. (From Bramlette, T. T., Green, R. M., Bartel, J. J., Ottesen, D. K., Schafer, C. T., and Brumleve, T. D., Rep. SAND 75-8063, Sandia Laboratories, Albuquerque, New Mexico, March 1976.

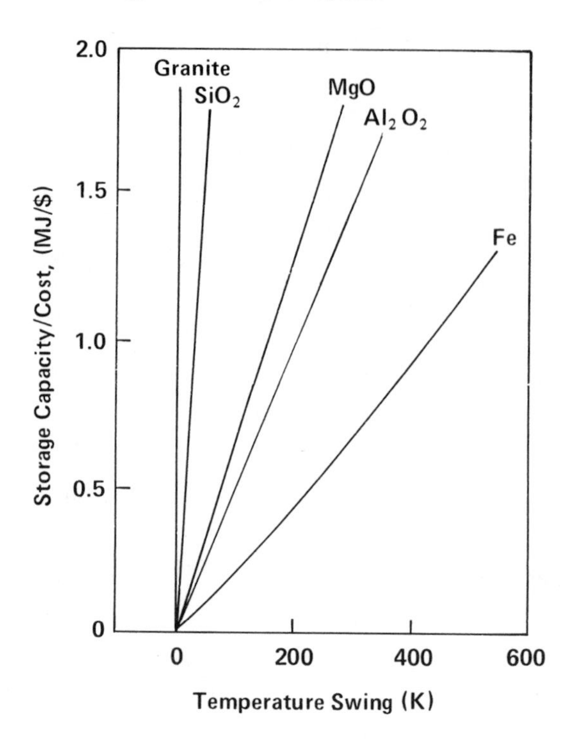

FIGURE 14. The energy stored per unit storage media cost for selected solid sensible heat storage media. (From Bramlette, T. T., Green, R. M., Bartel, J. J., Ottesen, D. K., Schafer, C. T., and Brumleve, T. D., Rep. SAND 75-8063, Sandia Laboratories, Albuquerque, New Mexico, March 1976.

ciated with finding a fluid with low vapor pressure, high heat capacity, and low cost are similar to those for storage in a liquid, but less severe. The properties of many sensible-heat-storage materials have been tabulated.[17]

A steam accumulator schematic for high temperature energy storage is shown in Figure 15. This device is really a pressure vessel to contain liquid water and is currently used in Europe to meet fluctuating steam demands.[19] It can be charged by a base load light-water reactor or solar thermal facility capable of supplying excess steam from the turbine during periods of low demand.[18] This steam is dumped into a vessel containing pressurized water. The vessel pressure may be charged up to as high as 20 atm and discharged down to 2 atm. This gives an energy storage density of about 100 kW_th/m^3.

The net electric power delivered from the steam accumulator depends on two more factors: the thermal turn-around efficiency, η_t and the peak power train efficiency, η_g'. The first factor is a measure of the fraction of the energy stored that can be recovered from a storage tank. The second factor is simply a measure of the effectiveness of the thermal storage plant to convert the stored energy into work. Golibersuch et al.[18] show that losses by sensible heat transfer to the accumulator walls are negligible. They also show that for a discharge time of 15 hr and an ambient temperature of 70°F, an insulated tank would have a turn-around efficiency of 97% while an uninsulated tank would have an 83% efficiency. Depending on which cycle is chosen, a power cycle efficiency of the thermal storage plant of between 20 and 25% seems reasonable.

Based on the above assumptions and calculations, energy storage is likely to cost between $7 and $11 per kW_th based on an accumulator cost of $700 to $1060/m³. If

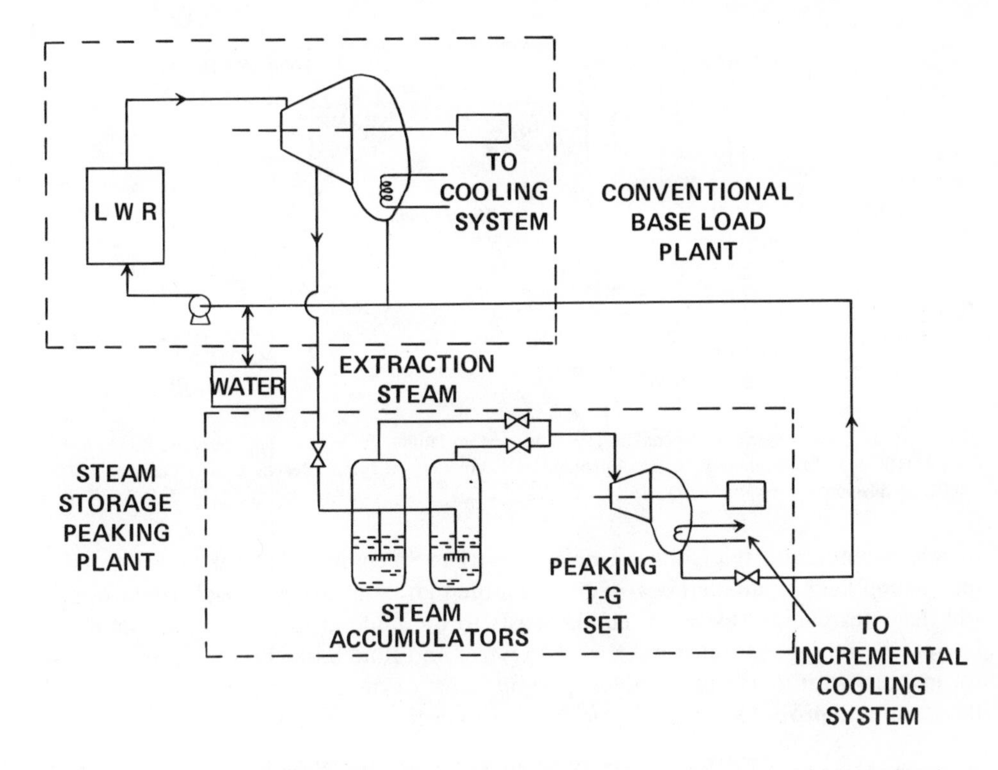

FIGURE 15. Steam storage peaking plant. (From Golibersuch, D. C., Bundy, F. P., Kosky, P. G., and Vakil, H. B., Rep. No. 75 CRD 256, General Electric Technical Information Exchange, Schenectady, New York, December 1975. This figure was originally presented at the Fall 1975 Meeting of The Electrochemical Society, Inc., held in Dallas, Texas.)

underground caverns at a cost of 17 to $\$70/m^3$ could be utilized, the storage costs could be reduced substantially to 20 to 70 cents per kW,h storage capacity.

The use of above ground accumulator tanks presents a significant safety hazard. In order to store enough energy to provide 4000 MWh of electrical energy, about $3 \times 10^5 m^3$ of accumulators filled with saturated water at 20 atm is required. If even the minimum volume flashes to steam during a rupture, the released energy is about 8.5×10^6 MJ (2.36 billion Wh).[18]

Another sensible heat storage device for high temperature applications is being developed for commercial scale solar power plants.[15] The storage tank is filled with 25 mm river gravel and 1.5 mm No. 6 silica sand in a 2 to 1 ratio. Then a heat transfer oil, Caloria HT43, is added to the vessel to fill the 25% void fraction. The sand and gravel are added to reduce the quantity of more expensive organic oil used for storage. Moreover, the solids also prevent natural circulation of the oil in the vessel, and temperature stratification is possible. To charge the unit shown in Figure 16, oil is circulated from the bottom of the storage vessel, through a heat exchanger to pick up heat from the source, and back to the top of the tank. A fairly sharp temperature transition or thermocline will occur naturally between the incoming hot fluid and the cold fluid in the bed, and this thermocline will move downward through the bed during charging. During extraction of heat from the bed, the direction of oil flow is reversed, and the thermocline moves upward through the bed as shown in Figure 17a. Essentially a constant outlet temperature is thus provided during both charging and discharging of the system until the unit is almost completely charged or discharged (Figure 17b).[15]

A number of other storage devices have been proposed for sensible heat storage of thermal energy. In general, inexpensive ways of containing the materials are sought,

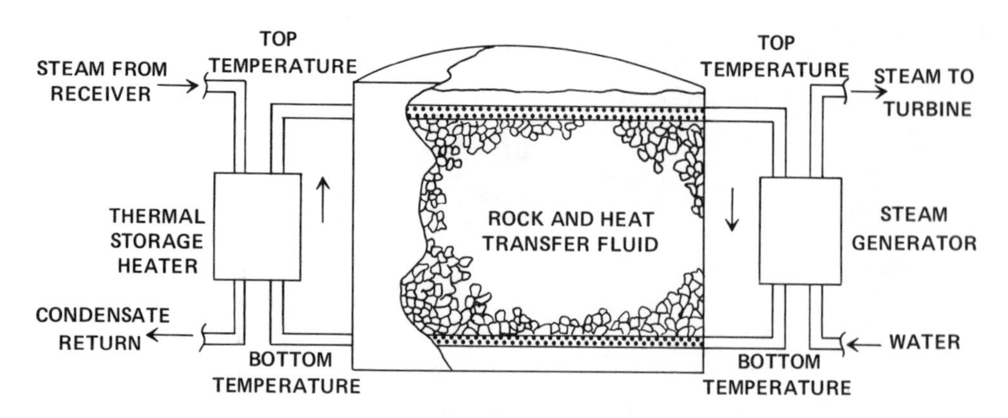

FIGURE 16. Dual medium thermal storage unit. (From Hallet, R. W., Jr. and Gervais, R. L., Rep. SAN/1108-8/5, McDonnell Douglas Astronautics Company, Redondo Beach, Calif., October 1977. With permission.)

since containment often is the major cost barrier. Steam accumulator tanks buried in the ground have significant cost reduction potential. Another proposal would use the soil itself for energy storage at similar costs to the underground steam accumulator. Use of aquifers to store thermal energy in water and sand has very low projected costs for low temperature storage. Table 7 presents a list of these candidates and others with the status and projected costs appropriate for 1975.[12]

B. Latent Heat Storage

A substantial absorption or release of energy generally accompanies a phase change such as from a solid to a liquid or from a liquid to a gas at a particular characteristic temperature. The potentially high energy storage densities over a relatively narrow temperature range make phase change materials attractive for thermal energy storage. Since a high volumetric energy storage density is essential, only solid-liquid or possibly solid-solid transitions are of practical interest. The volumes required to store a fixed amount of energy for heat of fusion materials are usually less than those for sensible heat materials (Figure 18), especially for small storage temperature swings. In Figure 18, sodium hydroxide undergoes a solid-solid phase change within about 25°C of the solid-liquid transformation. Some penalty must usually be assessed against phase change and solid sensible heat storage volumes to allow for passage of a heat transfer fluid.

The literature on selection of low temperature latent heat storage materials is voluminous with summaries available in References 20 and 21. High temperature fused salt storage for residential applications has also been extensively studied.[22] Material requirements include low cost, high heat of transition, high density, appropriate transition temperature, low toxicity, and long-term performance. Paraffin waxes[23] and salt hydrates[20] have been favored for low temperature storage applications although the former is very flammable while the latter is prone to subcooling without crystallization.

For higher temperature uses, some generalizations are possible.[12] Carbonates and possibly carbonate-chloride systems are serious candidates because their good corrosion characteristics make them relatively inexpensive to contain and their cost is reasonable. Nitrates and nitrites are good choices for applications below 500°C since they are relatively noncorrosive and fairly inexpensive. Chloride systems are cheap enough to be attractive but are more corrosive than the previous compounds. Hydroxides as a group tend to be more expensive and corrosive. Fluorides offer relatively high

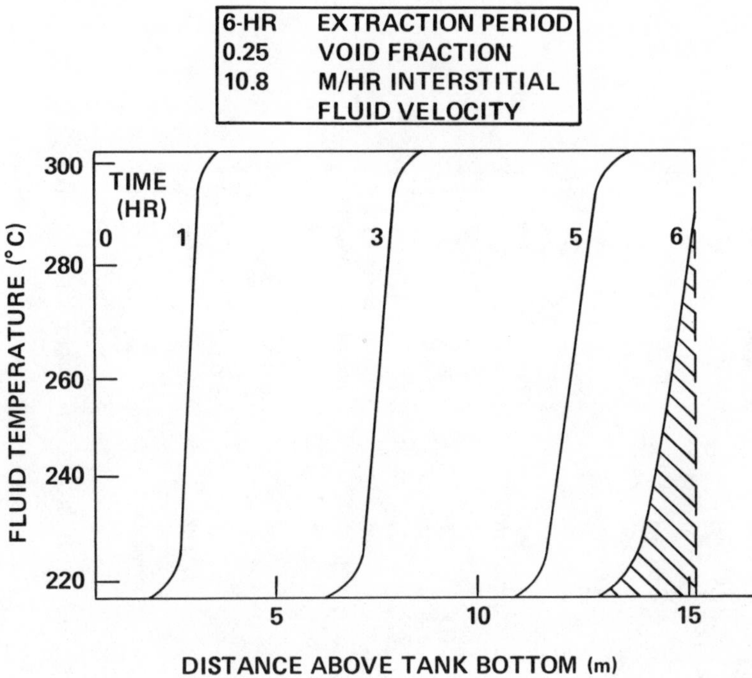

A.

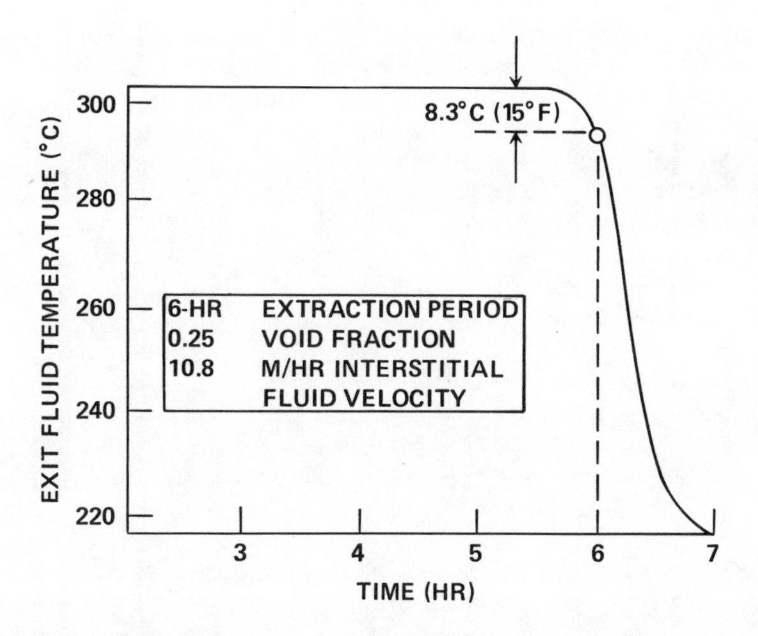

FIGURE 17. (A) Thermoclines during energy extraction from a 10-MW, dual medium thermal storage unit. (B) Fluid temperature at the exit from the dual medium thermal storage unit during six hour extraction. (From Hallet, R. W., Jr. and Gervais, R. L., Rep. SAN/1108-8/5, McDonnell Douglas Astronautics Company, Redondo Beach, Calif., October 1977. With permission.)

TABLE 7

Sensible Heat Storage Systems

Storage configuration	Storage medium	Status	T_{MAX}, °C	T_{MAX}-T_{MIN}, °C	Capacity MW,h	Cost $/kW,h
Aboveground tank	Water	Engineering design	210	87	4.1	8.0
			300	87	4.1	22.0
Aboveground tank	Water	Preliminary testing	232	56	0.41	—
	Therminol		343	56		
Steam accumulator	Water	Engineering design	200		14	3.0
			300		37	6.0
Undergroundtank	Water	Preliminary design	217	141	4370	0.4
Acquifers	Water and sand	Conceptual	170	110	42,000	0.003
Aboveground tanks	Therminol 55	Engineering design	315	55	226	62
(other fluids)	Therminol 66		315	55	226	27
	Caloria HT43		302	83		11
	HITEC		500	300		4
Solid storage materials	Cast iron	Operational	750	480	0.75	~6.00[a]
			700	430	0.64	~6.30[a]
Packed beds	Granite	Preliminary design	302	84	195	5.13
	Caloria HT43					
Fluidized bed	Sand	Conceptual	800	400	4000	—
	Fly ash					
Underground	Soil	Preliminary design	100	85	500	0.4—0.8
Rock bed	Rocks	Operational air heating	200	150	—	2.00—3.20

[a] Based on total system cost of $0.20/lb and the temperature swing Tmax-Tmin shown, not in reference cited.
[b] Not in reference cited.

From Bramlette, T. T., Green, R. M., Bartel, J. J., Ottesen, D. K., Schafer, C. T., and Brumleve, T. D., Rep. SAND 75-8063, Sandia Laboratories, Albuquerque, N.M., March 1976.

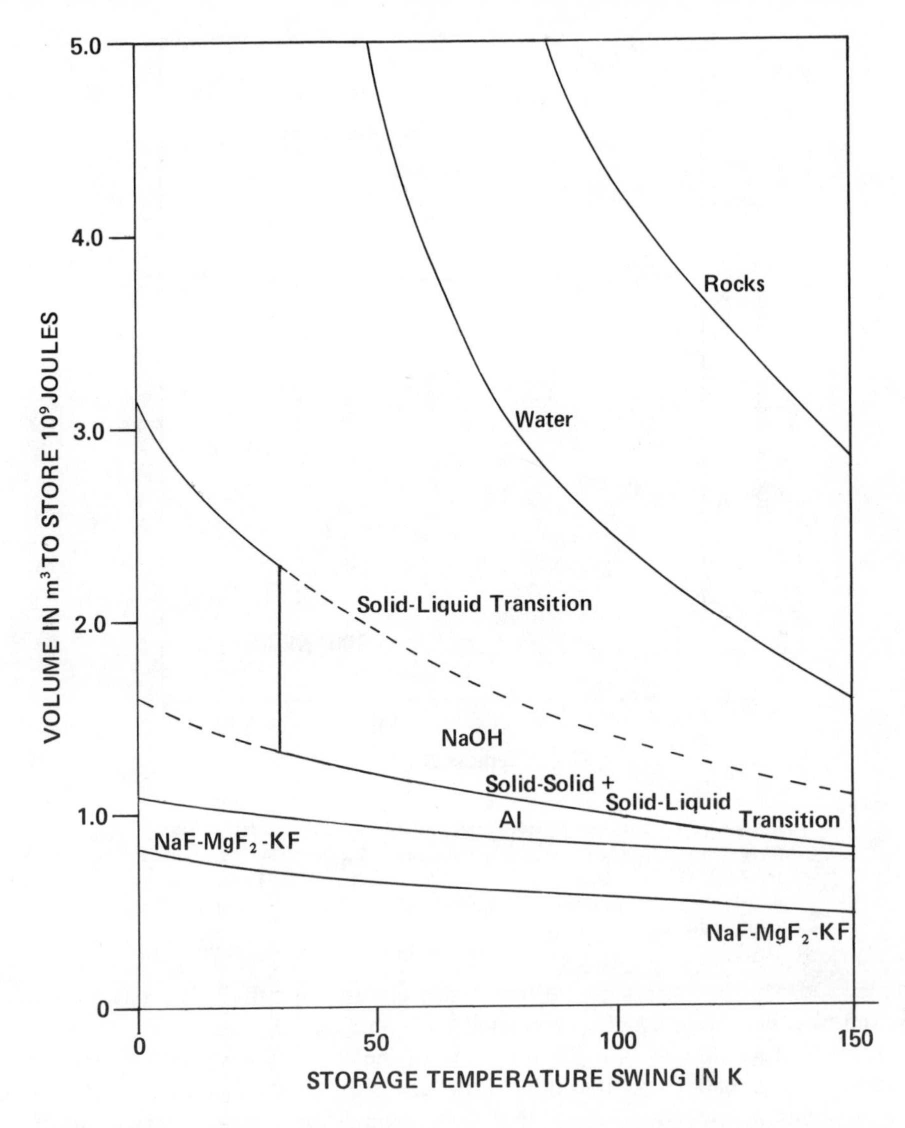

FIGURE 18. Approximate volumes of selected heat of fusion and sensible heat storage materials required to store one gigajoule of energy.

heat storage densities, but their usually high price and corrosive nature make the fluorides more expensive to use than the other salt systems discussed above. Figure 19 clearly shows the storage capacity per dollar of material cost for selected latent heat of fusion storage media. More details on these salts as well as others are in References 12 and 22.

A conceptual study has been carried out on the use of a latent heat storage system in conjunction with a high temperature gas reactor.[18] The system was designed to meet the following specifications: primary coolant: He, 48 atm, 400 to 780°C; storage capacity: 7200 MW_th; charge/discharge capacity: 600 MW_t; peak electrical generating capacity: 200 MW_e for 12 hr. The system is presented schematically in Figure 20. About 38 Gg of 70 Gg NaF/30 FeF_2 eutectic would be theoretically needed to store 7200 MW_th. In order to keep the mass reasonably fluid as a slurry, the total amount needed would be on the order of 70 Gg requiring a container about 37 m in diameter and 34 m high.

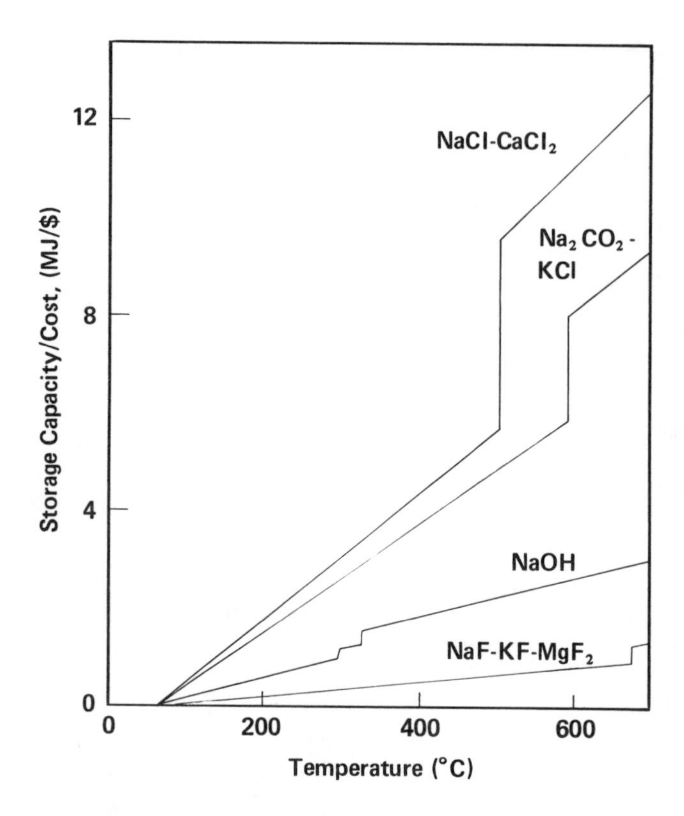

FIGURE 19. The energy stored per unit cost for selected latent heat of fusion storage media. (From Bramlette, T. T., Green, R. M., Bartel, J. J., Ottesen, D. K., Schafer, C. T., and Brumleve, T. D., Rep. SAND 75-8063, Sandia Laboratories, Albuquerque, New Mexico, March 1976.)

example. Such direct contact heat transfer also circumvents the high cost of containment and heat exchange usually associated with phase change materials. During discharge, the lead would be "rained" in the top of the slurry at about 370°C and would be heated to near 680°C as the globules of lead sink to the bottom. Similar direct contact systems which rely on an oil that rises through the storage reservoir are being developed for low temperature uses,[24] and a diagram of this system is shown in Figure 21.

Bundy et al.[17] concluded that all medium to high temperature storage systems using latent heat of fusion materials in conventional metal heat exchanger-containers tend to be quite expensive, and in many cases, there are serious chemical and mechanical problems associated with containment of the storage material. There are basic problems in achieving adequate and efficient heat transfer at reasonable cost.[12,17,25] In addition, at higher temperatures, small-sized units appear to suffer from excessive thermal leakage. The realistic general conclusion is that thermal energy storage as the latent heat of fusion is currently difficult to apply commercially in a competitive, efficient manner. Direct contact heat exchange devices such as the example promise one solution to the high price of containment and heat exchange, and plastic containment may be effective for lower temperature heating and cooling uses. Table 8 summarizes the 1975 projected costs of some latent heat storage systems.

C. Reversible Chemical Reaction Storage

Thermal energy may also be stored as the bond energy of a chemical compound by

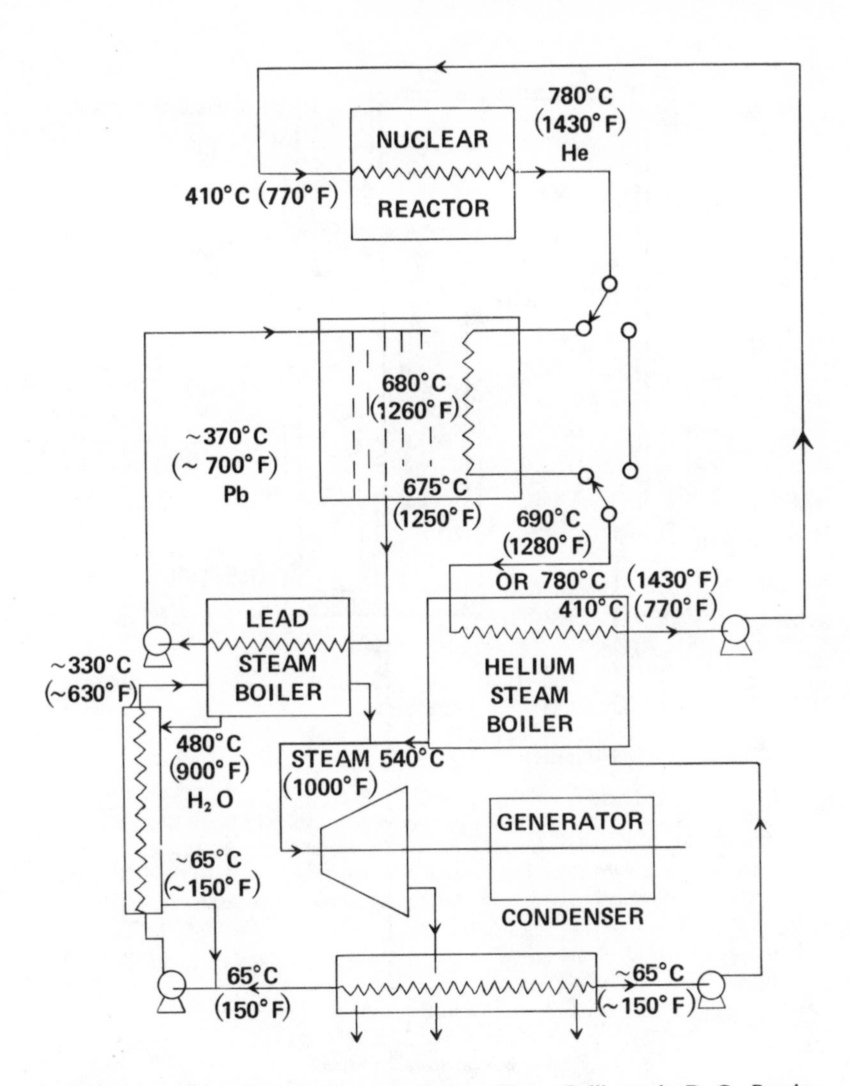

FIGURE 20. Latent heat steam storage system. (From Golibersuch, D. C., Bundy, F. P., Kosky, P. G., and Vakil, H. B., Rep. No. 75 CRD 256, General Electric Technical Information Series, Schenectady, New York, December 1975. This figure was originally presented at the Fall 1975 Meeting of The Electrochemical Society, Inc., held in Dallas, Texas.)

means of reversible chemical reactions. An endothermic forward reaction absorbs energy from the source under conditions which favor significant conversion to a high enthalpy chemical species. The reaction can only proceed until the equilibrium concentrations are reached. Then, to release the stored energy, the conditions are altered to favor high conversion by the exothermic reverse reaction to the low enthalpy species. The equilibrium concentrations of the species may be altered by: (1) changing the concentration (or pressure) of the chemical species and/or (2) changing the temperature of these species.

The energy storage density by reversible chemical reactions is generally higher than for phase change storage. Chemical storage also has significant cost potential since some of the materials are available for pennies a pound. Chemical storage has the added advantage in that significant energy storage densities are possible even at ambient conditions.[26] However, careful heat exchange between products and reactants is required to minimize sensible heat losses and provide efficient storage of energy.

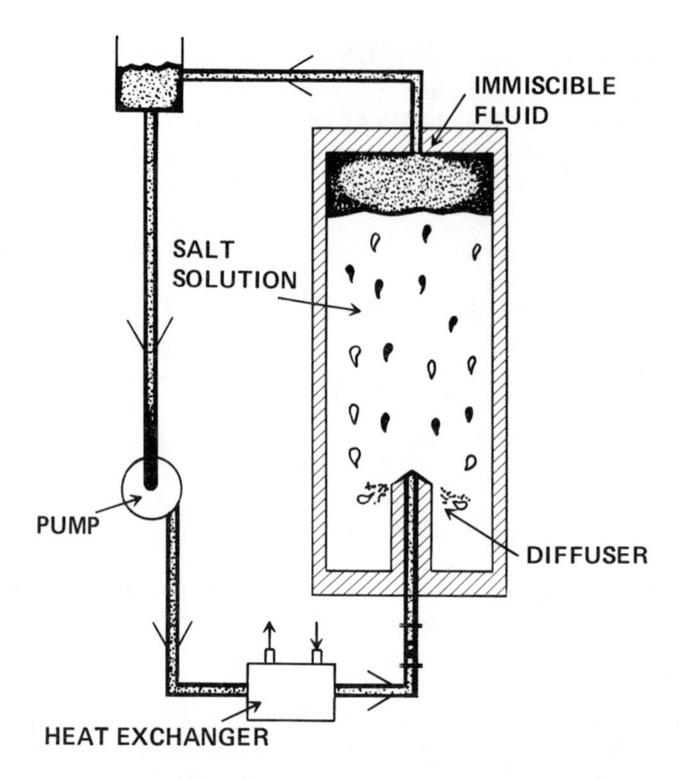

FIGURE 21. Heat of fusion storage system using an immiscible fluid for heat transfer. (From Edie, D. D., Melsheimer, S. S., and Mullins, J. C., Proc. Second Annual Thermal Energy Storage Contractors' Information Exchange Meeting, Gatlinburg, Tennessee, September 1977.)

TABLE 8

Heat of Fusion Storage Systems

Storage configuration	Storage medium	Status	T_{MAX}, °C	Capacity kW$_t$h	Cost \$/kW$_t$h
Annulus	LiH	Lab scale experiment	688	1.87	—
Cylinder	LiH	Lab scale experiment	680	0.281	—
Cylinder	Lif/LiOH Eutectic	Lab scale experiment	427	3.51	—
Cylinder	NaOH	Lab scale experiment	510	40	
Cylinder	NaOH	Operational units	482	193	4.60
Rectangular models	NaOH	Operational units	482	117	5.10
Cylinder	NaF/FeF_2 Eutectic	Conceptual	680	9.6×10^6	21

From Bramlette, T. T., Green, R. M., Bartel, J. J., Ottesen, D. K., Schafer, C. T., and Brumleve, T. D., Rep. SAND 75-8063, Sandia Laboratories, Albuquerque, N.M., March 1976.

Despite all the promise offered by chemical reaction storage, the technology is at such an early stage of development that systems can not be generally used for commercial applications. Indeed, relatively little experience has even been compiled in the laboratory to date; and significant scale demonstration units have not been run. Research and development are therefore needed to show:

- Reversibility of reactions with minor degradation of the chemical species and catalysts, when required
- Satisfactory kinetics, specificity, and conversions of the reactions
- Acceptable heat transfer rates in cost-effective containers and/or heat exchangers
- Easy storage and transportation (when required) of the chemical species
- No excessive corrosion
- Sufficient energy storage densities
- Acceptable storage efficiencies

Due to the generally immature information on the engineering design of reversible chemical reaction storage, this discussion will focus only on the operating concepts presently considered for storage of thermal energy. Reasonable cost projections will have to await further research to clarify the design of storage systems.

Table 9 presents some possible reactions for thermochemical storage of thermal energy.[27,28,29] Since a high energy storage density is essential at low ion cost in most applications, only reversible reactions with reactants and products which can be stored as liquids or solids are of practical interest. For example, solid calcium hydroxide (slaked lime) will endothermically decompose to solid calcium oxide (quicklime) and water vapor if it is heated to 520°C at one atmospheric pressure. The water vapor is condensed for storage. When heat is to be supplied from storage, water and the calcium oxide are mixed and the exothermic reverse reaction of the two species produces energy.[30-32]

The "turning temperature" T* in Table 9 is defined as the temperature for which the equilibrium constant is one and is approximated by the ratio of the standard enthalpy change to the standard entropy change for the reaction:[27]

$$T^* = \frac{\Delta H^\circ}{\Delta S^\circ} \tag{3}$$

When T > T*, the endothermic storage reaction is favored; while for T < T*, the exothermic reaction dominates.

Figure 22 illustrates the chemical heat pump mode of operation in which a dilute sulfuric acid solution is concentrated by using solar energy (or any other energy source) to evaporate water. The water vapor is condensed for storage, and the heat of condensation is given off to the load if it can be used at the condensation temperature or to the environment if it cannot. When heat is demanded from storage, energy from the atmosphere evaporates the liquid water and, provided the temperatures of the water and acid solution are properly regulated, the water vapor will condense in the solution. Consequently, the heat of condensation as well as the heat of mixing is released for the load, pumping energy from the environment. If all the energy is useful, more energy can be supplied in principle to the load than was captured by the sun. By interchanging the load and environment positions in Figure 22, the chemical heat pump may be used for air conditioning as well. Table 10 presents a number of reactions suitable for chemical heat pumping.[29,32,33] The temperatures shown are those typically

TABLE 9

Thermochemical Storage Reactions

Reaction	Heat of reaction, $\Delta H°$, kJ	Turning temperature, T*, K
$NH_4F(s) \rightleftharpoons NH_3(g) + HF(g)$	149.3	499
$Mg(OH)_2(s) \rightleftharpoons MgO(s) + H_2O(g)$	81.04	531
$MgCO_3(s) \rightleftharpoons MgO(s) + CO_2(g)$	100.6	670
$NH_4HSO_4(l) \rightleftharpoons NH_3(g) + H_2O(g) + SO_3(g)$	337	740
$Ca(OH)_2(s) \rightleftharpoons CaO(s) + H_2O(g)$	109.2	752
$LiOH(l) \rightleftharpoons ½ LI_2O(s) + ½ H_2O(g)$	56.7	1000
$CaCO_3(s) \rightleftharpoons CaO(s) + CO_2(g)$	178.4	1110

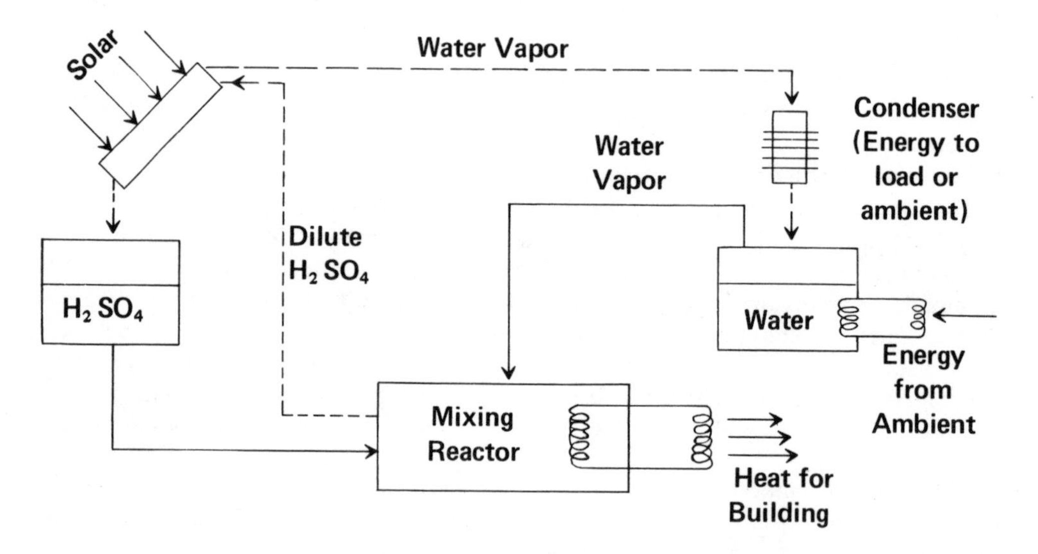

FIGURE 22. Chemical heat pump storage for sulfuric acid concentration/dilution with the charging cycle shown as dashed (- -) lines and discharging cycle by solid (—) lines.

considered. For ammonia systems, a second salt or liquid which reacts with ammonia at low temperatures such as $CaCl_2$ is often used to store the vapor.

IV. HYDROGEN

Batteries can be considered to be a special case of chemical energy storage where the functions of the initial conversion of electric to chemical energy, storage of this energy, and its reconversion to electric energy are combined in a single device. Thermal energy storage in reversible chemical reactions transforms thermal energy to chemical bond energy for storage and converts it back to heat at a later time by reversing the original reaction. One potentially important example of chemical energy storage with a separate process to release the stored energy is the use of hydrogen as a storage medium. Hydrogen may be produced either electrochemically or thermally.

Hydrogen is not a primary fuel. As such, its available energy is less than the energy from the source used to produce it due to inefficiencies. Its value lies in its ability to store and transport energy and in its potential to supplant oil and natural gas for those types of services for which they are most noted. Hydrogen has received a good deal

TABLE 10

Chemical Heat Pump Reactions

Reaction	Temp., K
Ammoniated salt pairs	
$CaCl_2 \cdot 8NH_3(s) \rightleftharpoons CaCl_2 \cdot 4NH_3(s) + 4NH_3(g)$	305
$NH_4Cl \cdot 3NH_3(l) \rightleftharpoons NH_4Cl(s) + 3NH_3(g)$	320
$MnCl_2 \cdot 6NH_3(s) \rightleftharpoons MnCl_2 \cdot 2NH_3(s) + 4NH_3(g)$	364
$MgCl_2 \cdot 6NH_3(s) \rightleftharpoons MgCl_2 \cdot 2NH_3(s) + 4NH_3(g)$	408
$MnCl_2 \cdot 2NH_3(s) \rightleftharpoons MnCl_2 \cdot NH_3(s) + NH_3(g)$	521
$MgCl_2 \cdot 2NH_3(s) \rightleftharpoons MgCl_2 \cdot NH_3(s) + NH_3(g)$	550
Hydrated salts	
$MgCl_2 \cdot 4H_2O(s) \rightleftharpoons MgCl_2 \cdot 2H_2O(s) + 2H_2O(g)$	380
Concentration-dilution	
$H_2SO_4 \cdot nH_2O(l) \rightleftharpoons H_2SO_4 \cdot (n - m)H_2O(l) + mH_2O(g)$	<600

of attention in the context of the "hydrogen economy" in which, for example, electricity generated by base-load nuclear power plants could be used to electrolyze water to hydrogen and oxygen. Then the hydrogen could be stored or transported to appropriate sites via pipelines; and at the load site the hydrogen could be used directly as fuel, or could be converted back to electricity via fuel cells or high temperature turbines. However, it is possible to conceive of a less grand scheme that would permit hydrogen to act as a storage medium both for utility load leveling applications and as a fuel for vehicles.

Production of hydrogen for use in central-station or transportation applications might be done by electrolyzing water into its constituents: hydrogen and oxygen. Another possibility is the thermochemical splitting of water using sources of high temperature process heat. At present, water electrolysis is a well established technology but is handicapped by a modest efficiency and high capital costs. Some observers[5] believe that with the development of advanced technology, the conversion efficiency might approach 100% with capital costs of $40 to $70/kW.

Unlike electrolysis systems which are available essentially as off-the-shelf items, the development of thermal processes for water splitting is still in the conceptual stage. Overall efficiencies and economics of thermal splitting might be superior to those offered by electrolysis, particularly if sources of high temperature heat — such as high temperature, gas cooled reactors — were available. It should be recognized that the establishment of technically and economically feasible processes for thermal splitting of water will undoubtedly require very extensive development efforts.

Hydrogen can be stored in metals. When pressurized hydrogen gas comes in contact with the metal surface, it readily diffuses into the metal and forms a metal hydride compound. A number of metals, usually alloys, have been and are being investigated. The hydride formation is exothermic, and the heat generated during the reaction must be removed. Later, heat must be applied to the metal hydride to evolve the hydrogen. These processes both occur at close to the ambient temperature. Storage of hydrogen in liquefied or compressed forms also has been proposed.

Hydride storage of hydrogen could be used for storing electric power if off-peak power were used to electrolyze water into hydrogen and oxygen. At times of peak demand, the hydrogen could be recovered and used to power fuel cells. One difficulty with hydrogen for storing electrical energy is that the overall efficiency of systems based on available fuel cells is only about 40%. Fuel cells at present are expensive. However, improvements in fuel cells operating on pure hydrogen are projected to permit system efficiencies of 65%. The alternative of converting hydrogen to electrical power via high temperature turbines is also awaiting advances in technology.

V. MECHANICAL STORAGE OF ENERGY

A. Pumped-Storage Hydroelectric Plants

The potential energy stored in water by virtue of its elevation can be converted into electricity with a high efficiency. In recent years a technique has been developed that allows the construction of hydroelectric plants that have some of the characteristics of storage batteries. They are called pumped-storage plants. At present they are the only practical way for large-scale storage of electrical energy.

Pumped-storage plants generally operate by transferring large amounts of water from a river or lake up to a reservoir at higher elevation. Power from base-load plants is used to drive the pumps during off-peak hours when these plants normally have excess capacity. During hours of peak demand, the water in the reservoir is allowed to fall back through the pump to the river or lake below. However, the pump now serves as a water driven turbine. The power from that turbine turns the electric pump motor backwards, and the motor is designed to act as an electrical generator when reversed. Thus excess electrical capacity available during off-peak hours can be stored until needed at peak demand.

The power used during off-peak hours is relatively cheap. All efficiencies considered, it requires about three cheap kilowatt-hours of off-peak electricity to produce two kilowatt-hours at peak demand. Thus, the system has an overall efficiency of about 67%. If peak pricing were employed, this type of storage would be even more attractive since the two kilowatt-hours sold during a time of high demand would command premium prices.

Pumped-storage units also have the ability to respond to rapid changes in load. A spinning turbine can be fully loaded in minutes, and the newer installations can be converted from pumping to generating in 5 to 10 min. Pumped-storage units can be combined with conventional hydroelectric plants or can be pure pumped-storage plants. The need for this kind of system is clear from the growth now projected; in 1970 there was 3,600 MW of pumped storage, by 1980 there will be 27,000 MW, and by 1990, 70,000 MW.

B. Compressed Gas Energy Storage

There are relatively few geographical locations that have a topography suitable for pumped-storage plants. However, by storing energy in air instead of water, the topographical problem can be overcome. An air-storage power plant employs a conventional gas turbine modified so that the compressor and turbine sections may be uncoupled and operated separately as shown in Figure 23. During off-peak, low-load periods, the turbine clutch is disengaged and the compressor is driven by a motor with power from base-load power plants in the system. The compressed air is stored for use during peak-load periods, at which time it is mixed with fuel for combustion, burned, and then expanded through the turbine. During these periods, the compressor clutch is disengaged, and the entire output of the turbine is used to drive the motor to generate electricity.

Storing air for this purpose in fabricated containers would probably be too expensive for the large storage volumes needed. The least expensive storage volumes would include dissolved-out salt caverns, porous-ground reservoirs, depleted gas and oil fields,

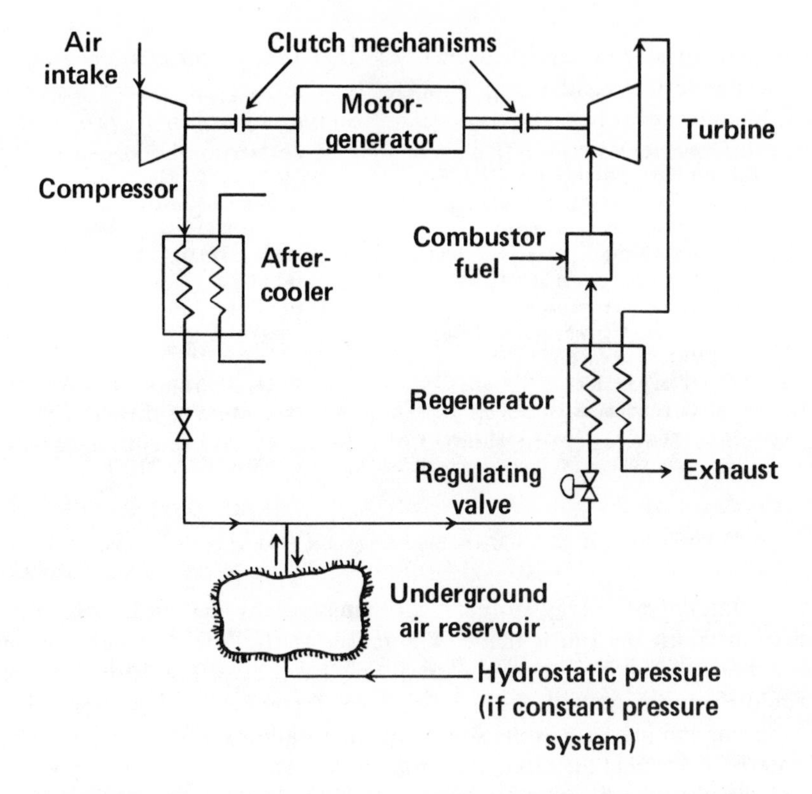

FIGURE 23. A compressed-air energy storage system. During charging, the motor drives a compressor which stores air under relatively high pressure in an underground reservoir. During discharge, the air is used to burn a fossil fuel which drives a gas turbine connected by means of a clutch to a generator.

and abandoned mines. Olsson[34] suggests that the air in abandoned mine storage regions could be kept under more or less constant pressure by hydrostatic pressure of water from a nearby lake or reservoir. Wide variations in stored gas pressure, however, can be tolerated with some penalty in performance.

The cavern volume required for a certain mass flow of compressed air is inversely proportional to the air pressure and proportional to the required number of hours of peak-load operation per day. Technical data for a 220 MW air storage plant is given in Table 11. Olsson asserts that such a plant could be built for $50/kW (1970 prices) excluding the cost of land, roads, and taxes.[34] Though no such plant is presently in operation, the technology for constructing such a plant is available.

C. Energy Storage In Flywheels

Flywheels have been used as energy storage devices for centuries. Today they are used widely in the internal combustion engines of automobiles, trucks, and diesel locomotives to carry the rotation of the engine between pulses of energy delivered by the pistons. Until recently, it was thought that employing flywheels to store energy in a wider range of applications was out of the question because of cost and because not enough energy could be stored for a given flywheel weight to satisfy the foreseeable needs. However, this picture has now been changed by recent advances in materials technology. These advances have come largely as a result of research and development in the aerospace industry.

Energy storage in a flywheel is governed by the mass of the rim and by how fast the wheel is spinning; the stored energy varies as the square of the rotation speed. The

TABLE 11

Technical Data for a 220-MW Air-Storage Power Plant

Station rating	220 MW
Air flow to storage at 5°C (41°F)	351 kg/sec (775 lb/sec)
Maximum storage pressure	43.5 atm (640 psia)
Air temperature in cavern	15°C (59°F)
Cavern depth	435 m (1425 ft)
Compressor power at 5°C (41°F)	161 MW
Turbine inlet temperature	800°C (1470°F)
Continuous power at 5°C (41°F)	73 MW
Efficiency/heat rate	
(1) at peak load	71.5%/4770 Btu/kWh
(2) continuous	27%/12,650 Btu/kWh
Power ratio: off peak kWh/peak kWh	0.76
Cavern volume per hour/day peak load operation	27,400 m^3 (970,000 ft^3)

After Olsson, E. K. A., *Mech. Eng.*, November 1970. With permission.

limit to the amount of energy stored is ultimately set by the tensile strength of the material from which the rim is made. The tensile strength must be great enough to withstand the so-called "hoop stress" resulting from centrifugal forces or else the wheel would fly apart. As with the energy stored, these forces are proportional to the mass of the rim and increase as the square of the rotation speed. Thus two properties of the material determine the amount of energy that can be stored in a flywheel: mass density which provides kinetic energy and tensile strength which resists centrifugal forces.

Quantitatively the above description can be expressed as:

$$E = \frac{1}{2} I \omega^2 \qquad (4)$$

or

$$E\,(\text{kJ}) = 5.0 \times 10^{-4}\, I\,[\text{kg–m}^2]\ \omega^2\ [\text{rad/sec}]^2 \qquad (5)$$

where I is the moment of inertia and ω is the angular velocity of rotation. An equivalent way of picturing the energy stored in a flywheel is as the energy in the "spring" formed by the tension created in the rim of the flywheel by the centrifugal force, which slightly expands the diameter of the flywheel.

The theoretical maximum specific energy that can be stored in a flywheel is fixed by the strength-to-density ratio of the material from which it is made.[35]

$$\frac{E}{W}\left(\frac{\text{kJ}}{\text{kg}}\right) = 1.31 \times 10^{-3} K_W\, \sigma(\text{MPa})/\delta(\text{Mg/m}^3) \qquad (6)$$

where σ and δ are the stress-level and density of the structural material and the numerical factor derives from the choice of units. K_w expresses the efficiency with which the particular design utilizes the material's strength and is a maximum if the stress is distributed uniformly throughout. In an optimum design for isotropic materials, both radial and tangential stresses would be equal and uniform, and K_w can approach a value of one. In designs optimized for materials such as fiber-reinforced composites, only one stress direction can be utilized; and the maximum value of K_w is 0.5. With either class of materials, the absolute maximum value of K_w is reached only in very slender or "flimsy" configurations with vanishing energy per unit volume of enclosure.

In order to compare various types of flywheel configurations it is convenient to define a volumetric specific energy as the maximum energy stored per unit volume of the cylinder enclosing the flywheel's maximum axial height and its maximum radial dimension.[35]

$$\frac{E}{V}\left(\frac{kJ}{m^3}\right) = 1.0 \times 10^4\ K_v\ \sigma\ (MPa) \tag{7}$$

Again the numerical factor derives from the choice of units. K_v expresses the efficiency with which the particular design fills the cylindrical volume as well as utilizing the material's strength; for a uniform density material, it equals K_w times the fraction of the cylindrical volume occupied by the flywheel.

Figure 24 illustrates how the weight efficiency factor K_w and the volumetric efficiency factor K_v are related for several classes of high-performance flywheel designs. Table 12 summarizes the estimated realizable properties of some candidate materials for flywheel construction.[35] The normalized parameters provide an indication of the relative energy storage performance (strength/density) and cost effectiveness [strength/(density-cost)] expected for rotors made from different materials using equivalent designs.

It is unlikely that flywheel energy storage can displace existing battery technology on the basis of weight, volume, or cost savings for a given amount of energy stored. On the other hand, power levels, and probably service life, may far exceed the capabilities of common battery systems.

VI. EVALUATING STORAGE METHODS

A. Governing Relationships

In this section a first order economic analysis of energy storage systems is presented. This analysis, though based on a conventional utility's needs, can clearly be generalized to a nonutility system. This work follows that which has been outlined by Golibersuch et al.[18]

In this analysis it is assumed that electrical energy being produced during nonpeak demand hours by a thermal central-station plant is being stored for use during peak hours. The costs of heat and electricity during peak and off-peak periods are summarized as follows:

$$w_o = q_o/\eta_g \tag{8}$$

$$w_e = f_e/t + w_o/\eta_e \tag{9}$$

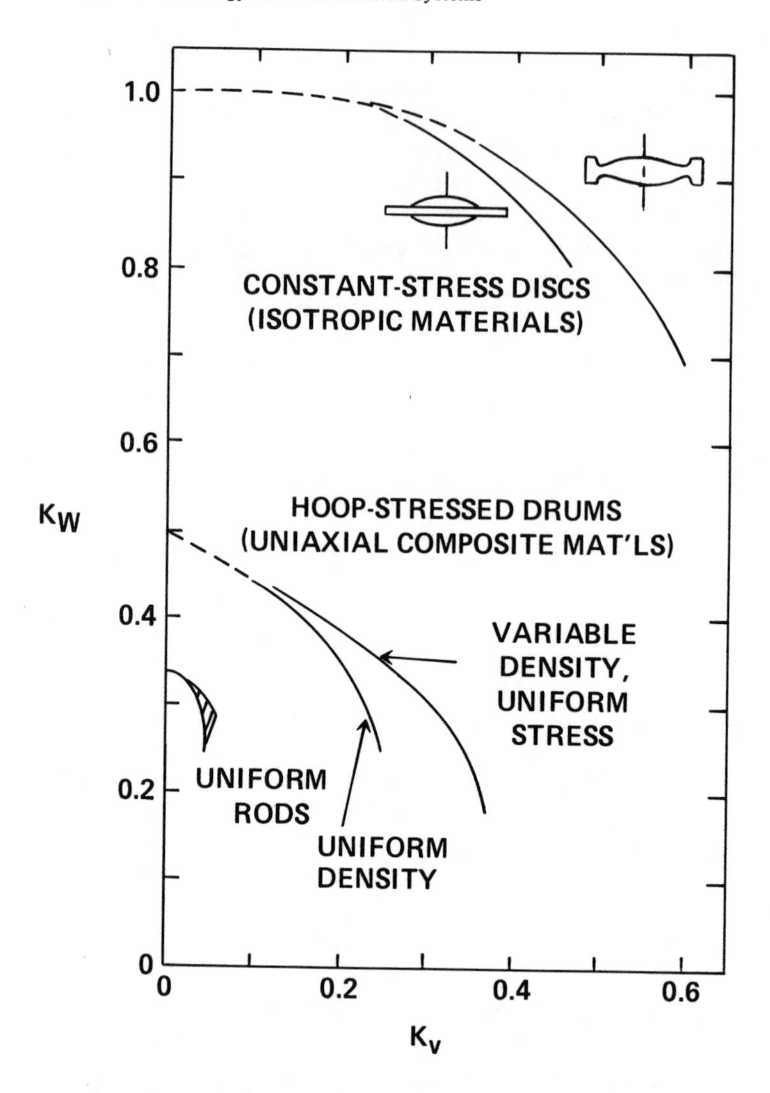

FIGURE 24. Relationship of weight-efficiency factor K_w and volumetric-efficiency factor K_v for high-performance flywheel designs. (From Fullman, R. L., Rep. No. 75 CRD 051, General Electric Technical Information Series, Schenectady, New York, April 1975; Proc. Tenth Intersociety Energy Conversion Engineering Conference. Permission granted by the Institute of Electrical and Electronic Engineers, Inc., New York.)

$$q_{th} = f_{th}/t + q_o/\eta_{th} \tag{10}$$

$$w_{th} = f_{g'}/t + q_{th}/\eta_{g'} \tag{11}$$

where f_e = annual fixed cost, nonthermal storage (mills/kW$_e$-year); $f_{g'}$ = annual fixed cost, thermal storage power plant (mills/kW$_e$-year); f_{th} = annual fixed cost, thermal storage (mills/kW$_e$-year) t = hours of peak operation per year; q_o = cost of off-peak heat (mills/kW$_{th}$-hr); q_{th} = cost of on-peak heat (mills/kW$_{th}$-hr); w_e = cost of peak electricity, nonthermal storage (mills/kW$_e$-hr); w_o = cost of off-peak electricity (mills/

TABLE 12

Properties of Candidate Flywheel Materials

	Density (mg/m³)	Cost ($/kg)	Cycles	Working stress (MPa)	Stress density (km)	Stress Density × Cost (kg·kg/$)
Maraging Steel	8.000	6.60	10^4	696	8.9	1.3
(18 Ni-#250)			10^5	340	4.3	0.65
4340 Steel	7.83	1.30	10^4	430	5.6	4.3
			10^5	280	3.6	2.8
Ti-6-4	4.43	6.60	10^4	630	14	2.1
			10^5	280	6.4	0.97
Al 2024-T3	2.77	1.10	10^4	230	8.5	7.7
			10^5	120	4.4	4.0
60 v/o S-glass/	1.96	1.80[a]	10^4	1000	52	29
epoxy			10^5	760	39	22
60 v/o E-glass/	1.99	1.10	10^4	830	42	38
epoxy			10^5	620	32	29
62 v/o Graphite/ epoxy	1.69	33[a]	$\leqslant 10^6$	830	50	1.5
63 v/o Kevlar®[b] 49 epoxy	1.36	7.70[a]	$\leqslant 10^6$	1000	75	9.7

[a] Based on anticipated fiber price reductions.
[b] ®DuPont trademark.

From Fullman, R. L., Rep. No. 75 CRD 051, General Electric Co., Schenectady, N.Y., April 1975; Proc. Tenth Intersociety Energy Conversion Engineering Conference. Permission granted by the Institute of Electrical and Electronics Engineers, Inc., New York.

kW_e-hr); w_{th} = cost of peak electricity from thermal storage plant (mills/kW_e-hr); η_e = nonthermal storage efficiency; η_g = overall power cycle efficiency base-load plant; η_g' = overall power cycle efficiency thermal storage plant; η_{th} = thermal turn-around efficiency.

For purposes of comparing electrical generation and thermal storage systems, it is convenient to relate the cost of peak electricity to off-peak electricity for both cases. Combining Equations (8), (10), and (11) one obtains:

$$w_{th} = (f_{g'} + f_{th}/\eta_{g'})/t + \frac{\eta_g}{\eta_{g'}} \frac{w_o}{\eta_{th}} \qquad (12)$$

The fraction η_g/η_g' takes into account the generally lower grade of thermal energy available from storage than from primary sources and the correspondingly different efficiencies of the two types of power plants (conventional power plant and thermal storage plant).

The annual fixed cost is determined by multiplying the installed cost by the capital recovery factor. The fixed cost for the storage plant, f_{th}, includes two terms: one determined by the cost per unit of installed power capacity and a second determined by the cost per unit of installed energy storage capacity times the number of hours of storage required for the particular duty cycle (Equation 1). Operating and maintenance costs have been neglected as they are generally small and difficult to estimate.

Figure 25 summarizes the cost range estimates for a number of particular peak power thermal energy storage plants based on the equations in this section[18]. Also shown is

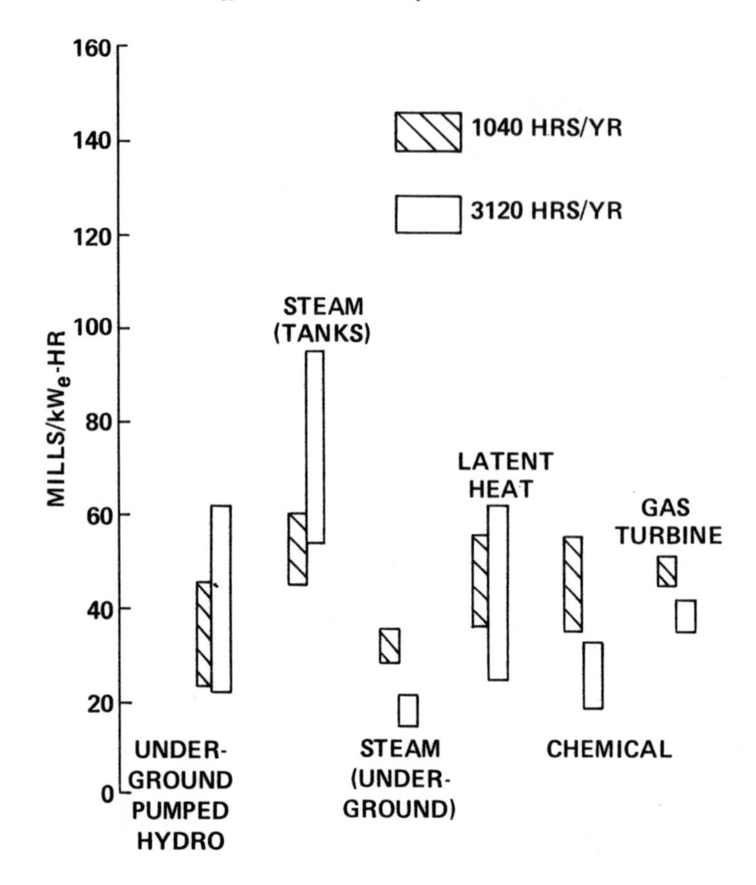

FIGURE 25. Peak power costs using various storage and direct production options. (From Golibersuch, D. C., Bundy, F. P., Kosky, P. G., and Vakil, H. B., Rep. No. 75 CRD 256, General Electric Technical Information Series, Schenectady, New York, December 1975; This figure was originally presented at the Fall 1975 meeting of The Electrochemical Society, Inc., held in Dallas, Texas.)

the cost of peak energy based on simple cycle turbine generation with an installed cost of \$100/kW_e, thermal efficiency of 28%, and a fuel cost of about \$3/GJ (\$3 per million Btu). The cost of off-peak electricity, w_o, is taken to be 5 mills/kW_e-hr. This is equivalent to a nuclear fuel cost of about 50¢/GJ (50 cents per million Btu). The assumed cost of off-peak electricity has little impact on the relative economic comparisons of various storage options.

B. The Cost of Stored Energy

To conceptualize the impact of energy storage on delivered energy costs, the energy in a storage device is arbitrarily broken down into fictitious elements according to usage. A solar installation is considered for which the demand is regular and somewhat independent of the season. The final conclusions should apply to any system, however.

Initial blocks of storage added to the system are used nights and during periods of cloud cover, and substantial amounts of energy pass through these elements of storage. Eventually, a point is reached where additional blocks of storage are only useful nights. Once enough storage is available to last the shortest night, additional storage elements are only discharged during longer nights in the year and cloudy periods which follow sunny days. Storage blocks beyond those required to last the longest night have lower utilization since cloudy and sunny days rarely alternate in a regular pattern. This di-

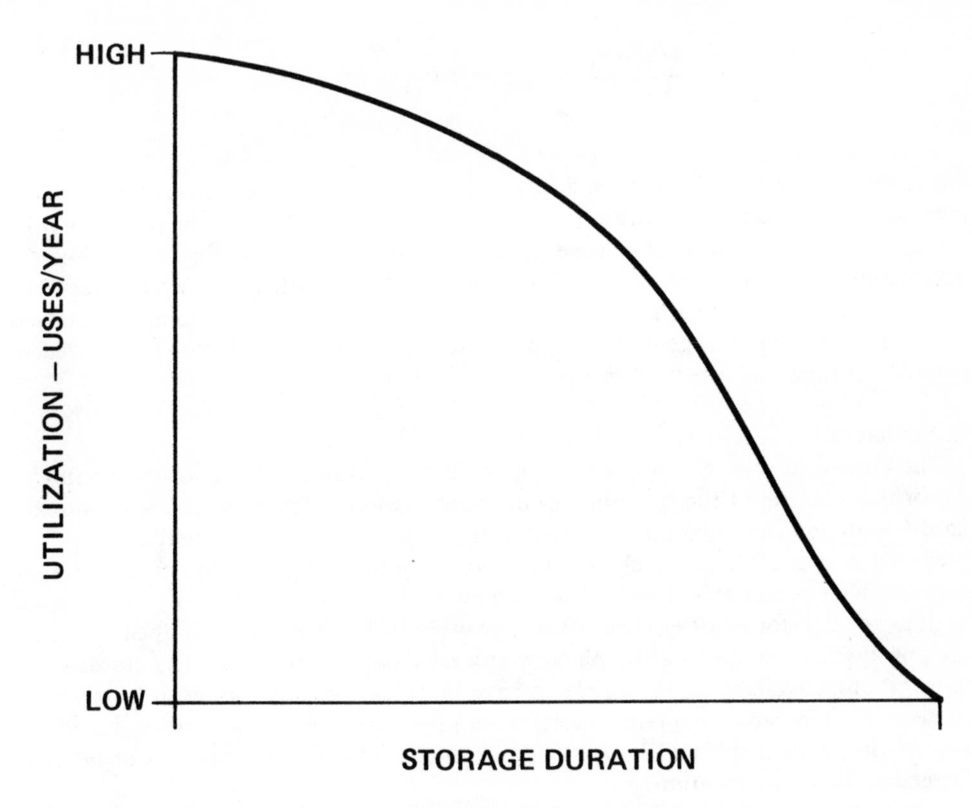

FIGURE 26. Utilization of last element of storage added to system vs. the storage duration to that point.

minishing utilization of additional storage elements continues until enough storage is added to provide a 100% solar powered system. Any storage beyond that amount is not used at all. Figure 26 presents an idea of how the storage element utilization might change as elements are added.

The degree of utilization of storage discussed above has important consequences on the price of the energy delivered from the storage subsystem. For example, from Table 7, a price of \$6.00 per kW_t-hr of capacity would be a reasonable projected cost for a sensible heat storage system. If that unit were amortized at 10% interest over 20 years, the yearly cost of the unit would be about \$0.70/kW_t-hr. If the device were used 300 times per year to meet evening loads and periods of cloud cover, the cost of the energy from the storage unit would be about:

$$\frac{\$0.70/\text{kW}_t\text{-hr-year}}{300\ \text{uses/year}} = \$0.0023/\text{kW}_t\text{-hr delivered}$$

This cost neglects the price of collectors, hardware, etc., required to charge the unit. The cost is low enough that storage is not a major cost barrier for such high utilization.

If the storage block added to meet extended cloudy periods is examined, its utilization is far less. Perhaps this fictitious block of storage would be used only four times per year. With the same capital recovery factor as above, the cost of energy delivered from this element is approximately:

$$\frac{\$0.70/kW_t\text{-hr}}{4 \text{ uses/year}} = \$0.175/kW_t\text{-hr delivered}$$

Now, when the costs of collectors, hardware, etc., are added, the delivered energy cost is certainly not promising. Furthermore, for a utility, the costs must be divided by the efficiency of thermal to electric generation to obtain the cost for electricity. Thus for applications where the utilization of storage is low, it is critical to develop very low cost storage subsystems such as the aquifers in Table 7. In all cases, a complete analysis of the system with storage and an appropriate capital recovery factor is required to determine the cost of delivered energy.

C. Summary

The energy storage systems considered here are at varying levels of development. Accordingly, some of the economic evaluations projected for these systems represent hardly more than speculation. However, it is not unreasonable to state that even these preliminary economic and technical projections indicate that several energy storage methods have reasonable potential for achieving application. Table 13 presents some options for the storing of electric energy for utilities. It is clear that pumped storage is the one application that is available now at a reasonable cost. (Lead-acid storage batteries, though available now, appear to have too high a cost per kilowatt-hour stored to make them useable on a commercial basis.) Because each of these storage methods has specific advantages and limitations, the choice of a storage device will be quite dependent on the application.

TABLE 13

Storage Options for Electric Utilities

	Round trip efficiency (%)	Capital ($/kW)	costs ($/kWh)	Energy density (kWh/m³)	Development stage	Potential application
Mechanical						
Pumped hydro	67—75	100—140		1.4	Existing application: engineering studies for underground	Central energy storage for peak shaving and load leveling
Compressed air-gas turbine system	65—75	120—150	3—10	3.5—17.5	First commercial demonstration 1977	Central energy storage for peak shaving and load leveling reserve generating capacity
Flywheels	70—85	80—120	50—100	17—70	Initial development	Distribute energy storage: power factor correction emergency generating capacity
Thermal						
Steam (pressure vessel)	70—80	150—250	15—25	up to 35	Historical installations, engineering studies of modern systems	Central energy storage integrated with baseload steam generation
Hot oil	65—80	150—250	10—50			
Batteries						
Lead-acid	60—75	60—100	25—50	35—70	State-of-the-art	Distributed energy storage for daily peak shaving: stand-by and emergency generating capacity; vehicle propulsion; energy storage in solar energy systems
Advanced aqueous	60—75	60—100	15—50	35—100	Small prototypes	
High-temperature	70—80	60—100	15—35	70—170	Laboratory cells	
Redox	60—70	100—200	5—15	17—70	Conceptual and laboratory studies	

From Kalhammer, F. R., *Energy Storage: Applications, Benefits and Candidate Technologies,* Proc. Symp. Energy Storage, Berkowitz, J. B. and Silverman, H. P., Eds., The Electrochemical Society, Princeton, N.J., 1976. With permission.

REFERENCES

1. **Löf, G. O. G. and Tybout, R. A.**, Cost of house heating with solar energy, *Sol. Energy*, 14, 253, 1973.
2. **Tybout, R. A. and Löf, G. O. G.**, Solar house heating, *Nat. Resour. J.*, 10, 268, 1970.
3. **Klein, S. A., Beckman, W. A., and Duffie, J. A.**, A design procedure for solar heating systems, *Sol. Energy*, 18, 113, 1976.
4. **Brown, J. T. and Cronin, J. H.**, Battery Systems for Peaking Power Generation, 9th Intersociety Energy Conversion Engineering Conference, American Society of Mechanical Engineers, New York, 1974.
5. **Kalhammer, F. R. and Zygielbaum, P. S.**, Potential for Large-Scale Energy Storage in Electric Utility Systems, Paper No. 74-WA/Ener-9, American Society of Mechanical Engineers, New York, 1974.
6. **Asbury, J. G. and Kouvalis, A.**, Electric Storage Heating: The Experience in England and Wales and in the Federal Republic of Germany, Rep. ANL/ES-50, Argonne National Laboratory, Argonne, Ill., May 1976.
7. **Gross, S.**, Review of Candidate Batteries for Electric Vehicles, Proc. Battery Council Int. Golden Anniversary Symp., London, May 1974.
8. **Stockel, J. F., Von Ommering, G., Swette, L., Gains, L.**, A Nickel-Hydrogen Secondary Cell for Synchronous Orbit Application, 7th Intersociety Energy Conversion Engineering Conference, American Chemical Society, Washington, D.C., 1972.
9. **Anon.**, Storage batteries: the case and the candidates, *EPRI J.*, No. 8, October 1976.
10. **Mitoff, S. P. and Bush, J. B., Jr.**, Characteristics of a Sodium-Sulfur Cell for Bulk Energy Storage, 9th Intersociety Energy Conversion Engineering Conference, American Society of Mechanical Engineers, New York, 1974.
11. **Weininger, J. L. and Secer, F. W.**, Nonaqueous lithium-bromine secondary galvanic cell, *J. Electrochem. Soc.*, 121, March 1974.
12. **Bramlette, T. T., Green, R. M., Bartel, J. J., Ottesen, D. K., Schafer, C. T., and Brumleve, T. D.**, Survey of High Temperature Thermal Energy Storage, Rep. SAND 75-8063, Sandia Laboratories, Albuquerque, March 1976.
13. **Silverman, M. D. and Engel, J. R.**, Survey of Technology for Storage of Thermal Energy in Heat Transfer Salt, Report ORNL/TM-5682, Oak Ridge National Laboratory, Tenn., January 1977.
14. **Freid, J. R.**, Heat-transfer agents for high-temperature systems, *Chem. Eng.*, 80, 89, 1973.
15. **Hallet, R. W., Jr. and Gervais, R. L.**, Central Receiver Solar Thermal Power System, Phase 1. CDRL Item 2, Pilot Plant Preliminary Design Report, Vol. 5, Thermal Storage Subsystem, Rep. SAN/1108-8/5, McDonnell Douglas Astronautics Company, Redondo Beach, Calif., October 1977.
16. Technical and Economic Assessment of Phase Change and Thermochemical Advanced Thermal Energy Storage (TES) Systems, Rep. EPRI EM-256, Vol. 1, Electric Power Research Institute, Palo Alto, Calif., December 1976.
17. **Bundy, F. P., Herrick, C. S., and Kosky, P. G.**, The Status of Thermal Energy Storage, Rep. No. 76 CRD 041, General Electric Technical Information Series, Schenectady, N.Y., 1976.
18. **Golibersuch, D. C., Bundy, F. P., Kosky, P. G., and Vakil, H. B.**, Thermal Energy Storage for Utility Applications, Rep. No. 75 CRD 256, General Electric Technical Information Series, Schenectady, N.Y., 1975.
19. **Goldstern, W.**, *Steam Storage Installations*, Pergamon Press, New York, 1970.
20. **Telkes, M.**, Solar energy storage, *ASHRAE J.*, 16, 39, 1974.
21. **Hale, D. V., Hoover, M. J., and O'Neill, M. J.**, Phase Change Materials Handbook, National Aeronautics and Space Administration CR-61363, September 1971.
22. **Glenn, D. R.**, Technical and Economic Feasibility of Thermal Energy Storage, Rep. C00-2558-1, U.S. Energy Research and Development Administration, Washington, D.C., 1976.
23. **Lorsch, H. G., Kauffman, K. W., and Denton, J. C.**, Thermal energy storage for solar heating and off-peak air conditioning, *Energy Convers.*, 15, 1, 1975.
24. **Edie, D. D., Melsheimer, S. S., and Mullins, J. C.**, Imiscible Fluid Heat of Fusion Heat Storage System, Proc. Second Annual Thermal Energy Storage Contractors' Information Exchange Meeting, Gatlinburg, Tenn., September 1977.
25. **Turner, R. H.**, *High Temperature Thermal Energy Storage*, Franklin Institute Press, Philadelphia, 1978.
26. **Offenhartz, P. O'D.**, Chemical methods of storing thermal energy, *Sharing the Sun! Solar Technology in the Seventies*, 8, 48, 1976.
27. **Wentworth, W. E. and Chen, E.**, Simple thermal decomposition reactions for storage of solar thermal energy, *Sol. Energy*, 18, 205, 1976.

28. **Wentworth, W. E., Batten, C. F., and Chen, E. C. M.,** Thermochemical Conversion and Storage of Solar Energy, Int. Symp. on Energy Sources and Development, Barcelona, Spain, October 19—21, 1977, proceedings to be published.
29. **Mar, R. W. and Bramlette, T. T.,** Thermochemical Energy Storage Systems — A Review, Rep. SAND 77-8051, Sandia Laboratories, Albuquerque, February 1978.
30. **Ervin, G.,** Solar Heat Storage Based on Inorganic Chemical Reactions, Proc. Workshop on Solar Energy Storage, University of Virginia, Charlottesville, April 1975, 91.
31. **Bauerle, G., Chung, D., Ervin, G., Guon, J., and Springer, T.,** Storage of solar energy by inorganic oxide/hydroxides, *Sharing the Sun! Solar Technology in the Seventies,* 8, 192, 1976.
32. **Ervin, G.,** Solar heat storage using chemical reactions, *J. Solid State Chem.,* 22, 51, 1977.
33. **Howerton, M. T.,** private communication, 1978.
34. **Olsson, E. K. A.,** Air storage plant, *Mech. Eng.,* 92, 20, 1970.
35. **Fullman, R. L.,** Energy Storage by Flywheels, Rep. No. 75 CRD 051, General Electric Company, Schenectady, N.Y., April 1975.
36. **Kalhammer, F. R.,** *Energy Storage: Applications, Benefits and Candidate Technologies,* Proc. Symposium Energy Storage, Berkowtiz, J. B. and Silverman, H. P., Eds., The Electrochemical Society, Princeton, New Jersey, 1976.

NOMENCLATURE

The symbols in this list are used in several chapters. When additional symbols are used in a chapter, they are defined in the text.

- A — Area m^2 (ft^2)
- $\overline{B}$ — Monthly-averaged, daily beam radiation, kWh/$m^2 \cdot$day (Btu/day$\cdot ft^2$)
- B — Daily total beam radiation, kWh/$m^2 \cdot$day (Btu/day$\cdot ft^2$)
- C — Cost, $
- C_{se} — Average annual cost of delivered solar enrgy or energy savings by conservation, $/GJ ($/MBtu)
- c_P — Specific heat at constant pressure, kJ/kg$\cdot$K, (Btu/lb°F)
- c_v — Specific heat at constant volume, kJ/kg$\cdot$K, (Btu/lb°F)
- COP — Coefficient of performance of a heat pump or refrigeration system
- CR — Concentration ratio
- CRF — Capital recovery factor
- d — Diameter, m (ft)
- d_H — Hydraulic diameter, m (ft)
- D — Daily total diffuse (scattered) radiation, kWh/$m^2 \cdot$day (Btu/day$\cdot ft^2$)
- $E_{b\lambda}$ — Spectral emissive power of a black body at λ, W/$m^2 \cdot \mu$ (Btu/h$\cdot ft^2 \cdot \mu$)
- e — Specific internal energy; eccentricity of earth orbit
- E — Energy
- F — Fin efficiency; force
- F' — Plate efficiency of a flat plate collector
- F'' — Flow factor of a flat plate collector
- F_{ij} — Radiation shape factor between surfaces i and j
- F_R — Heat removal factor of a flat plate collector
- f — Fanning friction factor; frequency
- f_s — Fraction of energy demand delivered by solar system
- g — Gravitational acceleration, m/s^2 (ft/sec^2)
- g_c — Inertia proportional factor occurring in relation Force = (mass × acceleration)/g_c
- Gr_L — Grashof number based on length dimension L
- G — Mass flow per unit area (= ϱv), kg/$m^2 \cdot$h (lb/$ft^2 \cdot$h)
- $\overline{H}$ — Monthly-averaged, total radiation on a horizontal surface on earth kWh/$m^2 \cdot$day (Btu/$ft^2 \cdot$day)
- $\overline{H}_o$ — Monthly-average, total radiation on a horizontal surface outside the atmosphere kWh/$m^2 \cdot$day (Btu/$ft^2 \cdot$day)
- h — Specific enthalpy, kJ/kg (Btu/lb); Planck's constant; altitude
- h_c — Convection heat transfer coefficient between a surface and a fluid, W/$m^2 \cdot$K (Btu/h$\cdot ft^2$°F)
- I — Insolation, defined as the instantaneous, hourly or annual solar radiation on a surface, W/m^2 (Btu/h$\cdot ft^2$)
- I_o — Solarrar constant, W/m^2 (Btu/h$\cdot ft^2$)
- i — Interest rate, percent; incidence angle, rad (deg)
- K — Thermal conductivity W/m$\cdot$k (Btu/h$\cdot$ft°F); Boltzmann's constant
- L — Length, m (ft); latitude; rad (deg); thermal load or demand
- l — Length, m (ft)
- m — Mass, kg (lb_m)
- $\dot{m}$ — Mass flow rate, kg/s (lb_m/h)
- n — Index of refraction; number

index

NTU — Number of transfer units (dimensionless)

Nu_L — Nusselt number based on length dimension L

P — Power, W (hp)

p — Pressure, Pa = N/m² (lb/in²)

Pr — Prandtl number

PV — Present value

PWF — Present worth factor

q — Rate of heat flow, W (Btu/hr)

Q — Quantity of energy or heat, kJ or kWh (Btu)

r — Radius, m (ft)

R — Thermal resistance, (K·m²/W) (°F·ft²·h/Btu); tilt factor; gas constant in pv = RT

Ra — Rayleigh number

Re_L — Reynolds number based on length dimension L

S — Surface area, m² (ft²)

s — Specific entropy, kJ/kg·K (Btu/lb°R)

T — Temperature, K or °C (°F or °R); tax or tax rate

t — Time, s (h); thickness, m (ft)

U — Overall heat transfer coefficient, W/m²·K (Btu/h·ft²°F)

u — Specific internal energy, kJ/kg (Btu/lb_m)

V — Volume, m³ (ft³); Voltage, V

v — Velocity, m/s (ft/sec); specific volume, m³/kg (ft^3/lb_m)

w — Width, m (ft)

W — Humidity ratio, kg water/kg dry air, (lb water/lb dry air); Work, kWh ($ft\ lb_f$)

z — Zenith angle, deg; altitude above mean sea level, m (ft)

Subscripts

a — air

ab — absorbent

amb — ambient conditions

ann — annual

atm — atmospheric

aux — auxiliary

b — beam

c — collector; convection

d — diffuse

eff — effective

f — fluid; fin; fuel

h — horizontal surface

i — incident

in — inlet or inside

k — conduction

max — maximum

min — minimum

n — normal

o — reference or standard; extraterrestrial

opt — optimum

out — outlet or outside

r — reflected

rad — radiation

s — solar; surface

terr — terrestrial

T — total

u — useful

w — wall; water

y — yearly

z — zenith

∞ — environmental conditions

Greek Symbols (May also be used as Subscripts)

α — absorptance; solar altitude angle, deg

β — volumetric thermal expansion coefficient collector tilt angle from horizontal plane

γ — specific heat ratio

δ — boundary layer thickness, m (ft); declination, deg

Δ — difference

ε — emittance

E — heat-exchanger effectiveness

η — efficiency; effectiveness

θ,ϕ — angles

λ — wavelength; mean free path, m (ft)

μ — dynamic viscosity, kg/m·s (lb_m/ft·h)

ν — frequency; kinematic viscosity, m^2/s (ft^2/s)

ϱ — reflectance; density, kg/m^3 (lb_m/ft^2)

σ — Stefan-Boltzman constant

τ — transmittance; shear stress, N/m^2 (lb_f/m^2)

ω — solid angle

$\dot{\omega}$ — angular velocity, s^{-1}

Prefixes (factor, prefix, symbol)

10^{12} tera T

10^{9} giga G

10^{6} mega M

10^{3} kilo k

10^{-2} centi c

INDEX

A

B

C

D

E

F

G

H

I

J

L

M

N

O

P

R

S

T

U

W

Z